Polymer–Polymer Miscibility

Polymer–Polymer Miscibility

OLAGOKE OLABISI
and
LLOYD M. ROBESON
Chemicals and Plastics
Union Carbide Corporation
Bound Brook, New Jersey

MONTGOMERY T. SHAW
Department of Chemical Engineering and
Institute of Materials Science
The University of Connecticut
Storrs, Connecticut

1979

ACADEMIC PRESS
New York   London   Toronto   Sydney   San Francisco
A Subsidiary of Harcourt Brace Jovanovich, Publishers

COPYRIGHT © 1979, BY ACADEMIC PRESS, INC.
ALL RIGHTS RESERVED.
NO PART OF THIS PUBLICATION MAY BE REPRODUCED OR
TRANSMITTED IN ANY FORM OR BY ANY MEANS, ELECTRONIC
OR MECHANICAL, INCLUDING PHOTOCOPY, RECORDING, OR ANY
INFORMATION STORAGE AND RETRIEVAL SYSTEM, WITHOUT
PERMISSION IN WRITING FROM THE PUBLISHER.

ACADEMIC PRESS, INC.
111 Fifth Avenue, New York, New York 10003

United Kingdom Edition published by
ACADEMIC PRESS, INC. (LONDON) LTD.
24/28 Oval Road, London NW1 7DX

Library of Congress Cataloging in Publication Data

Olabisi, Olagoke.
   Polymer--polymer miscibility.

   Includes bibliographical references.
   1. Polymers and polymerization. 2. Solution
(Chemistry) I. Robeson, Lloyd M., joint author.
II. Shaw, Montgomery T., joint author. III. Title.
QD381.8.063   547'.84   79–6942
ISBN 0–12–525050–9

PRINTED IN THE UNITED STATES OF AMERICA

79 80 81 82   9 8 7 6 5 4 3 2 1

To our wives (Gail, Saundra, and Stephanie) and our children. Their perseverance and encouragement aided the transformation of our dream into a reality.

Contents

Preface   xi

Chapter 1
Introduction

    1.1 Two-Component Systems: Definitions   3
    1.2 Historical Background with Commercial Examples   13
          References   17

Chapter 2
Thermodynamics of Polymer–Polymer Miscibility

    2.1 General Thermodynamic Principles   19
    2.2 Phase Separation Phenomena   31
    2.3 Hildebrand Approach: The Solubility Parameter   47
    2.4 Flory–Huggins Approach: The Lattice Theory   64
    2.5 Equation-of-State Approach   75
    2.6 Thermodynamics of Block Copolymer Systems   104
          References   111

Chapter 3
Methods for Determining Polymer–Polymer Miscibility

    3.1 Criteria for Establishing Miscibility   117
    3.2 Glass Transition Temperature   119
    3.3 Microscopy   136
    3.4 Scattering Methods   140
    3.5 Ternary-Solution Methods   157
    3.6 Miscellaneous   174
          References   189

Chapter 4
Methods of Enhancing Miscibility

4.1	Minor Modifications of Structure	196
4.2	Block and Graft Copolymer Formation	200
4.3	Interpenetrating Network Formation	203
4.4	Cross-Linking	204
4.5	Introduction of Interacting Groups	206
4.6	Miscellaneous	210
	References	211

Chapter 5
Comprehensive Survey of Miscible Polymer Systems

5.1	Criteria for Selection	215
5.2	Referenced Listing	217
5.3	Discussion	267
	References	268

Chapter 6
Properties of Miscible Polymer Systems

6.1	Thermal and Thermomechanical	277
6.2	Mechanical	287
6.3	Electrical	292
6.4	Rheological and Viscoelastic	295
6.5	Transport	300
6.6	Crystallization	306
6.7	Degradation	314
	References	316

Chapter 7
Utilization of Miscible Polymers

7.1	Industrial Examples	321
7.2	Mechanical Compatibility versus Miscibility in Polymer Blends	339
	References	353

Appendix 1   Nomenclature   359

Appendix 2	Abbreviations for Polymer Names	363
Index		365

Preface

Miscibility of polymeric mixtures was once viewed as an interesting but very rare occurrence. However, the number of miscible polymer blends reported in the literature has increased by an order of magnitude in the past decade with the number of technical papers in this area increasing accordingly. Polymer miscibility has become a major area of research in many industrial and academic laboratories. Nevertheless, this subject has only been covered in books on polymer blends as a special phenomenon. We believe the number of literature references and investigators active in this area clearly justifies a book specifically dealing with polymer miscibility.

This book elucidates both the theoretical as well as practical aspects of polymer miscibility. The historical background (Chapter 1) and commercial utility (Chapter 7) will give the reader a basis for appreciating the importance of this subject. The treatment of the thermodynamics of miscible polymer systems (Chapter 2) is considered to be a major part of this book even though future work in this area, as will be apparent to the reader, is definitely necessary. The experimental methods for assessing miscibility in polymer blends are reviewed in Chapter 3. Methods for enhancing miscibility are discussed in Chapter 4. A detailed review of miscible blends, presented in Chapter 5, includes not only simple blends but also block copolymers, graft copolymers, interpenetrating networks, and polyelectrolyte complexes. The properties of miscible polymer systems are compared with immiscible systems in Chapter 6. These chapters will provide a helpful reference for future investigation in this important field of polymer science.

Hopefully, our individual areas of expertise are collectively broad enough to encompass the subject matter presented. We believe therefore that a book on this subject coauthored by several investigators is advantageous. With full recognition that there are other investigators with at least equal qualifications to also present this subject in comprehensive form, we encourage additional books on this subject.

Our acknowledgments must start with Dr. C. N. Merriam, whose foresight (over a decade ago) concerning the importance of polymer miscibility in polymer science and application, provided the catalysis for our involvement in this interesting subject. Equally important acknowledgments are due to Drs. T. T. Szabo and L. M. Baker, who as Directors of Research in the Union Carbide laboratory at Bound Brook, also recognized the importance of this area and were actively involved with the experimental and theoretical programs. The review of the organization of this book and/or specific chapters by Drs. J. W. Barlow, J. P. Bell, J. V. Koleske, R. Koningsveld, W. J. MacKnight, M. Matzner, J. E. McGrath, L. P. McMaster, C. N. Merriam, and J. M. Whelan was particularly helpful in the evolution of this manuscript to its final form. The advance copies of seven manuscripts to be published from work at the University of Texas kindly forwarded by Dr. D. R. Paul provided important additions to the comprehensive review of miscible polymer systems.

The transformation of rough drafts into a net manuscript required the talents of many secretaries including Linda Blocker, Kathi DeFeo, Nancy Holder, Eloise Hall, Kathryn Kuchta, Annette Marshall, Judy Mastroserio, Mary Roche, Betty Stonesifer, and Carol Urbaniak. Their perseverance with our many corrections, long and detailed equations, and questionable penmanship is greatly acknowledged. The reproduction of many figures by J. Nicoll and J. Reinert and photography by S. Crisafulli was also a necessary element for completion of the manuscript.

Without the support and encouragement of our research management at Union Carbide in writing this book, we never would have embarked on this endeavor.

Olagoke Olabisi
Lloyd M. Robeson
Montgomery T. Shaw

Chapter 1

Introduction

In the initial stages in the development of polymer science, studies in polymer physics were involved with the understanding of the basic properties of homopolymers. Variations of the basic polymer structure leading to copolymers and graft and block copolymers were quickly followed by the detailed studies of the physical nature of these variations. More recently, detailed studies of the physics of polymer blends have been emphasized in both academic and industrial research laboratories. Many books have been written covering the general as well as specific aspects of these variations in the field of polymer physics.

An important case where the polymer–polymer mixture exhibits miscibility or solubility has provoked recent interest as miscible polymer blends are increasingly reported in the technical polymer literature. Until now, polymer–polymer miscibility has been treated as a special case in the field of polymer blends or alloys; thus, a detailed review in the form of a book has not been forthcoming. With the recent commercialization of new polymer–polymer miscible blends combined with very definite advances in the thermodynamics of polymer–polymer phase behavior, the necessity for a treatise specifically related to polymer–polymer miscibility has therefore evolved.

Several reviews of miscible polymer mixtures have been published in the past decade with listings of up to 50 mixtures exhibiting some degree of miscibility [1–4]. The most comprehensive reviews have been authored by Krause [3, 4] and provide an excellent listing of both miscible and immiscible blends. The literature search for this treatise has revealed over 100 different mixtures with the necessary criteria to be considered miscible polymer mixtures. Polymer miscibility is not only important in the case of simple polymer mixtures, but also determines the physical nature of block and graft copolymers, interpenetrating networks, and thermosetting networks of polymer mixtures; thus, these topics are covered in this treatise.

Before proceeding, clarification of the use of the term "miscibility" to describe single-phase, polymer–polymer blends is necessary. Prior studies and reviews have generally used the term "compatible" to describe single-phase behavior. However, "compatibility" has been used by many other investigators involving various studies of polymer–polymer blend behavior to describe good adhesion between the constituents, average of mechanical properties, behavior of two-phase block and graft copolymers, and ease of blending. The term "solubility," which is more descriptive and exact than "compatibility," could be another choice for describing molecular mixing in polymer–polymer blends. For single-phase, solvent–solvent and polymer–solvent mixtures, *solubility* is the accepted term. With polymer–polymer blends, ideal or random molecular mixing may not adequately describe the true nature of the blend even though the physical parameters of the blend would suggest true solubility. After much deliberation and discussions with many investigators involved with polymer–polymer blend research, the term *miscibility* has been chosen to describe polymer–polymer blends with behavior similar to that expected of a single-phase system. The term miscibility used in this treatise does not imply ideal molecular mixing but suggests that the level of molecular mixing is adequate to yield macroscopic properties expected of a single-phase material.

The level of molecular mixing existing in polymer blends that exhibit macroscopic properties indicative of single-phase behavior is commanding considerable attention without widespread agreement. For specific blends, recent studies have been able to provide experimental evidence of the level of molecular mixing. Until recently, only microscopic (i.e., electron microscopy) techniques were utilized to provide further insight into the level of mixing. In some blends, heterogeneous structure was observed at high levels of magnification even though macroscopic properties implied single-phase behavior. Heterogeneous structures (domains), however, have been observed in amorphous homopolymers (i.e., atactic polystyrene), thus confusing the interpretation of polymer blend miscibility [5]. These conflicting conclusions from macroscopic and microscopic experiments have resulted in research directed toward answering a specific question [6, 7]: How large is the size of a domain required to be in order to yield macroscopic properties (i.e., glass transition temperature) distinctly different from other domains of different molecular structures?

In earlier investigations of the thermodynamics of polymer mixtures, serious deficiencies were observed with the application of methods used successfully for predicting solvent–solvent or solvent–polymer solubility. The single-value solubility parameter approach was found to be quite unsuccessful in predicting polymer–polymer miscibility, and new techniques

1.1. Two-Component Systems: Definitions

that offered more promise were investigated. Equation-of-state thermodynamics has been applied recently to polymer phase behavior and qualitatively agrees with the experimentally observed phase behavior of polymer mixtures. This approach reveals definite differences from those one would expect via extrapolation of techniques commonly applied to solvent–solvent and solvent–polymer mixtures. Indeed, the thermodynamics of polymer–polymer mixtures requires much more rigorous treatment than the previously accepted methods for solvent–solvent and solvent–polymer mixtures, as will be detailed in this treatise.

Although the miscible blends of poly(vinyl chloride) and butadiene–acrylonitrile copolymers have been commercial since the 1940s [8, 9], recent interest has been provoked by the commercialization of a blend of poly(2,6-dimethyl-1,4-phenylene oxide) (PPO) and polystyrene [10] under the trade name Noryl, as well as the need for more permanent (i.e., polymeric) plasticizers for poly(vinyl chloride) [11, 12]. The best commercial advantages of a miscible polymer blend can best be summarized by the word "versatility." With a specific polymer, the number of possible variations in usable properties is limited without resorting to composition changes. Random, block, and graft copolymerization, polymer blends, and composites offer significant property diversification. With polymer blends exhibiting two-phase behavior, definite advantages can be derived [e.g., impact polystyrene and acrylonitrile–butadiene–styrene (ABS)] if mechanical compatibility can be achieved. With miscible polymer mixtures, mechanical compatibility is assured and a property compromise between the constituents is therefore achieved. Thus, with a miscible polymer–polymer blend, a range of price/performance characteristics between the component polymers can lead to a large number of potentially useful and different products. This versatility places miscible polymer blends into a unique situation with potential commercial importance.

This book is designed to cover both the theoretical and the practical aspects of polymer–polymer miscibility. A comprehensive review of miscible polymer blends listed in this book will provide the reader with information not previously available, other than in reviews that are not as comprehensive and/or are presently outdated due to the recent surge of new miscible polymer blends documented in the open literature.

1.1 TWO-COMPONENT SYSTEMS: DEFINITIONS

1.1.1. Blends—Two Phase and Single Phase

Although the principal subject of this book is the uncommon, single-phase, polymer blend, the two-phase system must be defined and contrasted with miscible systems to delineate the two subjects and to establish the criteria for excluding what is certainly the vast majority of polymer–polymer mixtures. The boundary line will not satisfy all participants—that would take the skills of a master politician—and in many instances some trespassing will occur.

In very general terms, if a blend in question possesses properties analogous to those expected for a single-phase material, the blend will be considered proper subject matter for this treatise. In most instances the critical property will be the glass transition; a blend with a single glass transition will be classified as miscible. Blends of components having similar glass transition temperatures will provide ambiguous cases and other techniques must be employed. But experience has shown that immisicibility in polymer blends is rarely well concealed, revealing itself as opacity, delamination, double glass transition, or combinations of these properties.

From a thermodynamic point of view, every polymer has some solubility in every other polymer, but the magnitude in most cases is exceedingly low. For example, if polystyrene is fluxed on a mill with poly(methyl methacrylate), a *two-phase* mixture results, no matter how long or intensive the mixing. One could, in principle (but with difficulty in practice), separate the two phases, analyze the composition of each, and arrive at values for the mutual solubilities. In this case, as with hexane in water, the solubilities would be less than 1%.

On the other hand, if one fluxes polystyrene on a mill with poly(2,6-dimethyl-1,4-phenylene oxide) as the second component, one phase results. It is thermodynamically stable because no matter how slowly the mixer turns or how long one waits there is still only one phase.

Consider a third example: polystyrene (PS) plus poly(vinyl methyl ether) (PVME). If PS is fluxed with an equal amount of PVME on the mill at 80°C, a clear, one-phase mixture results. However, if the temperature is raised to 140°C, two phases appear. A return to 80°C restores one phase. This behavior has been summarized for a range of compositions by the experimental cloud-point curves shown in Fig. 1.1.

These three examples provide an excellent basis on which to build a definition of polymer miscibility. The first represents an example of an immiscible blend, the second a miscible blend; the third illustrates that miscibility and immiscibility can be exhibited by the same mixture depending on the ambient condition. Furthermore, the third example demonstrates that the driving forces for the transition from the one-phase (miscible) to the two-phase (immiscible) state are thermodynamic in origin and do not depend, for example, on the extent or intensity of mixing.

In spite of the seemingly unquestionable behavior described in these examples, an exact definition of miscibility in polymer mixtures is a subject of considerable debate because it represents different characteristics to dif-

1.1. Two-Component Systems: Definitions

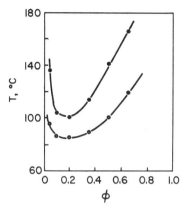

Fig. 1.1. Plot of initiation temperatures (○) and completion temperatures (●) of phase separation for several polystyrene concentrations, ϕ, of polystyrene–poly(vinyl methyl ether) mixtures at a heating rate of 0.2°C/min. [Reprinted by permission from T. Nishi, T. T. Wang, and T. K. Kwei, *Macromolecules* **8**, 227 (1975). Copyright by the American Chemical Society.]

ferent investigators. For the investigator interested in macroscopic properties useful in practical industrial problems, a miscible polymer mixture is that which exhibits a single glass transition temperature (T_g), and it is irrelevant whether or not changes of state occur during the preparation of the sample or during measurement. Here, miscibility implies homogeneity of the mixture up to a scale whose dimension is similar to the segmental size responsible for the major glass transition. However, for the investigator interested in statistical thermodynamics, miscibility implies homogeneity on a scale equivalent to the range of intermolecular forces. Miscibility in this sense is not necessarily satisfied by the T_g criterion. Miscibility is *not* even absolutely satisfied by the usual qualitative criterion given in several texts that the Gibb's free energy of mixing, ΔG^M, must be negative for a system to be thermodynamically stable. Such a criterion is only a degenerate form of Gibb's criteria because it neither distinguishes unstable from stable states nor gives any understanding of the concept of metastability, which is so important in phase separation phenomena. Furthermore, it is known [13] that mixtures are often unstable at negative ΔG^M (*not* relative to the pure constituents but to some intermediate composition) and that the ΔG^M becomes more negative as the mixtures phase-separate further.

Thermodynamic miscibility is governed by the subtle details of the composition dependence of the Gibb's free energy of mixing, namely, $\partial^2 G/\partial \Phi^2 > 0$, where Φ is the volume fraction of component 2. Such data can be measured easily only in the liquid state with measuring techniques such as light scattering and neutron scattering or through proper analysis of the location and

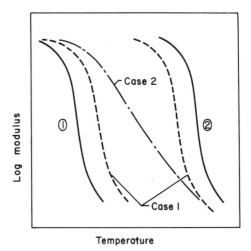

Fig. 1.2. Schematic of modulus–temperature behavior for blends exhibiting partial miscibility. The curves labeled 1 and 2 are for pure components.

shape of miscibility gaps. An example of an experimental miscibility gap is illustrated in Fig. 1.1 by the cloud-point curve of polystyrene–poly(vinyl methyl ether) mixtures [14]. In this figure, the system is completely miscible at temperatures below the lower curve (open circles) and completely immiscible above the upper curve (filled circles). The space between the two curves reflects the kinetic nature of the phase transformation; it also reflects the fact that a phase transition point can be observed only after big enough clusters have formed creating sufficient refractive index differences. This point will be discussed further in Section 3.4; it is sufficient for now to note that an equilibrium miscibility is, in fact, very difficult to discuss in distinct terms when dealing with the solid forms of macromolecules because of the high viscosity of these forms.

While the present text will discuss in detail the foregoing points of view about polymer–polymer miscibility, the "macroscopic," or T_g, definition has been adopted in selecting most of the examples of miscible polymer blends cited in this book. The "macroscopic" point of view is practical; in any case, all single-phase polymer mixtures would, on close scrutiny, reveal areas rich in one component or another—a condition necessitated by the size of the polymer molecules and the geometrical restraints imposed by covalent linking in the chainlike macromolecules. In a truly miscible mixture, such regions would not grow in size even if given every incentive to do so; that is, the mixture would be stable to reasonable time–temperature excursion. On the other hand, in an immiscible mixture, such regions will grow rapidly with time depending on the ambient conditions.

1.1. Two-Component Systems: Definitions

At present, the appearance of a single glass transition for polymer blends is not universally accepted as evidence of mixing at the macromolecular scale, to say nothing of the segmental scale [15]. Acceptable experimental evidence supporting unambiguously either side of this question is lacking, and the level of mixing required to produce a single T_g is actively being sought. The best studies to date [16] tend to show that the level of mixing in blended systems with a single T_g is essentially the same as that in a pure component, where each macromolecule tends to exclude measurably its neighbors. Thus, it does not appear likely that mixing on a segmental scale occurs, except in the special, but not uncommon, circumstance of strong specific interactions.

Microscopic evidence of two-phase behavior has been found in several cases where bulk properties imply single-phase behavior [17, 18]. However, electron microscopic examination of pure amorphous homopolymers (e.g., atactic polystyrene) has revealed domain structures as well [5]. Small-angle neutron scattering studies, on the other hand, show only random chain conformation in amorphous polymers. The latter technique is currently being applied to polymer–polymer systems and should provide important insight into the structure of single-phase and two-phase polymer systems [19].

Numerous *experimental* studies of phase behavior using bulk, or "macroscopic," properties such as glass transition have demonstrated that many polymer blends exhibit neither true two-phase behavior nor single-phase behavior. From these intermediate cases, two classes of behavior can usually be constructed:

(1) Two-phase structure where both phases contain different and finite concentrations of each component, as revealed by T_g values shifted significantly from the pure-component values. This behavior is analogous to that found with low molecular weight systems.
(2) Multiphase, or interphase, behavior where the glass transition is broadened over that commonly observed in single-phase systems.

These intermediate cases, shown schematically in Fig. 1.2, are quite common and will be covered in this treatise (Chapters 3, 5, and 6).

The vast majority of polymer pairs form two-phase blends after mixing as can be surmised from the small entropy of mixing for very large molecules. These blends are generally characterized by opacity, distinct thermal transitions, and poor mechanical properties. However, special precautions in the preparation of two-phase blends can yield composites with *superior* mechanical properties. These materials play a major role in the polymer industry, in several instances commanding a larger market than either of the pure components.

A typical example is impact-modified polystyrene (IPS), shown in Fig. 1.3. In this material, the polybutadiene phase (dark) is bonded to the matrix

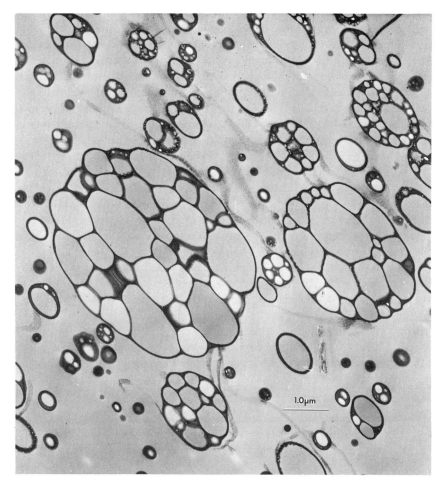

Figure 1.3. Structure of impact modified polystyrene containing 6% polybutadiene. [Reprinted by permission from E. R. Wagner and L. M. Robeson, *Rubber Chem. Technol.* **43,** 1129 (1970). Copyright by the American Chemical Society.]

phase by chemical grafting. In addition, it is cross-linked to maintain its integrity under processing and is sized to optimize the impact performance [20].

Other unique characteristics of two-phase blends have been exploited. For example, plastic paper [21] can benefit from the opacity of an immiscible system, while the pearlescence of partially miscible systems is exploited for decorative effects. Two-phase interconnected structures can be freed of one component by selective extraction, leaving a feltlike substance with unique

1.1. Two-Component Systems: Definitions

properties [22]. Other properties often improved in two-phase blends are stress–crack resistance, processability, friction and wear resistance, and flame retardance.

The characteristics of two-phase blends have been covered in detail in a series of reviews and books [23–29]. Single-phase behavior has not been covered to this depth, although two reviews do offer considerable information [3, 4].

The distinction between miscible and immiscible systems has been introduced in this section mainly for the purpose of establishing the scope of the material to be covered. More comprehensive examinations of these concepts will be found in Sections 2.1.1 and 3.1, where the perspectives are theoretical and experimental, respectively.

1.1.2. Block and Graft Copolymers

Block and graft copolymers, by virtue of covalent bonding between polymeric structures of differing compositions, will be considered separately from simple polymer mixtures. This is necessary because the thermodynamics of phase separation of the constituent polymers of the block and graft copolymers is different from that of common polymer mixtures. The existence of covalent bond(s) has a profound effect on the phase behavior (due primarily to entropic effects); and, as will be discussed in later sections, covalent bonding markedly improves the miscibility for two dissimilar polymeric species.

Block copolymers are characterized by the existence of two (or more) structurally dissimilar polymers in a single linear polymer molecule. Several variations exist, as illustrated below (A and B are monomer units):

AAAAAA......AAABBBBBBBB....BBBBBB (A_x–B_y)	AB Block Copolymer
A_x–B_y–A_x	ABA Block Copolymer
$(A_x$–$B_y)_n$	$(AB)_n$ Block Copolymer

Commercial examples of block copolymers include polystyrene–polybutadiene AB block copolymers [30], polystyrene–polybutadiene–polystyrene ABA block copolymers [31], poly(butylene terephthalate)–poly(tetrahydrofuran) $(AB)_n$ block copolymers [32], bisphenol A polycarbonate–silicone rubber $(AB)_n$ block copolymers [33], and thermoplastic polyurethanes [34], which are classified as $(AB)_n$ block copolymers. The chemistry and physical behavior of block copolymers have been reviewed in several comprehensive books and reviews [35–39] which should be consulted for more detailed information.

Graft copolymers have been an integral part of polymer science for many years although commercialization of specific pure graft copolymers has been limited. However, graft copolymers are an important constituent in blends in which at least one of the graft copolymer components exists as an ungrafted species at a significant concentration. These blends include, of course, impact polystyrene and ABS (acrylonitrile/butadiene/styrene), which are composed of a matrix of ungrafted polymer and a dispersed phase of matrix polymer grafted onto a lightly cross-linked rubber network [40]. Similar cases exist commercially for impact-modified poly(methyl methacrylate) and poly(vinyl chloride) (PVC). With poly(vinyl chloride), the polymer grafted on the dispersed rubber phase is not PVC but rather a structure that exhibits mechanical compatibility with PVC, that is, styrene/acrylonitrile copolymers or poly(methyl methacrylate) [41, 42].

A graft copolymer is characteristically different in structure from a block copolymer in that the individual polymer segments in a single molecule do not form a linear chain; instead, they consist of a linear chain of one composition from which side chains or another composition are attached, as diagrammed below:

```
                           B
                           .
                           B
                           B
       AAAAA......A......AAAAAAAA..........AAAABAAAAAAA
         B           B                          B
         B           B                          B
         B           B                          B
         B           :                          B
                     :                          B
                     B                          B
                     B
```

A special case of graft copolymers, having a high concentration of low molecular weight grafts, has been recently termed comb copolymers [43, 44]:

```
.....AAAAAAAAAAAAAAAAAAAAAAAAAAAAAAAAAAAAAAAAAAAAAAAA.....
  B B   B      B B     B       B B    B      B B
  B B   B      B  ·    B       B B    B      B B
  · B   ·      · B     ·       · B    ·      B ·
  B ·   B      B       B       B ·    B      · B
        B                             B             B
        B                             B
```

Another recent variation of block, graft, and comb copolymers is the star or radial block copolymer [45, 46]. In this case, the structure has a central

1.1. Two-Component Systems: Definitions

position from which three or more linear block copolymer molecules are covalently attached:

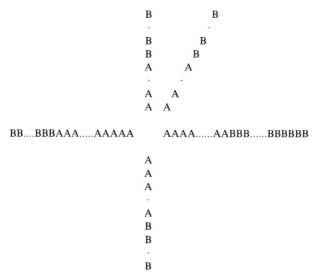

BB....BBBAAA......AAAAA   AAAA......AABBB......BBBBBB

The center position is a low molecular weight polyfunctional entity to which the individual linear blocks can be covalently bound. The termination of the anionic polymerization of a polystyrene–polybutadiene AB block copolymer with divinyl benzene will yield a star or radial polymer [45]. Also the termination of the same block copolymer with silicon tetrachloride will yield a four-membered star copolymer under ideal conditions of stoichiometry and reaction completion [47]. Radial block copolymers of styrene and isoprene have been touted to achieve lower viscosity at equivalent mechanical properties than the linear block copolymer counterparts [48, 49]. This advantage has resulted in several commercial products used in adhesive and packaging application areas.

1.1.3. Interpenetrating Networks

Interpenetrating networks (IPNs) can be considered a unique subset of polymer blends for which many variations and methods of preparation can be devised. Interlocking rings which exhibit no chemical bonds are termed catenanes [50]:

In polymer blends, structures exhibiting this type of interlocking ring structure have been termed interpenetrating networks.

One of the more common methods for the production of IPNs consists of swelling a cross-linked polymer A with a second monomer B_i containing appropriate cross-linking agents, followed by polymerization of the second monomer. If polymers A and B exhibit a high level of miscibility, a continuous interpenetrating network throughout the specimen can be envisioned. Under many situations where polymer A and polymer B do not constitute miscible polymer pairs, phase separation during the polymerization of monomer B_i will occur. The level of interpenetration will then be markedly lower and observed primarily at the phase boundaries. Because interpenetration of the respective networks at the phase boundaries is possible using the above-described blend preparation technique, mechanical compatibility can, therefore, be achieved. At higher levels of interpenetration, a shift in the phase behavior from immiscibility for the simple, un-cross-linked blends to partial miscibility for the IPN system may be observed.

Other methods of producing IPNs have been described in the literature [51–53] and are summarized here:

(1) Mixing of linear polymers (melt or solution) followed by cross-linking *in situ*

(2) Mixing latices of linear polymers containing cross-linking agents and removal of the carrier (i.e., H_2O) followed by cross-linking

(3) Mixing linear prepolymers and their respective cross-linking agents and curing the resultant blend

(4) Mixing monomers and their respective cross-linking agents and curing the resultant blend

Procedures (3) and (4) are referred to as simultaneous interpenetrating networks (SINs).

These techniques have been experimentally investigated for many blend combinations. These methods of polymer blend preparation are considered important in the discussions of polymer–polymer miscibility due to the improved miscibility shown experimentally for many IPN systems over that achieved with simple polymer mixtures of the same constituents. Pure IPNs can be considered those cases in which no or an insignificant chemical reaction occurs between the constituent polymers during preparation. In many cases, however, chemical reaction can occur yielding IPNs composed of grafted polymer (A–g–B) as well as interpenetration of the networks. Donnatelli *et al.* [54] termed semi-IPNs as a graft copolymer in which one of the polymers is cross-linked while the other retains its linear structure. This semi-IPN is, of course, characteristic of the grafted matrix on cross-linked rubber in impact polystyrene and ABS.

Of course, there are many other variations of the IPNs. Possible arrangements and classifications of these systems have been reviewed elsewhere [55, 56]. The interested reader should consult these references for more detail.

1.2 HISTORICAL BACKGROUND WITH COMMERCIAL EXAMPLES

The advantages one can envision for miscible polymer systems can probably be best described by the word "versatility." By simply varying the concentration of the constituents of a miscible blend, an innumerable variety of materials, each with a unique set of properties, can be obtained. Under certain conditions of widely differing price/performance characteristics of the constituents, the blend can offer a set of properties that may give it the potential of entering application areas not possible with either of the polymers comprising the blend. This is indeed the case with the first two cases discussed in this section, namely, poly(vinyl chloride) (PVC)-butadiene/acrylonitrile copolymer (NBR) and polystyrene-poly(2,6-dimethyl-1,4-phenylene oxide) (PPO) blends. These miscible polymer blends, as well as others to be discussed in this section and in Chapters 5 and 7, provide excellent examples for the justification of industrial interest in miscible polymer systems.

The interest of the academic community in studying the phenomena of miscible polymer blends is also easily explained by the fact that the combinations of polymer blends are all but innumerable, whereas the known miscible systems are relatively few. The thermodynamic arguments for phase separation or, conversely, phase miscibility for polymer mixtures are relatively new and open for considerable investigation as opposed to solvent-solvent and solvent-polymer thermodynamics. As is generally the case in a new area of research activity, many unanswered questions exist and contrasting views on the nature of polymer-polymer miscibility have arisen. A certain polarization of various disciplines has occurred with the question of molecular mixing. This is indeed the case with the first historical example we will discuss (PVC-NBR). Certain conclusions will be drawn concerning these areas of disagreement, but hopefully within the context of presenting both sides of the question fairly. PVC-NBR, being the most-investigated miscible polymer blend, offers an excellent example for discussion in this regard.

1.2.1 Poly(vinyl chloride)-Butadiene/Acrylonitrile Copolymers

The unique and useful properties of blends of poly(vinyl chloride) and butadiene/acrylonitrile copolymer rubber have been known for over 35 years [57, 58]. Two separate product areas have evolved from these blends: the addition of PVC (generally with low molecular weight plasticizers) to nitrile rubber, followed by vulcanization, and the utility of nitrile rubber as a permanent plasticizer for PVC. Emmett [59] showed that addition of plasticized PVC to nitrile rubber vulcanizates resulted in improved ozone and sunlight resistance, flex cracking, thermal aging, and chemical resistance. Lower tensile strength, abrasion resistance, and higher compression set were found to be limitations of the blend versus the nitrile rubber controls. These blends were evaluated via mixing in typical rubber processing equipment followed by vulcanization of the nitrile rubber. The utility of these cross-linked blends has been described by other investigators [60] for applications including printing roll covers, gaskets, fuel hose covers, cable jackets, and conveyor belt covers.

The addition of nitrile rubber to PVC to produce a nonvulcanized, permanently plasticized PVC has resulted in a significant number of specialty plasticized vinyl applications for many years. While low molecular weight plasticizers offer definite cost, efficiency of plasticization, and processing advantages over the PVC–NBR blends, the inferior chemical resistance (ability to be leached out by swelling non-solvents), poor permanence (due to migration to the surface and subsequent volatilization), and biological degradation prevent their use in more demanding applications. Until recently, no other high molecular weight polymers ($>10,000\ M_n$) have been available for plasticizing PVC [12]. Oligomers (aliphatic polyesters) with molecular weights generally less than 4000 M_n have been used to improve the permanence of plasticized PVC for applications requiring intermediate performance. Thus, nitrile rubber has held a unique position for many years as the only available permanent plasticizer for PVC. This has resulted in a number of applications, including jacket and low-voltage primary insulations for the wire and cable industry. Permanence is required for jacket insulation, because low molecular plasticizers migrated into the primary insulation and altered the electrical properties. In the shoe industry [60], the PVC–NBR blend was considered useful because it would not be affected by shoe bonding cement. As fatty foods and oils (i.e., margarine) can leach out plasticizers resulting in taste and film property changes, the PVC–NBR blend allowed for a commercial solution to this problem [61]. Similar performance leads to utility in oil containment vessels, including pollution control pond liners used for oil containment.

1.2.2. Poly(2,6-dimethyl-1,4-phenylene oxide)–Polystyrene Blends

As will be illustrated by data shown later in this treatise, poly(2,6-dimethyl-

1.2. Historical Background with Commercial Examples

1,4-phenylene oxide) (PPO) –(C₆H₂(CH₃)₂–O)ₙ– and polystyrene are miscible polymers. PPO has a glass transition temperature, T_g, of 210°C and polystyrene has a T_g of 100°C with blends exhibiting intermediate T_g's. This allows compositions that exhibit short-term load-bearing capabilities between 90° and 200°C (heat distortion temperatures, ASTM D-635, for amorphous polymers are generally 10° to 15° lower than the glass transition temperatures). PPO, because of higher monomer cost and a more difficult polymerization procedure, is significantly higher in cost than polystyrene. Thus, combinations of the two polymers allow a balance in the price/performance profile, with PPO yielding advantages in higher T_g and better impact strength while polystyrene offers a very important price advantage.

The combination of these two polymers has resulted in a commercial system under the trade name Noryl offered by General Electric Company [10]. Note that, with this system, many price/performance profiles are available [62, 63]. This versatility, of course, is one of the major attributes of miscible polymers and is well demonstrated by Noryl and the product line available under that trade name.

The plastics industry uses Noryl in the manufacture of electronic components, appliance housings, television yokes, business machine housings, automobile dashboards, and pump components. With the recent requirements for improved flame-resistant plastics in appliances, Noryl has enjoyed a specific advantage because the PPO–polystyrene blend can be modified with low-cost phosphorus-containing additives (i.e., triphenyl phosphate [64]) at levels which do not significantly alter usable properties. This advantage has been one of the primary reasons for the growth of Noryl markets since its commercial introduction.

1.2.3. High-Heat ABS

Styrene–acrylonitrile copolymer grafted onto a polybutadiene or styrene–butadiene copolymer rubber is commonly referred to as ABS (acrylonitrile–butadiene–styrene). The glass transition of the styrene–acrylonitrile matrix (100°C) determines the upper use temperature of the product (thus, maximum heat distortion temperatures of 90°C). To extend the potential of ABS in applications requiring improved short-term load-bearing capabilities at elevated temperature, a miscible polymer mixture can be employed.

α-Methylstyrene can be substituted for styrene in the polymerization to yield a higher heat distortion temperature product as one possible route.

Another approach, believed commercially utilized, is to produce an α-methylstyrene (MS)–acrylonitrile (AN) copolymer (generally at the azeotropic composition, 69/31 by weight α-MS/AN) [65]. This copolymer has a T_g of 125°C and is added to ABS. The polymer blend exhibits glass transition temperatures intermediate between the styrene–acrylonitrile matrix of ABS (100°C) and the α-methylstyrene–acrylonitrile copolymer. The miscibility of this pair was first noted by Slocombe [66].

1.2.4. Coating Resins

Perhaps the earliest examples of polymer–polymer systems can be drawn from the history of the coatings industry. Even before the commercialization of coatings, people experimented with the mixing of resins to obtain better properties. Gardner [67] reports that the art of making varnish was described as early as the eleventh century. Varnish is perhaps the first example of an IPN or, considering the relatively low molecular weight of its components, an example of a copolymer. The method for making varnish is as follows: Linseed oil, a triglyceride of unsaturated fatty acids, is heated with a brittle resin, usually an unsaturated C_{20} monocarboxylic acid [68]. The resin itself could be used as a coating, but the linseed oil is incorporated to reduce the brittleness of the product. On the other hand, the linseed oil by itself forms an oily, slow-to-harden coating. Thus, the principle of property averaging was recognized a long time ago. The resin contains unsaturation and, presumably during the heating with the oil and the subsequent oxidative cross-linking, is incorporated covalently. Nevertheless, no segregation occurs and clear, tough, films of high molecular weight result. Many modern varnishes use phenol–formaldehyde resins rather than natural resins. Modification of the basic phenol structure is required in order to obtain solubility of the resin in the linseed oil. Para substitution with *tert*-butyl, phenyl, and other hydrocarbon groups increases the aliphatic character (reduces solubility parameter) enough to achieve solubility.

In the nonconvertible coatings (lacquers) area, the importance of polymer–polymer systems was recognized as soon as it was found that cellulose nitrate required plasticization to obtain adequate toughness for coatings. Shellac, a natural resin obtained from insects, was an early example of a polymeric additive to cellulose nitrate. The mixture forms clear, tough films and "the two appear to act as mutual plasticizers" [69]. Later it was observed that poly(vinyl acetate), poly(vinyl methyl ether), and poly(vinyl ethyl ether) performed the same function. As cellulose acetate made inroads into cellulose nitrate markets, it was found that not all the plasticizers for the latter were suitable for the former. Phenolic resins, poly(α-methylstyrene) and poly-(vinyl acetate), gave clear coatings with cellulose acetate. The coatings chemists also discovered the concept of using a third component to improve the miscibility of the two others. Phenolic resin, shellac, and styrene–co-α-methylstyrene was an example of such a ternary mixture [69].

REFERENCES

1. D. J. Buckley, *Trans. N.Y. Acad. Sci.* [2] **29,** 735 (1967).
2. L. Bohn, *Rubber Chem. Technol.* **41,** 495 (1968).
3. S. Krause, *J. Macromol. Sci., Rev. Macromol. Chem.* **7** (21), 251 (1972).
4. S. Krause, *in* "Polymer Blends" (D. R. Paul and S. Newman, eds.), Chapter 2. Academic Press, New York, 1978.
5. P. H. Geil, *Ind. Eng. Chem., Prod. Res. Dev.* **14,** 59 (1975).
6. P. R. Couchman and F. E. Karasz, *J. Polym. Sci., Polym. Phys. Ed.* **15,** 1037 (1977).
7. D. S. Kaplan, *J. Appl. Polym. Sci.* **20,** 2615 (1976).
8. R. A. Emmett, *Ind. Eng. Chem.* **36,** 730 (1944).
9. M. C. Reed, *Mod. Plast.* **27,** 117 (1949).
10. E. P. Cizek, U.S. Patent 3,383,435 (1968) (assigned to General Electric Co.).
11. P. E. Graham, *33rd ANTEC SPE* p. 275 (1975).
12. C. F. Hammer, *Am. Chem. Soc., Div. Org. Coat. Plast. Chem., Pap.* **37**(1), 234 (1977).
13. R. Konigsveld and L. A. Kleintjens, *Lect., 5th Discuss. Conf. Phases Interfaces Macromol. Syst., 1976.*
14. T. Nishi, T. T. Wang, and T. K. Kwei, *Macromolecules* **8,** 227 (1975).
15. J. Stoelting, F. E. Karasz, and W. J. McKnight, *Polym. Eng. Sci.* **10,** 133 (1970).
16. F. B. Khambatta, F. Warner, T. Russell, and R. S. Stein, *J. Polym. Sci., Polym. Phys. Ed.* **14,** 1391 (1976).
17. R. W. Smith and J. C. Andries, *Rubber Chem. Technol.* **47,** 64 (1974).
18. M. Matsuo, C. Nozaki, and Y. Jyo, *Polym. Eng. Sci.* **9,** 197 (1969).
19. W. A. Kruse, R. G. Kirste, J. Haas, B. J. Schmitt, and D. J. Stein, *Makromol. Chem.* **177,** 1145 (1976).
20. C. G. Brigaw, *Adv. Chem. Ser.* **99,** 86 (1971).
21. J. R. Stell, D. R. Paul, and J. W. Barlow, *Polym. Eng. Sci.* **18,** 496 (1976).
22. W. A. Miller, M. A. Spivack, F. R. Tittman, and J. S. Byck, *Textile Res. J.* **43,** 728 (1973).
23. L. H. Sperling, ed., "Recent Advances in Polymer Blends, Grafts, and Blocks." Plenum, New York, 1974.
24. L. E. Nielsen, "Mechanical Properties of Polymers and Composites." Dekker, New York, 1974.
25. E. M. Fettes and W. N. Maclay, *Appl. Polym. Symp.* **7,** 3 (1968).
26. D. R. Paul and S. Newman, eds., "Polymer Blends." Academic Press, New York, 1978.
27. P. A. Marsh, A. Voet, and L. D. Price, *Rubber Chem. Technol.* **40,** 359 (1967).
28. N. A. J. Platzer, ed., "Multicomponent Polymer Systems," Adv. Chem. Ser. No. 99. Am. Chem. Soc., Washington, D.C., 1971.
29. P. F. Bruins, ed., "Polyblends and Composites," Appl. Polym. Symp. No. 15. Appl. Polym. Symp., New York, 1970.
30. G. Kraus and H. E. Railsback, *in* "Recent Advances in Polymer Blends, Grafts and Blocks" (L. H. Sperling, ed.), p. 245, Plenum, New York, 1974.
31. G. Holden, E. T. Bishop, and N. R. Legge, *J. Polym. Sci., Part C* **26,** 37 (1969).
32. G. K. Hoeschele and W. K. Witsiepe, *Angew. Makromol. Chem.* **29/30,** 267 (1973).
33. R. P. Kambour, *J. Polym. Sci., Part B* **7,** 579 (1969).
34. C. S. Schollenberger, H. Scott, and G. R. Moore, *Rubber World* **137,** 549 (1958).

35. A. Noshay and J. E. McGrath, "Block Copolymers: Overview and Comprehensive Treatise." Academic Press, New York, 1977.
36. R. J. Ceresa, ed., "Block and Graft Copolymerization." Wiley, New York, 1973.
37. R. J. Ceresa, "Block and Graft Copolymers." Butterworth, London, 1962.
38. S. L. Aggarwal, ed., "Block Copolymers." Plenum, New York, 1970.
39. G. M. Estes, S. L. Cooper, and A. V. Tobolsky, *J. Macromol. Sci., Rev. Macromol. Chem.* **4** (2), 313 (1970).
40. E. R. Wagner and L. M. Robeson, *Rubber Chem. Technol.* **43,** 1129 (1970).
41. N. G. Gaylord, *Adv. Chem. Ser.* **142,** 76 (1975).
43. M. Pegoraro, *Pure Appl. Chem.* **1,** 291 (1976).
43. R. Milkovich and M. T. Chiang, U.S. Patent 3,832,423 (1974) (assigned to CPC International Inc.).
44. R. Milkovich and M. T. Chiang, U.S. Patent 3,786,116 (1974) (assigned to CPC International, Inc.); see also related patents: U.S. Patents 3,842,059 and 3,842,146 (1974).
45. L. K. Bi and L. J. Fetters, *Macromolecules* **8,** 98 (1975).
46. L. K. Bi, L. J. Fetters, and M. Morton, *Polym. Prepr., Am. Chem. Soc., Div. Polym. Chem.* **15** (2), 157 (1974).
47. C. Price and D. Woods, *Eur. Polym. J.* **9,** 827 (1973).
48. J. R. Haws, *Am. Chem. Soc., Div. Org. Coat. Plast. Chem., Pap.* **34** (1), 114 (1974).
49. L. M. Foclor, A. G. Kitchen, and C. C. Baird, *Am. Chem. Soc., Div. Org. Coat. Plast. Chem., Pap.* **34** (1), 130 (1974).
50. H. L. Frisch and E. Wasserman, *J. Am. Chem. Soc.* **83,** 3789 (1961).
51. K. C. Frisch, D. Klempner, H. L. Frisch, and H. Ghiradella, *in* "Recent Advances in Polymer Blends, Grafts and Blocks" (L. H. Sperling ed.), p. 395, Plenum, New York, 1974.
52. L. H. Sperling, T. W. Chiu, C. P. Hartman, and D. A. Thomas, *Int. J. Polym. Mater.* **1,** 331 (1972).
53. K. C. Frisch, D. Klempner, S. K. Mukherjee, and H. L. Frisch, *J. Appl. Polym. Sci.* **18,** 689 (1974).
54. A. A. Donatelli, A. D. Thomas, and L. H. Sperling, *in* "Recent Advances in Polymer Blends, Grafts and Blocks" (L. H. Sperling, ed.), p. 375. Plenum, New York, 1974.
55. L. H. Sperling, *in* "Recent Advances in Polymer Blends, Grafts and Blocks" (L. H. Sperling, ed.), p. 93. Plenum, New York, 1974.
56. L. H. Sperling and K. B. Ferguson, *Macromolecules* **8,** 691 (1975).
57. D. E. Henderson, U.S. Patent 2,330,353 (1943) (assigned to B. F. Goodrich Co.).
58. E. Badum, U.S. Patent 2,297,194 (1942).
59. R. A. Emmett, *Ind. Eng. Chem.* **36,** 730 (1944).
60. J. E. Pittenger and G. F. Cohan, *Mod. Plast.* **25,** 81 (1947).
61. M. C. Reed, *Mod. Plast.* **27,** 117 (1949).
62. G. Bommi and P. LeBlanc, *Mod. Plast. Encycl.* **53,** No. 10A, 42 (1976).
63. J. S. Eickert, *Mod. Plast. Encycl.* **50,** No. 10A, 56 (1973).
64. A. Katchman and G. F. Lee, Jr., Great Britain Patent 1,372,634 (1974) (assigned to General Electric Co.).
65. H. H. Irvin, U.S. Patent 3,010,936 (1961) (assigned to Borg-Warner Corp.).
66. R. J. Slocombe, *J. Polym. Sci.* **26,** 9 (1957).
67. W. H. Gardner, *in* "Protective and Decorative Coatings" (J. J. Matteillo, ed)., p. 1. Chapman & Hall, London, 1941.
68. C. L. Mantell, C. W. Kopf, J. L. Curtis, and E. M. Rogers, "The Technology of Natural Resins." Wiley, New York, 1942.
69. Oil and Color Chemists Association, "Paint Technology Manuals. Part One. Non-Convertible Coatings." Chapman & Hall, London, 1961.

Chapter 2

Thermodynamics of Polymer–Polymer Miscibility

Before the current thinking on the thermodynamics of polymer–polymer systems is discussed, it is beneficial to review briefly some fundamental solution thermodynamics. The reasons for this discussion are to establish a reference to which polymer–polymer solutions can be compared and to introduce a nomenclature system.

2.1 GENERAL THERMODYNAMIC PRINCIPLES

2.1.1 Definition of a Solution

A solution is defined in most texts as a mixture of two or more components constituting a single phase. But the concept of a phase is very difficult to define, particularly in the case of polymers, and the best that can be done is to use somewhat weak descriptions. Thus, a phase is a "uniform" piece of matter with "reproducible," "stable," properties depending only on thermodynamic variables. Time independence or thermodynamic equilibrium is implied. A certain amount of nonuniformity, such as small concentration and density fluctuations and surfaces, must be allowed, and in some cases the appearance and behavior of the phase can be changed significantly by these fluctuations. For example, a single phase near the critical point does not even look uniform to the naked eye, yet it meets the thermodynamic reproducibility requirement and thus is a single phase. The effect of subdivision on the activity of a solid is well known and can constitute an important energy contribution. In fact, one can imagine a situation where a system solid + liquid + surface, comprising a reproducible stable colloid, depends only on temperature T, pressure P, and composition ϕ and is in reversible equilibrium with the solid. In this case, the colloid, in spite of its lack of molecular dispersion, would have to be considered a single phase. Macromolecules are so large that they fall necessarily into the colloidal size range with the possibility of a single molecule existing simultaneously and stably in two phases, to say nothing of its ability to form a single-molecule phase. It is understandable, therefore, that a great many effects have been observed with mixtures of macromolecules and there is little in the form of rules except the basic ones of uniformity, reproducibility, and stability. The reproducibile single T_g is a possible criterion for polymer–polymer miscibility as discussed in Section 1.1.1. It is perhaps the most widely used and the most practical, and is fairly sensitive. As the segmental size responsible for the major T_g is estimated at about 10 to 50 repeat segments, the presence of 50- to 100-Å inhomogeneities in polymer–polymer solutions is not surprising and is not in itself a proof of the existence of two phases. Heterogeneities of this order have been observed in amorphous single-component polymer systems as well [1]. The question of intercomponent entanglements and the associated question of the perturbation of end-to-end distances of the components are gradually being answered.

As pointed out by Yu [2], the miscible polymer–polymer solution is a relative, rather than an absolute, state, with the uniformity requirements calling for a length scale and the reproducibility and stability requirements calling for a time and energy scale. It has been the authors' experience that polymer–polymer systems classified herein as miscible do not require great energy or time to phase-separate or mutually dissolve. For example [3], heating rates of 0.2°/min typically spread the observable cloud point by only a few degrees on cyclic heating and cooling, with no mixing applied.

It is useful to review briefly the types of liquid phase behavior that have been observed with *low molecular weight* compounds and relate these to the system thermodynamics. No new types of phase behavior have been observed with quasi-binary polymer systems when the added constraint of covalent bonds between phases, as in block copolymers, is taken into account.

The simplest phase behavior is represented by complete miscibility at all concentration ratios and at all temperatures between the melting and boiling points, as illustrated by phase diagram A in Fig. 2.1. Cyclohexane–carbon tetrachloride is an example of a completely miscible system [4]. In the diagrams of Fig. 2.1, T is the absolute temperature and ϕ_2 is the volume fraction* of component 2. The area identified by horizontal lines is the region of temperature and concentration where two phases coexist. Volume fraction is

---

* Volume fraction is defined as $\phi_2 = N_2 V_2/(N_1 V_1 + N_2 V_2)$, where N is the number of molecules and V is volume.

2.1. General Thermodynamic Principles

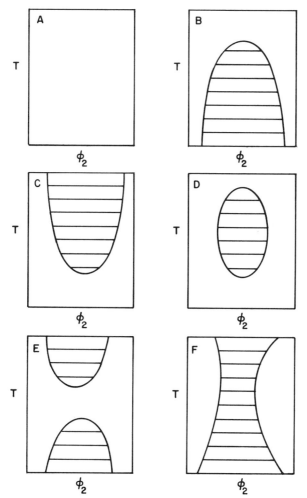

Fig. 2.1. Schematic of liquid–liquid temperature–composition phase diagrams. Shaded areas represent the temperature–composition regimes of a solution instability where phase separation occurs.

often used in polymer work because the ratio of segments occupying adjacent lattice sites, not the ratios of entire molecules, determines interaction energies. In polymer–polymer systems of equal molecular weights, the choice of the concentration variable is not particularly important.

Besides T and ϕ_2, the pressure P is also an independent variable. Usually it will be close to atmospheric, but because the polymer may be processed at high pressures it will be important to know the possible effect of pressure on the phase diagram [5].

Several other types of phase behavior in liquid systems have been observed.

A common one is diagram B (Fig. 2.1), but diagrams C, D, E, and F have been observed as well. Examples are water + n-butanol, water + triethylamine, and water + nicotine for B, C, and D, respectively [4]. Polystyrene in cyclohexane will give E or F depending on the molecular weight of the polystyrene [6].

There is little reason to exclude a priori any of these for polymer–polymer systems. In addition, with crystalline components, there are many other possible variations. (The reader is referred to books on metallurgy for these possibilities.) The glass transition should not, in principle, affect the phase behavior because the phase diagram represents the equilibrium condition and an indefinitely long wait should remove the glassy state as surely as an increase in temperature to above the glass transition temperature, T_g. In practice however, this is not so.

2.1.2 The Ideal Solution

Ideal solutions occur very rarely, CCl_4 + cyclohexane being a fairly good example [1], but it is useful for introducing the concept of a regular solution. An ideal solution is formed when two components mix to yield a free energy change that is completely determined by the entropy gained by each component due to the extra degrees of freedom created by the solution process. This is called the combinatorial entropy, which, for low molecular systems with equal-sized molecules, is given by

$$\Delta S^M = -k(N_1 + N_2)(x_1 \ln x_1 + x_2 \ln x_2) \quad (2.1)$$

where x_i is the mole fraction of component i. Ideal solubility precludes volume changes on mixing, and the enthalpy of mixing, ΔH^M, must be zero. This applies to a mixing of two materials that are effectively identical except for a tag on one. Changing the relative size will *reduce* the number of possible combinations and the entropy will be less than that above.

These ideas are depicted in Fig. 2.2, which illustrates the lattice picture (a concept to be heavily exploited in cases involving macromolecules) with a cell size equal to the volume occupied by these identical molecules. This conception aids in the derivation of Eq. (2.1), which is based on the Boltzman law for the entropy of mixing

$$\Delta S^M = k \ln \Omega \quad (2.2)$$

where Ω is the total number of ways of arranging N_1 and N_2 molecules on a regular lattice comprising $N = N_1 + N_2$ cells. This is just the combination of N things taken N_1 at a time.

$$\Omega = \frac{N!}{N_1! N_2!} \quad (2.3)$$

2.1. General Thermodynamic Principles

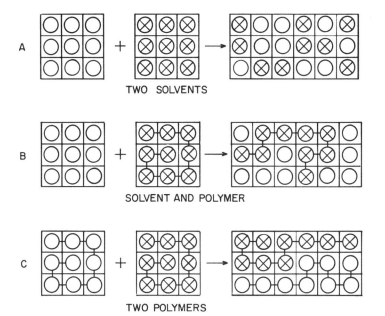

Fig. 2.2. Schematic illustration of the numbers of possible arrangements in a small molecule mixture (A), a polymer solution (B), and a polymer mixture (C). The polymer chains each contain nine segments of the size of the solvent molecules.

Recalling Sterling's approximation

$$\ln n! = n \ln n - n$$

and substituting Eq. (2.3) into (2.2) gives the result, Eq. (2.1), after some algebraic manipulation. The activity a_i of each component in the ideal solution is equal to its mole fraction; it will obey Raoult's law if the vapors form ideal gases. Because of the relationship,

$$\Delta G = \Delta H - T\Delta S \qquad \Delta H = 0, \text{ ideal} \tag{2.4}$$

the components forming an ideal solution will always be completely miscible.

2.1.3 Regular Solution

The concept of the regular solution was introduced by Hildebrand [7] and greatly simplifies the handling of solutions that deviate from ideality. The concept is that even if the molecules in the solution interact in some way, either favorably, producing lower activity, or unfavorably, producing higher activity than ideal, they will be so jostled by their kinetic energy that the combination of positions open to each will be the same as if the solution were ideal. Thus, the entropy of a regular solution will be given by the *same* expression as for the ideal solution. Again, this precludes volume changes on mixing and interactions strong enough to produce clustering.

In the regular solution the free energy of mixing is then the ideal ($-T\Delta S^M$) plus an enthalpy term. The latter can be developed most simply in terms of an exchange energy w,

$$w = \tfrac{1}{2}\varepsilon_{11} + \tfrac{1}{2}\varepsilon_{22} - \varepsilon_{12} \qquad (2.5)$$

where ε_{ij} is the energy of a contact between components i and j. Because the mixing must be close to random in order to have ideal entropy, the total enthalpy can be closely approximated by

$$\Delta H^M = (N_1 + N_2)zwx_1x_2 \qquad (2.6)$$

where z is a coordination number. Thus, the assumption of randomness has considerably simplified matters to give [7, 8]

$$\Delta G^M = (N_1 + N_2)zwx_1x_2 + kT(N_1 + N_2)(x_1 \ln x_1 + x_2 \ln x_2) \qquad (2.7)$$

For vaporizable liquids, this theory can be tested by measuring vapor pressures as a function of T and x_2 and by observing the phase behavior. If the exchange energy w is assumed independent of T and x_2, the phase diagram of Fig. 2.1B is predicted. This type of phase behavior is quite common, but quantitative agreement can be achieved only by permitting zw to be a function of composition and temperature. The exact nature of these dependencies can be predicted with fair accuracy by liquid-state theories to be discussed in more detail in Sections 2.4 and 2.5.

2.1.4 Polymer–Solvent Systems

The anomalous behavior of high molecular weight mixtures with low molecular weight solvents was discovered early [9]. Solvent vapor pressure was found to be *far* lower than predicted by the above free energy relation using mole fractions, even for solutions of polymers in liquids with very similar structures (e.g., polystyrene in ethylbenzene). Such solutions are expected to be nearly *athermal*; that is, w should be zero and the solvent vapor pressure above the polymer solution should be proportional to the solvent mole fraction multiplied by the pure-liquid vapor pressure (Raoult's law). This is far from what is observed. Indeed, in the case of a piece of rubber containing very few molecules, the vapor pressure of a swelling solvent is many orders of magnitude lower than predicted by Raoult's law. It seemed evident, therefore, that the mole fraction, or molecule/molecule interchangeability [the basis for the entropy part of Eq. (2.7)], would have to be replaced by some sort of fraction characteristic of chain molecular systems.

2.1. General Thermodynamic Principles

Thus arose the concept of a *lattice* where the sites of the lattice would represent the exchangeable units for the calculation of entropy. The polymer molecule, according to the concept, could occupy many adjacent sites of the lattice, e.g., r sites. This picture is schematically shown in Fig. 2.2B. Taking into account that the polymer segment must have at least two adjacent sites occupied by polymer segments yields the famous Flory–Huggins expression for the entropy of mixing of polymer with solvent:

$$\Delta S^M = -k(N_1 \ln \phi_1 + N_2 \ln \phi_2) \qquad (2.8)$$

Note that the *volume* fraction has replaced the mole fractions in the natural log terms of the ideal entropy of mixing expressed in Eq. (2.1).

The enthalpy of mixing must also undergo a similar modification because it is the *segments* of the polymers which are presumed to interact according to the expression of Eq. (2.5). The number of polymer-occupied lattice sites is N_2 (subscript 2 is commonly attached to the polymer) times the ratio of the size of the polymer chain to that of the solvent, i.e., V_2/V_1. This is equivalent to saying that the lattice size is determined by the solvent and the polymer occupies this lattice, as suggested by Fig. 2.2B. The enthalpy is then obtained from Eq. (2.6) simply by substituting

$$\phi_2 \text{ for } x_2, \qquad \phi_1 \text{ for } x_1, \qquad \text{and } (N_2 V_2)/V_1 \text{ for } N_2$$

to arrive at

$$\Delta H^M = \left(N_1 + \frac{N_2 V_2}{V_1}\right) zw \phi_1 \phi_2 \qquad (2.9)$$

Note that z should probably be reduced from the case of two equally sized components because the polymer segment has some surface blocked by other segments.

The free energy of mixing becomes

$$\Delta G^M = kT \left(N_1 + \frac{N_2 V_2}{V_1}\right) \left[\phi_1 \ln \phi_1 + \phi_2 \left(\frac{V_1}{V_2}\right) \ln \phi_2 + zw \phi_1 \phi_2 / kT \right] \qquad (2.10)$$

and is written in this form because the term in the square brackets is identifiable with the free energy per lattice site (in kT units), with its obvious decrease of (negative) entropy with V_2 but a reasonable constancy of enthalpy with molecular weight. Alternatively, the free energy can be calculated on a *volume* basis by dividing both sides of Eq. (2.10) by $N_1 V_1 + N_2 V_2$, which is the total volume V.

$$\frac{\Delta G^M}{V} = \frac{\Delta G^M}{N_1 V_1 + N_2 V_2} = kT \left[\frac{\phi_1}{V_1} \ln \phi_1 + \frac{\phi_2}{V_2} \ln \phi_2 + \frac{zw \phi_1 \phi_2}{V_1 kT}\right] \qquad (2.11)$$

The quantity zw/kT is often called the Flory interaction parameter χ, but there are other definitions as well. Careful experiments have shown χ to be a function of temperature, composition, and molecular weight distribution of the polymer [9]; thus it has lost its simplistic classification of "parameter."

2.1.5 Polymer–Polymer Systems

The conceptual changes on going from a solvent (1) + polymer (2) system to a polymer (1) + polymer (2) system can be either trivial or immense. The lattice concept cannot be rejected offhand although it will be difficult to use a lattice of "solvent" size without leaving many vacant sites, because both components must retain their chain character. But, as we have seen, lattice parameters do not enter directly and Eq. (2.8) can be written in terms of an interacting segment volume V_s which conveniently preserves the interaction energy w given in Eq. (2.5) at around the same value [6]. Then

$$\Delta H^M = Vzw\phi_1\phi_2/V_s, \qquad \Delta S^M = -k[N_1 \ln \phi_1 + N_2 \ln \phi_2] \quad (2.12)$$

Combination to give the free energy on a volume basis yields

$$\frac{\Delta G^M}{V} = +kT\left[\frac{\phi_1}{V_1} \ln \phi_1 + \frac{\phi_2}{V_2} \ln \phi_2\right] + \frac{zw}{V_s}\phi_1\phi_2 \quad (2.13)$$

Note that the magnitude of the combinatorial entropy is now decreased in *both* terms because of the polymer volume V_i and that it makes a negligible contribution to the free energy of mixing. The value of V_s as well may be quite large compared with typical solvent molecular volumes because the connectivity of polymer chains leads to the exclusion of neighboring polymer molecules from the domain of others, unless considerable organization is allowed. Organization usually occurs if the molecules are strongly interacting and leads to the qualitative conclusion that extremes in polymer–polymer solubility might be expected (and are indeed found). The short-range nature of intermolecular forces and the chainlike nature of polymers suggest that interacting *surfaces* and interacting segments, instead of volumes, would be more appropriate in polymer systems. This will be discussed in Sections 2.4 and 2.5. The origin of the molecular forces generating the interaction term w will be discussed briefly in the following section.

2.1.6 Molecular Forces of Attraction

a. Random Dipole-Induced Dipole. This type [7, 10, 11, 12] of interaction is possible between any two molecules, regardless of structure, because the

2.1. General Thermodynamic Principles

only requirement is the ground-state oscillation of charge in the molecule. The result of such oscillation is a temporary dipole moment, which immediately induces dipoles in all other neighboring molecules. These dipoles then interact. Polarizability, the susceptibility of a molecule to charge separation, is the key parameter in this type of interaction.

London is credited [7] with having described the magnitude of the interaction energy due to these temporary dipoles. His relationship is

$$U = -3I_1 I_2 \alpha_1 \alpha_2 / 2(I_1 + I_2) r^6 \qquad (2.14)$$

with zero as the reference at infinite separation. Here U is the energy, I is the ionization potential, α is the polarizability, and r is the distance between the molecules. Note that the interaction drops off rapidly with distance, a fact that makes energetics very sensitive to volume changes. This interaction is known commonly as the London or dispersion force.

The polarizability of a molecule can be estimated by the Lorentz equation,

$$\alpha = [(n^2 - 1)/(n^2 + 2)](3V/4\pi N) \qquad (2.15)$$

where n is the refractive index, V is the molar volume, and N is Avogadro's number. Aromatic rings, unsaturation, and halogen substitution tend to increase the polarizability. Polarizability is, in general, a tensor, and pairs of molecules with nonspherical symmetry will exhibit some angular dependent interaction energy.

b. Dipole-Induced Dipole. If one component of a solution has a permanent dipole moment, it will induce a dipole in neighboring symmetrical molecules, leading to an interaction. Debye derived the expression for this type of interaction, often called induction,

$$U = -(\alpha_1 \mu_2^2 + \alpha_2 \mu_1^2)/r^6 \qquad (2.16)$$

where μ is the dipole moment. Note again the rapid drop-off of potential with distance.

The effective dipole moment of a complex molecule is not easy to estimate because of its vectorial nature. Most often it is measured in solution or gas phase from the change with temperature of an electromagnetic dispersion of suitable frequency. McClellan [13] gives an extensive list of dipole moments. The dipole moments for polymers can be measured but the accuracy of the measurements is usually not good.

The magnitude of inductive energy is generally smaller than the dispersion and the electronic (dipole–dipole) interactions by about an order of magnitude [14].

c. Dipole–Dipole. The energy due to the interaction of randomly oriented dipoles is given by the following expression.

$$U = -2\mu_1^2 \mu_2^2 / 3kTr^6 \qquad (2.17)$$

Note again the dependence on the inverse sixth power of the separation but also the inverse temperature dependence. The distance dependence can be reduced to the third power if alignment of the dipoles should occur; they will obviously do so if temperature is sufficiently low, the dipole moments are strong, and there are no geometric restraints. From Eq. (2.17) it can be seen that dipole–dipole interactions can be in considerable error if assumptions concerning the magnitude of the dipole moment or its location are wrong.

At this point, some general comments about the $1/r^6$ (van der Waals) *interaction* is in order. The dipole–dipole and dispersion interactions are even functions of material parameters, except for the essentially constant ionization potential, indicating that the molecules with these types of interactions can be expected to obey the geometric mean rule:

$$\varepsilon_{12} = \sqrt{\varepsilon_{11}\varepsilon_{12}} \tag{2.18}$$

The inductive force is not in this category, being an odd function.

The $1/r^6$ dependence emphasizes a significance of the constant volume of mixing approximation in that the geometric mean rule will not hold if the interaction distance changes on mixing. However, positive or negative deviations of large magnitude are possible because of the strong dependence of the interaction energy on the spacing between the segments.

d. Ion–Dipole Interactions. This interaction is responsible for the solubility of polyacrylonitrile and polyamides in salt solutions. Ions can also induce a dipole in an isotropic molecule, leading to a weaker interaction. The energies involved are

$$U = -q_1\mu_2/r^2 \tag{2.19}$$

$$U = -q_1^2\alpha_2/2r^4 \tag{2.20}$$

Note that these forces are quite long range in comparison to the van der Waals attraction, but that the effective value of the charge q can be quite low because of the presence of counterions.

e. Hydrogen Bonding. The prerequisites for a hydrogen bond [15] of significant strength are:

(1) a hydrogen atom covalently bound to an electron-withdrawing atom
(2) a structure with donatable electrons as the acceptor must situate at about 180° with respect to the first bond

This geometric consideration is quite important, the energy falling off rapidly with angle. The strength of the bond is increased as these factors become more favorable to a maximum of about 13 kcal/mol, far more than the energies

2.1. General Thermodynamic Principles

discussed so far. Indeed, hydrogen bonding is perhaps responsible for more miscible systems than any of the other types of interactions so far discussed. As an example, the large number of miscible systems in which PVC is a component can probably be ascribed to the donating character of the Cl—C—H groups shown below in an interacting situation with polycaprolactone [16].

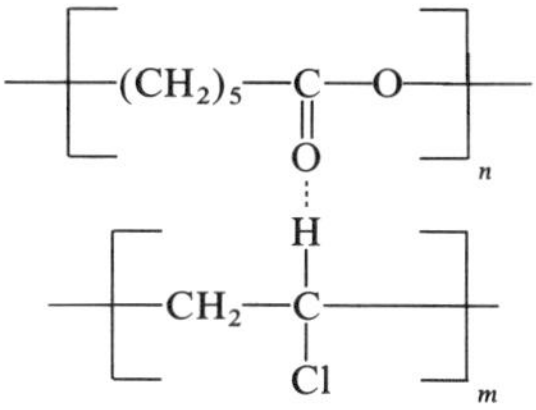

$$\left[-(CH_2)_5-\underset{\underset{O}{\|}}{C}-O-\right]_n$$

$$\left[-CH_2-\underset{\underset{Cl}{|}}{\overset{H}{C}}-\right]_m$$

The miscibility of urethane segments in the soft, polyester, or polyether phase of segmented polyurethane has been associated with hydrogen bonding involving the urethane NH and the ether or carbonyl groups of the soft segment [17]. The striking case of poly(ethylene oxide)–poly(acrylic acid) [18] is undoubtedly an example of multiple hydrogen bonding, depicted below.

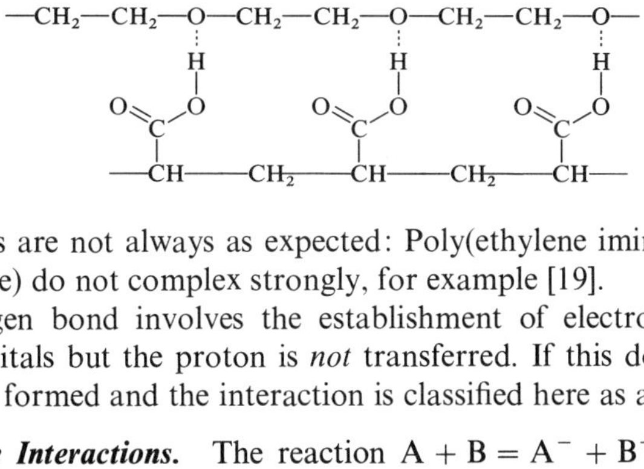

$$-CH_2-CH_2-O-CH_2-CH_2-O-CH_2-CH_2-O-$$

(structure showing H-bonds to three carboxylic acid groups: O=C(O)(OH) linked by —CH—CH₂—CH—CH₂—CH—)

But the results are not always as expected: Poly(ethylene imine) and poly(ethylene oxide) do not complex strongly, for example [19].

The hydrogen bond involves the establishment of electron-containing molecular orbitals but the proton is *not* transferred. If this does occur, an organic salt is formed and the interaction is classified here as acid–base.

f. Acid–Base Interactions. The reaction $A + B = A^- + B^+$ is common among the low molecular weight organic compounds. Adipic acid reacted with hexamethylene diamine to form the salt precursor to Nylon 6,6 is a famous example. Oddly enough, the occurrence of this type of reaction in polymer–polymer systems is not well documented. Complexes of acid and base polymers have been made, but the bonding has been hypothesized to be hydrophobic—the interaction of alkyl groups—or merely hydrogen bonding [20–23]. Clear cases of ion interaction are obtained if *neutralized* acid and base polymers or polyions are mixed in water.

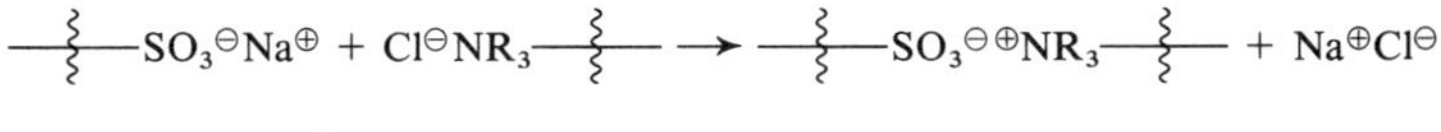

$$-SO_3^{\ominus}Na^{\oplus} + Cl^{\ominus}NR_3 - \longrightarrow -SO_3^{\ominus \oplus}NR_3- + Na^{\oplus}Cl^{\ominus}$$

An example [24] is:

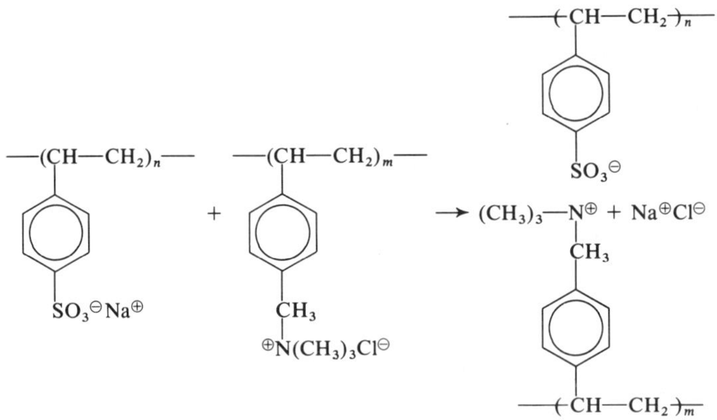

[Chemical reaction scheme showing sulfonated polystyrene sodium salt + quaternary ammonium chloride polystyrene → ion-paired copolymer + Na⊕Cl⊖]

It also suffices to mix a polyion with unneutralized acid or base:

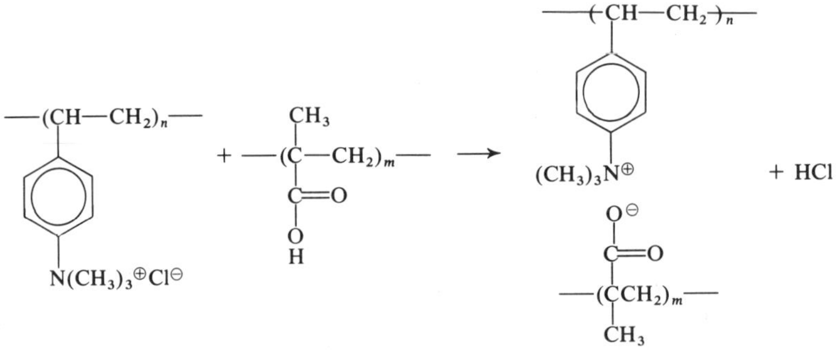

[Chemical reaction scheme showing quaternary ammonium chloride polystyrene + poly(methacrylic acid) → ion-paired product + HCl]

g. Charge Transfer. This term is used to describe many types or degrees of electronic delocalization but is generally considered to mean the existence of a *two-molecule* entity with complete removal of one electron from the orbitals of one molecule to those of the other. The interaction is thus electrostatic in nature. Usually color changes accompany the interaction and the complex exhibits paramagnetism.

Sulzberg and Cotter [25] have provided some examples of this type of interaction in polymer–polymer systems. One is shown below, where the nitro-containing polymer is the electron acceptor.

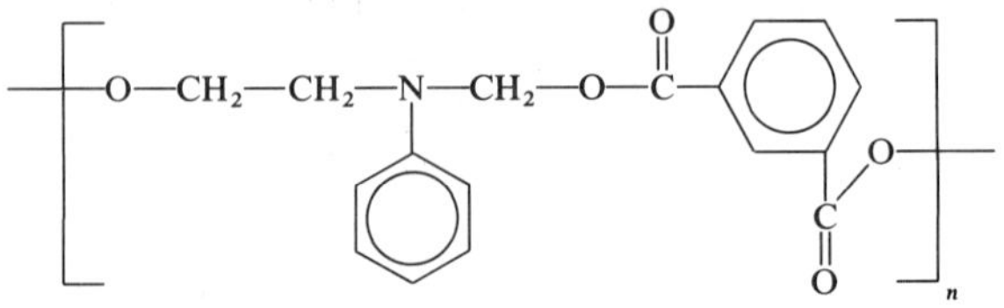

[Structure of polymer with repeat unit containing –O–CH₂–CH₂–N(phenyl)–CH₂–O–C(=O)–aryl–O–C(=O)– shown with subscript n]

2.2. Phase Separation Phenomena

$$\left[-\!\!-\!\!\mathrm{O}\!-\!(CH_2)_6\!-\!\mathrm{O}\!-\!\overset{\overset{\displaystyle O}{\|}}{C}\!-\!\!\underset{\underset{\displaystyle O}{\|}}{\overset{NO_2}{\underset{C}{\bigcirc}}}\!-\!\mathrm{O}\!-\!\!- \right]_m$$

After mixing these two polymers a colored complex was formed, characterized by a significant shift of the maximum ultraviolet (uv) absorption to higher wavelengths and enhanced mechanical properties. In addition, several of the mixtures showed higher electrical conductivities than the component polymers.

The reader must be cautioned here that the resolution of molecular interactions into the types mentioned above is not intended to apply exactly to the referenced polymer–polymer system; it is to aid in the discussion of probable reasons for miscibility. With polymers, as opposed to low molecular weight compounds, the classification of an interaction can be a severe problem due partly to the extra difficulty of obtaining optimum spectroscopic results. In addition, most interactions are not quite clear-cut; there is a gray area of intermediate cases, particularly when aromatic groups or conjugated double bonds are involved.

2.2 PHASE SEPARATION PHENOMENA

Phase separation phenomena in miscible polymer liquid are generally brought about by variations in temperature, pressure, and/or composition of the mixture. The two major classes of phase transitions in a dense phase are (i) liquid–solid phase transition and (ii) liquid–liquid phase transition. The first is exemplified by vitrification of one or all of the chemically different equilibrium liquid phases and by crystallization of one or all of the chemically different equilibrium liquid phases. In the second, the final stage of the system is dictated by the combined effects of the thermodynamics of the system (i.e., the extension, location, and nature of the miscibility gap in the temperature–composition behavior) and the flow properties of the different polymers. While the mechanism of liquid–solid phase separation is generally the classic one of nucleation and growth, for liquid–liquid the mechanism depends on the state of the thermodynamic stability of the system [26, 27]. In one region, nucleation and growth predominate; in another, spinodal decomposition is the mechanism of the phase transformation [26–31].

In order to gain some insight into the criterion of stability which ultimately determines the mechanism of the phase transition, let us consider a mechanical analog to the problem of phase stability. Figure 2.3 illustrates four

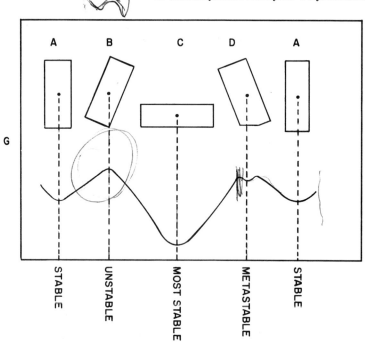

Fig. 2.3. Schematic illustration of four states of stability of a block on a table.

possible states of a block placed on a table. The potential energy curve traces the points corresponding to the center of gravity as the block passes through the successive states.

At state A, the block is stable because if it is subjected to a small finite displacement it returns back to state A as soon as the force producing the displacement is removed. At state B, the block is unstable because it does not return to state B if it is subjected to an infinitesimal displacement. It would instead proceed to the next more stable state of lower energy. State C is the state of lowest energy and, hence, the most stable state. The block is stable to a relatively larger displacement at state C than it is at state A.

State D is metastable. It is produced by cutting off the corner edge of the block and this leads to the dimple in the region of the maximum of the potential energy. A block at state D can survive finite displacements whose value depends on the height of the surrounding energy barriers; this is unlike state B, which is incapable of sustaining an infinitesimal displacement.

As we pass from this simple mechanical analog to the complex, homogeneous, two-component mixture, we consider the free energy–composition diagram in order to illustrate the stability of the system [31].

Using Figs. 2.4 and 2.5, consider the free energy of a mole of homogeneous

2.2. Phase Separation Phenomena

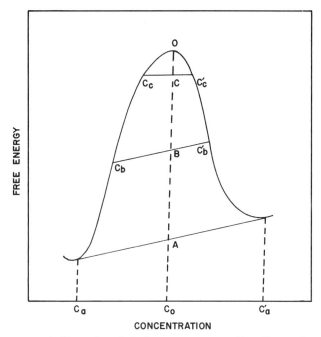

Fig. 2.4. Schematic illustration of the free energy–composition diagram for an unstable phase (nonactivated).

fluid of composition C_o. This should be compared to the free energy of a mole of fluid separated into two liquid phases differing in composition, but whose average composition is also C_o. The tie-lines C_a–C_a', C_b–C_b', and C_c–C_c' connect the points on the free energy curves representing the two compositions of the nonhomogeneous system whose average composition is C_o. The molar free energy of the mixture is the value of G where the tie-line crosses the average composition, C_o.

For Fig. 2.4, the free energy of the homogeneous solution of composition C_o is marked by O. Any small perturbation in composition about C_o resulting in any two phases also results in a lower free energy than the original homogeneous system. Hence, such a separation, like state B in Fig. 2.3, proceeds spontaneously and continuously through the successive phases represented by the tie-lines C_c–C_c' and C_b–C_b' until the lowest free energy is achieved at C_a–C_a'. Consequently, once the phase separation has occurred, the system cannot homogenize spontaneously because that would require an uphill climb from the free energy well.

These occurrences are made possible by the fact that the free energy curve around O is concave downward. The inflection points mark a change in the

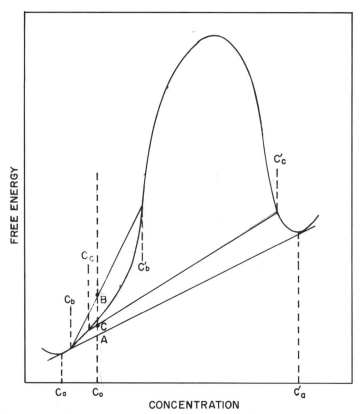

Fig. 2.5. Schematic illustration of the free energy–composition diagrams for a metastable phase (activated).

curvature of the free energy curve and, therefore, bound the range of spontaneous separation. This boundary is defined as the spinodal, and the spontaneous separation is referred to as spinodal decomposition.

In Fig. 2.5, any small perturbation in composition about C_o that results in the appearance of two phases also results in higher free energy. The composition change must be rather large in the right-hand direction before phase separation is accompanied by a free energy reduction. Thus, phase C_o, like state D in Fig. 2.3, is metastable and the minimum increase in free energy needed to render the system unstable is defined as its activation energy. This is generally the basis of the nucleation theory. The limit of metastability is the lowest free energy represented by the tie-line C_a–C_a', and it is defined as the binodal.

The consideration above is limited to fixed temperature and pressure. If for a given pressure the free energy–composition diagram is measured as a

2.2. Phase Separation Phenomena

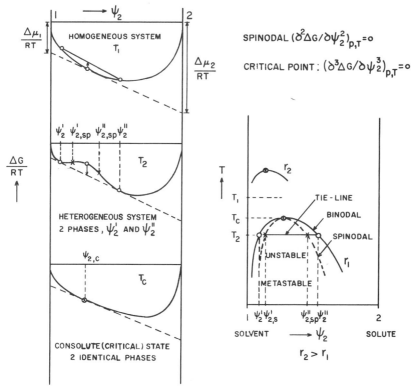

Fig. 2.6. Free energy of mixing as a function of concentration in a binary liquid system showing partial miscibility. [Reprinted by permission from R. Koningsveld, L. A. Kleintjens, and H. M. Schoffelers, *Pure Appl. Chem.* **39**, 1 (1974). Copyright by International Union of Pure and Applied Chemistry.]

function of temperature, the loci of the limits of stability and of metastability yield the spinodal and binodal curves shown in Fig. 2.6. Also, the figure schematically represents the essential concepts discussed above.

2.2.1 Nucleation and Growth

Nucleation is the process of generating within a metastable mother phase the initial fragments of a new and more stable phase [32]. This initial fragment is called a nucleus and its formation requires an increase in the free energy as illustrated in Fig. 2.5. From Fig. 2.6, this means that there must be a finite undercooling into the binodal region in order for a nucleus to develop. Hence, an analogy can be drawn to chemical reaction kinetics. Nucleation is an activated process which forms unstable intermediate embryos; the critical rate-determining embryo is called a nucleus.

A nucleus is different from an equal number of nearest neighbor molecules because it possesses an excess of surface energy which produces the aggregate as a new phase. This excess energy is, according to Gibbs [26], the work of forming a "fluid of different phase within any homogeneous fluid." There are two contributions to this work: work spent in forming the surface, and work gained in forming the interior mass. In the absence of any strong specific interaction, the total activation energy is given by Gibbs [26] as

$$W = \tfrac{1}{3}\gamma S \tag{2.21}$$

where γ is the interfacial tension and S is the surface area of the nucleus of the new phase produced.

Once these nuclei are formed, the system decomposes with a decrease in free energy, and the nuclei grow. This growth process as well as the corresponding phase structure can be represented by Fig. 2.7.

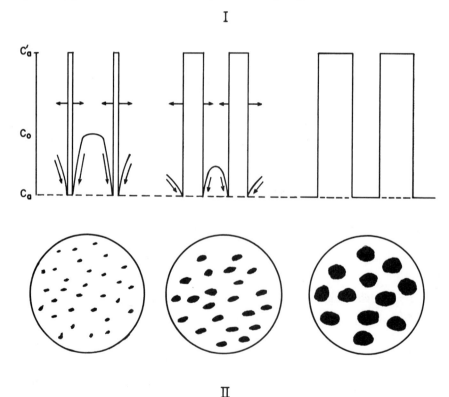

Fig. 2.7. Schematic illustration of a phase separation by the nucleation and growth mechanism: (I) one-dimensional evolution of concentration profiles; (II) two-dimensional picture of the resultant phase structure.

2.2. Phase Separation Phenomena

Given a homogeneous solution of composition C_o, if a nucleus of composition C_a' is formed the composition of the mother phase in the immediate vicinity of that nucleus would be C_a, in accordance with the thermodynamics of the two-phase system. A little distance away from the nucleus, the matrix concentration remains essentially at C_o. Because of the activated nature of nucleation itself, those individual molecules making up the nucleus are held strongly together and are unable to diffuse away. However, the individual

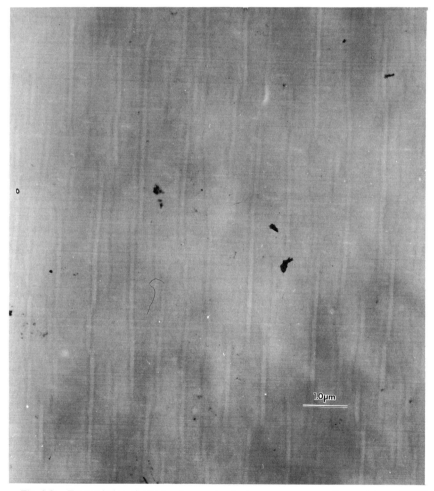

Fig. 2.8. Transmission electron micrograph of a homogeneous sample: a mixture of 25% PMMA and 75% styrene/acrylonitrile copolymer (28% acrylonitrile); $T = 265°C$. [Reprinted by permission from L. P. McMaster, *Adv. Chem. Ser.* **142**, 43 (1975). Copyright by the American Chemical Society.]

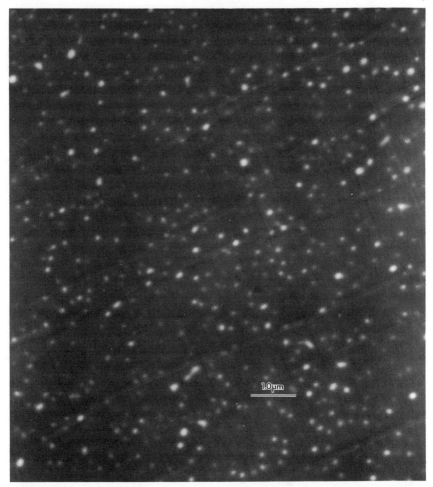

Fig. 2.9. Transmission electron micrograph of a dispersed phase structure: a mixture of 25% PMMA and 75% styrene/acrylonitrile copolymer (28% acrylonitrile) exposed to a constant electron beam intensity for 3.4 min. [Reprinted by permission from L. P. McMaster, *Adv. Chem. Ser.* **142**, 43 (1975). Copyright by the American Chemical Society.]

molecules within the mother phase with the concentration C_o would diffuse into the phase with the lower concentration C_a (i.e., downhill diffusion). The chemical potential gradient thus set up helps to feed the young nucleus. This is the growth mechanism. The concentration within the nucleus remains constant at C_a' and that of the second phase at C_a during the growth process; but the interfaces between the two phases move with time. This is illustrated in Fig. 2.7.

2.2. Phase Separation Phenomena

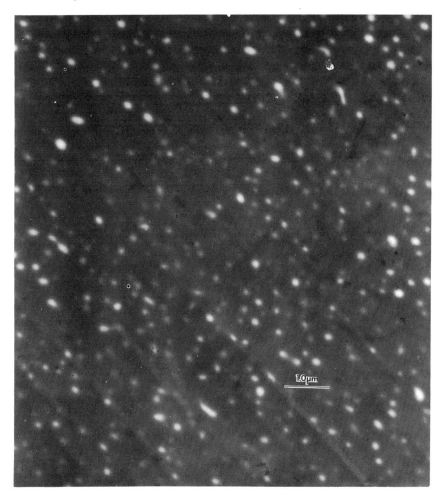

Fig. 2.10. Transmission electron micrograph of a dispersed phase structure: a mixture of 25% PMMA and 75% styrene/acrylonitrile copolymer (28% acrylonitrile) exposed to a constant electron beam intensity for 12.4 min. [Reprinted by permission from L. P. McMaster, *Adv. Chem. Ser.* **142**, 43 (1975). Copyright by the American Chemical Society.]

At the same period of time, other nuclei grow within the same mother phase and a finely dispersed two-phase system results. The final droplet sizes and the distances between them depend on the time scale of the experiment and on the rate of diffusion. Normally, the diffusion rate decreases with decreasing temperature while the rate of nuclei formation increases; the net result is a maximum rate of growth several degrees below the binodal. Assuming, therefore, that the diffusion rate is finite, the droplet increases in

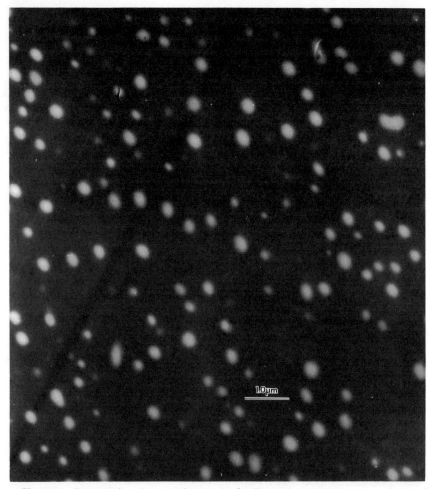

Fig. 2.11. Transmission electron micrograph of a dispersed phase structure: a mixture of 25% PMMA and 75% styrene/acrylonitrile copolymer (28% acrylonitrile) exposed to a constant electron beam intensity for 122.4 min. [Reprinted by permission from L. P. McMaster, *Adv. Chem. Soc.* **142**, 43 (1975). Copyright by the American Chemical Society.]

size initially by growth and later by coalescence, coarsening or ripening until there are two large phases, one of composition C_a' and the other of composition C_a.

2.2.2 Experimental Evidence in Miscible Polymer Mixtures

While phase growth in alloys of metals and in mixtures of inorganic glasses was studied in the early 1900s, the first semiquantitative study of phase growth in a miscible polymer system was carried out by McMaster [33].

2.2. Phase Separation Phenomena

He investigated the blend of styrene/acrylonitrile copolymer (28% AN) with poly(methyl methacrylate). Figures 2.8–2.11 are transmission electron micrographs of the dispersed phase structure of 75% SAN/25% PMMA blends at 265°C. The initial homogeneous blend appears in Fig. 2.8; succeeding photographs show the same composition after exposure times of 3.4, 12.4, and 122.4 min. The evolving structure is a dispersed phase of PMMA-rich droplets in the SAN-rich matrix. From studies of the experimental cloud-point curves, it was also established that at 265°C the blend is in the metastable region.

The time dependence of the volume average droplet size was correlated, with the Lifshitz–Slyozov's semiquantitative expression for Ostwald ripening [27, 34]:

$$d^3 = d_o^3 + 64\gamma(X_e V_m)Dt/9RT \qquad (2.22)$$

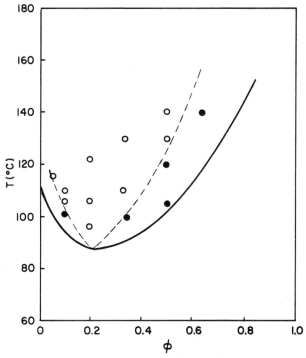

Fig. 2.12. Temperature–composition phase diagram for polystyrene–poly(vinyl methyl ether) mixtures. Phase separations by what appears to be spinodal mechanism (○) and nucleation and growth mechanism (●) are observed under a microscope. The dashed curve represents the line of demarcation between the two morphologies. [Reprinted by permission from T. Nishi, T. T. Wang, and T. K. Kwei, *Macromolecules* **8**, 227 (1975). Copyright by the American Chemical Society.]

where d_o is the droplet diameter after bulk thermodynamic equilibrium, γ is the interfacial tension between droplet and matrix phases, X_e is the equilibrium mole fraction of the droplet-rich component in the matrix phase, V_m is the molar volume of the droplet phase, and D is the diffusion coefficient for the matrix phase.

From the calculated diffusion coefficient and the experimentally measured coarsening rate, it was concluded that the mechanism of coarsening for the dispersed phase structure is Ostwald ripening whereby larger droplets grow at the expense of the smaller ones.

Another example of dispersed phase growth in polymeric systems was provided by Nishi *et al.* [35], who studied the thermally induced phase separation behavior of polystyrene–poly(vinyl methyl ether). Figure 2.12 shows several compositions and temperatures where phase separation was observed. The filled circles represent regions in which nucleation and growth are the mechanism of phase separation, whereas the open circles represent spinodal decomposition regions. The solid line is the cloud-point curve and the dashed curve represents the line of demarcation between the two modes of separation. The occurrence of the filled circles near the cloud-point curve, but away from the minimum, lends further credence to the existence of these two modes of phase separation phenomena in multicomponent polymeric mixtures.

2.2.3 Spinodal Decomposition

Spinodal decomposition is a kinetic process of generating within an unstable mother phase a spontaneous and continuous growth of another phase. The growth originates, not from nuclei, but from small amplitude composition fluctuations which statistically promote continuous and rapid growth of the sinusoidal composition modulation with a certain maximum wavelength [31, 36]. Thus, the decomposed system is characterized at some point by a high level of phase interconnectivity in both the minor and major phases—a structure which could possess some rather unusual mechanical and permeability characteristics. The absence of interconnectivity in a phase-separated system need not imply nucleation and growth as the mechanism of decomposition; an originally interwoven structure *can* coarsen to a somewhat dispersed structure. Likewise, the presence of interconnecting phases could be the result of shearing during the nucleation and growth mechanism. For this reason, the most acceptable demonstration of a spinodal decomposition mechanism is spinodal kinetics, namely, an observation of continuous growth of certain small amplitude composition modulation. This growth process and the corresponding phase structure are shown schematically in Fig. 2.13.

2.2. Phase Separation Phenomena

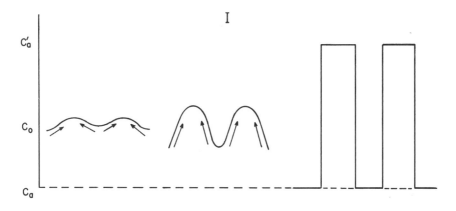

Fig. 2.13. Schematic illustration of phase separation by the spinodal mechanism: (I) one-dimensional evolution of concentration profiles; (II) two-dimensional picture of the resultant phase structure.

Consider a multicomponent system of composition C_o with a concentration fluctuation of component "X" where the individual molecules tend to join favorable permanent clusters. If the molecules of component "X" continue to diffuse uphill from the low concentration region surrounding the statistical fluctuation into the cluster, the system will spontaneously and continuously decompose into two phases of compositions C_a and C_a' as shown in Fig. 2.13. This system is in the unstable state. Neighboring clusters would develop because molecules of component "X" will move in the direction of the clusters since this high concentration region is a low energy region. By this mechanism, an oscillating concentration field is set up and the wavelength of the oscillation will be fixed by the scale of the phase structure which, in turn, is fixed by the minimization condition of the free energy. This scale is invariant during the growth process even though the amplitudes of the concentration fluctuation increase. The characteristic phase structure appears in Fig. 2.13. However, as the concentration gradient approaches its plateau, the system tends to minimize its interfacial free energy by minimizing the amount of interface area. Since a sphere contains the most volume and the least surface area, the interwoven structures coarsen by an interfacial–energy-driven viscous flow mechanism. Even though a spherically dispersed structure may not result, the previous regular order breaks down and, eventually, two layers, one on top of the other, may result if the viscosities of the two phases are sufficiently low. If the viscosities are relatively high, the final structure may bear some resemblance to the original interwoven structures. In summary, the essential characteristic features of the spinodal mechanism are [31, 36, 37]:

(1) It is an unstable process; it requires no activation energy.

(2) The order parameter which describes the spinodal system, namely, the local concentration, obeys a continuity equation and therefore is limited by diffusion, the diffusion coefficient being negative.

(3) It is an isothermal process.

(4) The process occurs coherently. For example, in a liquid crystalline system the two coexisting phases would exist on the same lattice. However, for amorphous isotropic materials, the interwoven structures of the two coexisting phases would be *uniform yet random*.

2.2.4 Kinetics of Spinodal Decomposition in Polymer Mixtures

Thermodynamic theories of nonhomogeneous solutions exist in two formulations given by Cahn [37] and by Debye [38]. Derived from quite different bases, both yield identical results for the free energy contribution due to concentration fluctuations in a nonhomogeneous system. Furthermore, each theory could be considered from two points of view. One relies purely on phenomenological, irreversible thermodynamics; the other is a statistical approach which takes cognizance of the statistical fluctuations caused by the interactions between the system and its thermal reservoir. In what follows, we consider the formalism of Cahn and Hilliard [37, 39].

The local chemical potential difference caused by concentration fluctuations in a homogeneous binary system is given by

$$(\mu_1 - \mu_2) = \partial G'/\partial \psi_1 \qquad (2.23)$$

where G' is the free energy density for the homogeneous system and ψ_1 is the segment fraction of component 1. In a nonhomogeneous binary sys-

2.2. Phase Separation Phenomena

tem, a finite concentration gradient necessarily exists and a gradient energy term must be added to the local chemical potential difference. This energy describes the extra energy associated with departures from uniformity:

$$\mu_1 - \mu_2 = (\partial G'/\partial \psi_1) - 2K\nabla^2 \psi_1 \qquad (2.24)$$

K is an empirical constant called the gradient energy coefficient because it multiplies the term which reflects the influence of the concentration gradient on the local free energy. This local free energy difference causes an interdiffusion flux j, and the associated thermodynamic driving force is

$$\Omega \nabla(\mu_1 - \mu_2) = -\mathbf{j} = \Omega\left\{\frac{\partial^2 G'}{\partial \psi_1^2}\nabla\psi_1 - 2K\nabla^3\psi_1\right\} \qquad (2.25)$$

where Ω is the Onsager-type phenomenological coefficient, a linear response proportionality coefficient between a flux and a gradient in chemical potential. When this flux relationship is substituted into the equation of continuity, the Cahn–Hilliard linearized result is obtained [37, 39]:

$$\frac{\partial \psi_1}{\partial t} = \nabla \cdot \Omega \nabla(\mu_1 - \mu_2) = -\nabla \cdot \mathbf{j} = \Omega\left\{\frac{\partial^2 G'}{\partial \psi_1^2}\nabla^2\psi_1 - 2K\nabla^4\psi_1\right\} \qquad (2.26)$$

This should be compared to the conventional diffusion equation:

$$\partial \psi_1 / \partial t = D\nabla^2 \psi_1 \qquad (2.27)$$

Thus, for relatively long diffusion distances, the diffusion coefficient in spinodal decomposition is equivalent to

$$D \sim \Omega \partial^2 G'/\partial \psi_1^2 \qquad (2.28)$$

a negative quantity in the unstable region since $\partial^2 G'/\partial \psi_1^2 < 0$. The general solution of the Cahn–Hilliard relation is

$$\psi_1 - \psi_{1,0} = \sum_{\text{all }\beta} \{\exp[R(\beta)t]\}\{A(\beta)\cos(\beta \cdot \mathbf{X}) + B(\beta)\sin(\beta \cdot \mathbf{X})\} \qquad (2.29)$$

where $\lambda_i = 2\pi/\beta_i$, and the growth rate $R(\beta)$ is given by

$$R(\beta) = -\Omega\beta^2\left\{\frac{\partial^2 G'}{\partial \psi_1^2} + 2K\beta^2\right\} \qquad (2.30)$$

Because K is always positive, $\partial^2 G'/\partial \psi_1^2$ governs the sign of the growth rate. Positive values yield negative values of $R(\beta)$ and all instabilities rapidly decay. As $\partial^2 G'/\partial \psi_1^2$ crosses zero and becomes negative, some wavelengths, notably the long wavelengths, will grow. As it becomes more and more negative, much shorter wavelengths become capable of growing. This property of the Cahn–Hilliard solution can be more easily ascertained if the growth rate is plotted as a function of wavelength. For certain values of $\partial^2 G'/\partial \psi^2$ and K, $R(\beta)$ exhibits a maximum and, because of the exponential nature of the Cahn–Hilliard solution, the scale of the spinodal decomposition is dominated by this most rapidly growing wavelength, given as

$$\lambda_m = \frac{2\pi}{\beta_m} = 2\sqrt{2}\pi\left[\left(-\frac{1}{2K}\right)\frac{\partial^2 G'}{\partial \psi_1^2}\right]^{-1/2} \quad (2.31)$$

It is this mode which Cahn and Hilliard have identified as characterizing the initial spinodal instability and determining the scale of the emerging precipitate.

In order to make a quantitative connection between this analysis and experiment, van Aarsten [40] utilized the Flory–Huggins theory to give

$$\frac{\partial^2 G'}{\partial \psi_1^2} = RT\left[\frac{1}{\psi_1 V_1} + \frac{1}{\psi_2 V_2} - \frac{2\chi_{12}}{V_1}\right] \quad (2.32)$$

By assuming that $\chi_{12} = \chi_{12,s} + \chi_{12,h}/T$, he simplified the expression to

$$\frac{\partial^2 G'}{\partial \psi_1^2} = \frac{2RT\chi_{12,h}}{V_1}\left(\frac{T - T_s}{T_s}\right) \quad (2.33)$$

where T_s is the temperature at the spinodal, V is the molar volume, $\chi_{12,s}$ is the entropic and $\chi_{12,h}$ the non-athermal contribution to the Flory–Huggins interaction parameter. From Debye's thermodynamic theory of nonhomogeneous solutions, the gradient energy coefficient is given by [38]:

$$K = (RT\chi_{12}/V_1)l^2/6 \quad (2.34)$$

where l is the range of molecular interaction which, for polymer–polymer mixtures, is given by $l = \bar{R}/\sqrt{3}$; $\bar{R}$ is the root mean square end-to-end distance of the polymer chain.

If $\bar{R}$ is assumed to be approximately the same for the component polymers and if the entropic contribution to χ_{12} is neglected, the scale of phase separation is related to the maximum wavelength through

$$\lambda_m/l = 2\pi[3|T - T_s|/T_s]^{-1/2} \quad (2.35)$$

Nishi, Wang, and Kwei [35] followed essentially the analysis above, coupled with morphological, nuclear magnetic resonance (nmr), and interaction parameter data, in order to estimate the parameters which describe the kinetics of spinodal decomposition for polystyrene–poly(vinyl methyl ether) mixtures. The values obtained were:

$$\beta_m = 5.24 \times 10^4 \quad cm^{-1}$$
$$D = -2.8 \times 10^{-13} \quad cm^2/sec$$

$$l = 580 \text{ Å}$$
$$K = 8.9 \times 10^{-11} \quad \text{cal-cm}^2/(\text{cm}^3 \text{ of PVME})$$
$$\partial^2 G'/\partial \psi_1^2 = -0.97 \quad \text{cal}/(\text{cm}^3 \text{ of PVME})$$

These should be compared to the typical values for metallic systems [41]:

$$D \sim -10^{-18} \text{ cm}^2/\text{sec}$$
$$K \sim 2.41 \times 10^{-13} \text{ cal/cm (atomic fraction)}^{-2}$$
$$\partial^2 G'/\partial \phi_1^2 \sim -2.41 \times 10^{-18} \text{ cal/cm}^3 \text{ (atomic fractions)}$$

It is to be noted, therefore, that the values are distinctly different while the signs are the same. A similar attempt at describing the kinetics of spinodal decomposition for polymers was made by McMaster [33]. He followed a similar analysis, as did Nishi *et al.* [35], to arrive at the following values for the binary styrene/acrylonitrile copolymer–poly(methyl methacrylate) system:

$$7.85 \times 10^4 \text{ cm}^{-1} \leq \beta_m \leq 3.14 \times 10^6 \text{ cm}^{-1}$$
$$40 \text{ Å} \leq l \leq 400 \text{ Å}$$
$$D \sim -10^{-14} \text{ cm}^2/\text{sec}$$

which are the same order of magnitude as those of Nishi *et al.* [35].

McMaster [33] went further and applied a modified form of the Tomotika stability analysis to study the coarsening mechanism of the spinodal decomposed structure of SAN–PMMA mixtures. He concluded that, while Ostwald ripening [27, 34] is the coarsening mechanism for the dispersed phase structures, interconnected structures coarsen by an interfacial tension-driven viscous flow mechanism.

2.3 HILDEBRAND APPROACH: THE SOLUBILITY PARAMETER

2.3.1 Historical Background

The solubility parameter was probably first identified by Hildebrand [7] as being a useful quantity for the characterization of the strength of interactions in simple liquids. Its usefulness has been proved experimentally dozens of times and it was only natural that the concept was extended to polymer–solvent systems, particularly by the coatings industry but also by the elastomer industry. The former was interested in providing "compatible" solvent systems for coatings resins, while the latter was more concerned with the unfavorable swelling of cured rubber by solvents. The extension to polymer–polymer systems was popularized by Bohn [8].

The solubility parameter is defined as the square root of the energy of vaporization per unit volume of material and is given the symbol δ [7].

$$\delta \equiv (\Delta E^v/V)^{1/2} \ (\text{cal}/\text{cm}^3)^{1/2} \tag{2.36}$$

Thus, δ is proportional to the cohesion of the material or the strength of attraction between the molecules making up the material. The range of δ for liquids at room temperature is from below 6 (fluorocarbon) to above 30 (mercury). The square of δ^2 is often called the cohesive energy density, for obvious reasons, and is abbreviated as CED. The concept of the solubility parameter, as applied to the properties of solutions, is that materials with a high solubility parameter require more energy for dispersion than is likely to be returned upon mixing with a material of low solubility parameter. This concept will be quantified below.

The attractiveness of the solubility parameter approach, and the source of its universal appeal, has been its ability to characterize a system using only component properties—no *system* parameter is needed. This, in principle, enables the practitioner to pick compatible components without any experiments.

2.3.2 Early Development of the Solubility Parameter Approach

The solubility parameter approach has taken many forms. The simplest presumes that the only source of a net unfavorable interaction is a difference in solubility parameter between the components. Because it is unfavorable, it must be minimized, for example, by selection of an alternate component yielding a lower solubility parameter difference. No attention need be paid to the possibility of a net favorable interaction because it is not a necessity for solution and is difficult to predict. This simplest form is a natural result of the geometric mean assumption, to be discussed in this section.

In its more sophisticated variations the solubility parameter approach can be used, with very few added assumptions, to aid in the calculation of the effects of pressure and temperature on the mixing free energy as well as the probable change of these with structure. Variations of this type will be discussed in turn.

Several general methods are available for determining the solubility parameter of a substance. For liquids which form ideal gases in the vapor phase, the solubility parameter is directly obtained by calorimetry through the relationship

$$\Delta E^v = \Delta H^v - PV = \Delta H^v - nRT \tag{2.37}$$

For polymers, this method is clearly impossible and methods of estimation

2.3. Hildebrand Approach: The Solubility Parameter

must be used. One approach is to calculate the solubility parameter through the internal pressure, P_i [42]:

$$\delta^2 = P_i = \left(\frac{\partial U}{\partial V}\right)_T \cong \frac{T\alpha}{\beta} \tag{2.38}$$

where α is the thermal expansion coefficient (°C^{-1}) and β is the compressibility (cm^3/cal). If the equation of state is known, δ can be calculated as a function of temperature and pressure using this internal pressure approximation. Based on their semiempirical equation of state for polymer melts, Olabisi and Simha [42] have developed a generalized relation for P_i which could be used to calculate δ according to Eq. (2.38).

A method often used for cross-linked polymers and applicable to partially crystalline material such as PVC is based on a maximum in swelling using series of solvents of varying and known solubility parameters. A typical result is shown in Fig. 2.14. The assumption is that the interaction and swelling will be maximum when the solubility parameter of the polymer matches that of the solvents. This may be an inaccurate assumption for systems with polar structures. The same reasoning and cautions are applicable to intrinsic viscosity determination in a series of solvents. The solu-

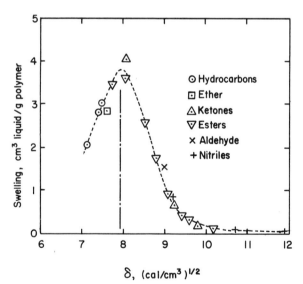

Fig. 2.14. Swelling of natural rubber in solvents with various solubility parameters, δ. The point of maximum swelling can be taken as the solubility parameter of the polymer. [From H. Mark and A. V. Tobolsky, "Physical Chemistry of High Polymeric Systems," p. 262. Copyright 1950 by Wiley (Interscience), New York.]

bility parameter is also calculated through the heats of solution [43, 44], directly or by chromatography, solution behavior [45, 46], and by extrapolation [47]. Figure 2.15 is an example of the latter technique applied to polyethylene. The extrapolated values are in good agreement and also agree with reported values for polyethylene, as shown in the figure.

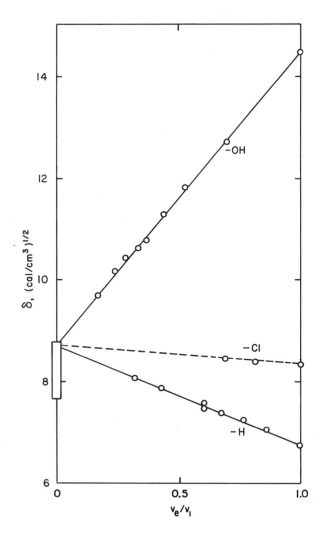

Fig. 2.15. Determination of the solubility parameter of polyethylene by extrapolation of homologous series of normal alcohols (–OH), 1-chloro compounds (–Cl), and alkanes (–H) to pure polymer. The abscissa is the volume of the functional end divided by the total volume of the molecule. [From B. A. Wolf, *Makromol. Chem.* **178**, 1869 (1977). Reprinted by permission of Huthig and Wepf Verlag, Basel.]

2.3. Hildebrand Approach: The Solubility Parameter

Methods for estimating the solubility parameter from group contributions have particular appeal for polymers because of the difficulty of direct determination, and will be discussed in Section 2.33.

The change of the solubility parameter with temperature is likely to be important in polymer systems. An estimate given by Hildebrand is $\partial \ln \delta / \partial T = 1.25\alpha$: A more elaborate dependence is obtainable from the Olabisi–Simha [42] generalized expression for internal pressure. The two relations (δ and its temperature coefficient) depend strongly on the thermal expansion coefficient, α, and this makes it a key property in determining phase behavior in systems with similar δ. An identical conclusion is reached via the equation of state approach to be discussed in Section 2.5.

Applied, in its simplest form, to the mixing of two polymers, the solubility parameter approach follows exactly the regular solution approach for low molecular weight material. The end result is

$$\frac{\Delta H^M}{V} = (\delta_1 - \delta_2)^2 \phi_1 \phi_2; \quad -\frac{T\Delta S^M}{V} = kT\left(\frac{N_1}{V}\ln \phi_1 + \frac{N_2}{V}\ln \phi_2\right) \quad (2.39)$$

where the sum of the left-hand sides is the free energy of mixing per unit volume; V is the total volume.

This relationship is very simple and convenient. It implies knowledge of phase behavior from a knowledge of the components' solubility parameters and their temperature dependencies. The essential assumptions leading from Eq. (2.12) to Eq. (2.39) can be appreciated from the following analysis.

From Eq. (2.12), the enthalpy of mixing is

$$\frac{\Delta H^M}{V} = (zw/V_s)\phi_1 \phi_2 \quad (2.40)$$

where z is the coordination number, w is the energy of interaction or exchange $\frac{1}{2}\varepsilon_{11} + \frac{1}{2}\varepsilon_{22} - \varepsilon_{12}$, and V_s is the volume of each interacting segment. The replacement of zw/V_s in Eq. (2.40) by $(\delta_1 - \delta_2)^2$ in Eq. (2.39) involves two steps.

Step 1: The geometric mean assumption states that the value of ε_{12} will be the geometric mean of ε_{11} and ε_{22}. According to Hildebrand [7], the assumption was made as early as 1898 by Berthelot but it is known commonly as the Scatchard geometric mean assumption. In many systems of small nonpolar or slightly polar molecules this assumption is very good. The consequence of the assumption when applied to polymer–polymer systems of similar polarity is unclear. In some cases it is expected to be *better* than with small molecules of equivalent polarity because geometric constraints discourage maximum dipole–dipole interaction [48]. However, in cases where orderly alignment of dipoles in low energy positions is encouraged by the chain structure, the assumption is expected to fail completely [49]. Such systems are very likely to be tightly bound complexes.

If the geometric mean assumption is made, a perfect square is formed:

$$w = \tfrac{1}{2}\varepsilon_{11} + \tfrac{1}{2}\varepsilon_{22} - \varepsilon_{12}$$
$$\varepsilon_{12} = (\varepsilon_{11}\varepsilon_{22})^{1/2} \text{ by geometric mean assumption} \quad (2.41)$$
$$w = \tfrac{1}{2}[(\varepsilon_{11}^{1/2})^2 - 2\varepsilon_{11}^{1/2}\varepsilon_{22}^{1/2} + (\varepsilon_{22}^{1/2})^2] = \tfrac{1}{2}(\varepsilon_{11}^{1/2} - \varepsilon_{22}^{1/2})^2 \quad (2.42)$$

Step 2: The remaining derivation consists of recalling that $\delta \equiv (E^v/V)^{1/2}$ and therefore of identifying $z(\sqrt{\varepsilon_{11}} - \sqrt{\varepsilon_{22}})^2/2V_s$ with $(\delta_1 - \delta_2)^2$. A more complete derivation of this argument is given by Hildebrand [7].

2.3.3 Refinements of the Solubility Parameter

a. Multidimensional Approach. Van Arkel [50] suggested an approach for explaining the deviations from Eq. (2.39) frequently observed in solutions of polar and nonpolar liquids. Recalling that dipole–dipole interactions are proportional to $\mu_1^2\mu_2^2$, whereas dispersive interactions go as $\alpha_1\alpha_2$, it seems reasonable that polar forces should follow a geometric mean rule. One writes the polar and dispersive interactions as:

$$\varepsilon_{11} = \alpha_1\alpha_1 + \mu_1^2\mu_1^2, \quad \varepsilon_{22} = \alpha_2\alpha_2 + \mu_2^2\mu_2^2, \quad \varepsilon_{12} = \alpha_1\alpha_2 + \mu_1^2\mu_2^2 \quad (2.43)$$

where the proportionality constants have been neglected. Introducing these relations into Eq. (2.41) yields:

$$w = \tfrac{1}{2}(\alpha_1^2 - 2\alpha_1\alpha_2 + \alpha_2^2) + \tfrac{1}{2}(\mu_1^4 - 2\mu_1^2\mu_2^2 + \mu_2^4)$$
$$= \tfrac{1}{2}(\delta_{d1} - \delta_{d2})^2 + \tfrac{1}{2}(\delta_{p1} - \delta_{p2})^2 \quad (2.44)$$

where

$$\delta_{di} \propto \alpha_i, \quad \delta_{pi} \propto \mu_i^2 \quad (2.45)$$

Hence, the heat of mixing becomes

$$\Delta H^M/(N_1V_1 + N_2V_2) = [(\delta_{d1} - \delta_{d2})^2 + (\delta_{p1} - \delta_{p2})^2]\phi_1\phi_2 \quad (2.46)$$

whereby

$$\delta_p = (\delta_{p1} - \delta_{p2}); \quad \delta_d = (\delta_{d1} - \delta_{d2}); \quad \text{and} \quad \delta^2 = \delta_p^2 + \delta_d^2 \quad (2.47)$$

b. Measuring Solubility-Parameter Components. Methods of determining the magnitude of the solubility parameter components in Eq. (2.47) were outlined by Gardon [14] based on the type of intermolecular forces discussed in Section 2.1.6. His method required an estimation of the dipole moment, a decided disadvantage.

2.3. Hildebrand Approach: The Solubility Parameter

Hansen [51, 52] took a different approach, allowing the dispersion contribution to be the same as that for a hydrocarbon of similar structure and volume. (This is known as the homomorph concept, and was also adopted by Blanks and Prausnitz in a two-dimensional approach [53].) This dispersion contribution was then subtracted from the total cohesive energy. The remainder was split into two parts—Hansen labeled them polar and hydrogen bonding, but they are in fact completely empirical—in a clever way. Several polymers were mixed with a large number of solvents, all with known total solubility parameters and homomorphs. The solubilities were measured roughly. The ratio between the hydrogen bonding and polar contributions was then optimized simultaneously for all the solvents to give the best picture of the solubility of all resins. Once the three parameters for the *solvents* were established, it was a simple matter to determine experimentally the solubility parameter components of any polymer by the use of characterized solvents.

Crowley *et al.* [54] took still another approach to the hydrogen bonding parameter, using spectroscopic data to estimate its strength. However, these authors, as well as Hansen, neglect the fundamental point of Small [55] that the hydrogen bonding effect requires at least two parameters, related to donating and accepting character, for its description. This was recently rediscovered by Nelson *et al.* [56], who also improved on the available spectroscopic data.

The solubility parameter components for polymers as measured by the various authors are listed in Table 2.1.

c. Correlation of the Solubility Parameter with Other Physical Properties.
The approximation $\delta = (T\alpha/\beta)^{1/2}$ has already been mentioned in Section 2.3.2. Bagley *et al.* [57] have used direct measurements of the thermal pressure coefficient $(\partial P/\partial T)_v = \alpha/\beta$ to evaluate the solubility parameter of low molecular weight liquids. The difference between their results and those from heat by vaporization was assigned to hydrogen bonding. Olabisi and Simha's [42] generalization is applicable to both polymers and oligomers but it describes only the dispersion contribution.

Correlations of solubility parameter with more readily available pure-component physical properties have been suggested and some have been tested for low molecular weight liquids. The properties include refractive index, surface tension, and the parachor.

Surface tension. Hildebrand [7] obtained the following empirical correlation of δ with surface tension γ:

$$\delta = 4.1(\gamma/V_m^{1/3})^{0.43} \tag{2.48}$$

Koenhen and Smolders [58] have associated surface tension with solubility parameter by a similar relationship:

$$\delta = (13.8\gamma/V_m^{1/3})^{1/2} \tag{2.49}$$

TABLE 2.1

Range of the Experimental Values of the Solubility Parameter and the Values of the Dispersive (d), Polar (p), and Hydrogen Bonding (h) Contributions to the Total Solubility Parameter[a]

Polymer	exp[b]	Components			Ref.
		d	p	h	
Polyethylene	7.7–8.4				
Polypropylene	8.2–9.2				
Polyisobutylene	7.8–8.1	8	1	3.5	[d]
Polystyrene	8.5–9.3	8.6	3.0	2.0	[d]
Poly(vinyl chloride)	9.4–10.8	9.4	4.5	3.5	[d]
Poly(vinyl bromide)	9.5				
Poly(vinylidene chloride)	9.9–12.2				
Poly(tetrafluoroethylene)	6.2				
Poly(chlorotrifluoroethylene)	7.2–7.9				
Poly(vinyl alcohol)	12.6–14.2				
Poly(vinyl acetate)	9.3–11.0	9.3	5.0	4.0	[d]
Poly(vinyl propionate)	8.8				
Poly(methyl acrylate)	9.7–10.4				
Poly(ethyl acrylate)	9.2–9.4	9.2	5.3	2.1	
Poly(propyl acrylate)	9.0				
Poly(butyl acrylate)	8.8–9.1				
Poly(isobutyl acrylate)	8.7–11.0				
Poly(2,2,3,3,4,4,4-heptafluorobutyl acrylate)	6.7				
Poly(methyl methacrylate)	9.1–12.8	9.2	5.0	4.2	[d]
Poly(ethyl methacrylate)	8.9–9.1	9.2	5.3	2.1	[d]
Poly(butyl methacrylate)	8.7–9.0				
Poly(isobutyl methacrylate)	8.2–10.5				
Poly(tert-butyl methacrylate)	8.3				
Poly(benzyl methacrylate)	9.8–10.0				
Poly(ethoxyethyl methacrylate)	9.0–9.9				
Polyacrylonitrile	12.5–5.4				
Polymethacrylonitrile	10.7				
Poly(α-cyanomethyl acrylate)	14.0–14.5				
Polybutadiene	8.1–8.6	8.8	2.5	1.2	[d]
Polyisoprene	7.9–10.0	8.5	1.5	1.5	[d]
Polychloroprene	8.2–9.2				
Polyformaldehyde	10.2–11.0				
Poly(tetramethylene oxide)	8.3–8.6				
Poly(propylene oxide)	7.5–9.9				
Polyepichlorohydrin	9.4				
Poly(ethylene sulfide)	9.0–9.4				
Poly(styrene sulfide)	9.3				
Poly(ethylene terephthalate)	9.7–10.7				
Poly(8-aminocaprylic acid)	12.7				
Poly(hexamethylene adipamide)	13.6				

2.3. Hildebrand Approach: The Solubility Parameter

TABLE 2.1 (Continued)

Polymer	exp[b]	Components			Ref.
		d	p	h	
Poly(dimethyl siloxane)	7.3–7.6[c]				
Cellulose diacetate	10.9–11.4[c]				
Ethyl cellulose	10.3[c]				
Cellulose nitrate (11.83% N)	10.5–14.9[c]				
Poly(vinyl butyral)	11[d]	8.5	4.3	5.5	[d]
Phenolic resin	11.3[d]	9.0	4.0	5.5	[d]

[a] Except as noted, adapted from D. W. Van Krevelen, "Properties of Polymers," 2nd ed., p. 143, Elsevier, Amsterdam, 1976; C. M. Hansen, *J. Paint Technol.* **39**, 104 (1967).
[b] D. W. Van Krevelen, "Properties of Polymers," 2nd ed., p. 143, Elsevier, Amsterdam, 1976.
[c] H. Burrel, in "Polymer Handbook" (J. Brandup and E. H. Immergut, eds.), Vol. IV, p. 337, Wiley, New York, 1975.
[d] C. M. Hansen, *J. Paint Technol.* **39**, 104 (1967).

Bonn and van Aartsen [59] suggested recently that the cube root of the molar volume should be replaced by the square root of the segment surface area when Eq. (2.49) is applied to polymers:

$$\delta \propto \gamma^{1/2}(1/\text{area})^{1/4} \tag{2.50}$$

The exponent 1/2 is claimed to be theoretically derivable. The area to be used is discussed by Siow and Patterson [60]. A good relationship between the critical surface tension and solubility parameter for a number of crystalline polymers has been noted by Gardon [61].

Refractive index. Bonn and van Aartsen [59] have derived a theoretical expression for the relationship between cohesive energy density and refractive index at infinite wavelength, n_∞. The use of infinite wavelength guarantees that all polarization will be included.

$$\delta \propto (I/v)^{1/2}[(n_\infty^2 - 1)/(n_\infty^2 + 2)] \tag{2.51}$$

These authors did not find this relationship to be very reliable, as it predicts the CED to within only 13% (standard deviation). Koenhen and Smolders [58] found that $\delta_d = 9.55 n_D - 5.55$ but recommend that v, the molar volume, be included to improve the results of this correlation.

Parachor. Siow and Patterson [60] have shown that the relationship between surface tension and parachor [P],

$$\gamma = ([P]/v)^4 \tag{2.52}$$

is quite good for predicting the temperature and molecular weight dependence of γ. This could be used in Eq. (2.49) or (2.50) above to give a method for calculating δ from group contributions to the parachor, found in older physical chemistry texts.

Dipole moment. The dipole moment, μ, has been used by several workers as a correlating property for the polar contribution to the solubility parameter, δ_p. Koenhen and Smolders [58] have found an empirical relationship

$$\delta_p = 50.1 \, \mu/v^{0.75} \qquad (2.53)$$

based on δ_p as estimated by Hansen [52]. Gardon [14] has pointed out a relationship between δ_p and the ratio of α/μ^2, where α is the polarizability. His equation is

$$\delta_p^2 = \delta^2/(1 + \tfrac{9}{8}I(\alpha/\mu^2)^2 kT + 3kT\alpha/\mu^2) \qquad (2.54)$$

where I is the ionization potential. Table 2.2 lists approximate dipoles of

TABLE 2.2

Approximate Dipole Moments of Polymer Groups as Estimated from Low Molecular Weight Compounds of Similar Structure

Group	μ, Debye	Group	μ, Debye
—SO₂—	5.0	—CH— | Cl	2.0
—CO₂—	1.7		
—CO₃—	1.0	—CH— | F	1.9
—C(=O)—O—C(=O)—	3.0	—CH— | C≡N	3.5
—O—	1.2		
—N(H)—C(=O)—	4.0	⟨Ph⟩—Cl	1.55
—N(H)—C(=O)—O—	2.6–4.0		
—C(=O)—N(H)—C(=O)—	2.5	⟨Ph⟩—O—⟨Ph⟩	1.2
—CH—CH— with O=C—N(H)—C=O ring	1.6	CH₃—⟨Ph⟩—O—⟨Ph⟩	0.94
—CH— | COOH	1.2		
—CH— | OH	1.7	CH₃—⟨Ph⟩(CH₃)—O—	0.74
—CH— | Br	1.9		

2.3. Hildebrand Approach: The Solubility Parameter

some common groups found in polymers. The importance of shielding effects can be seen.

Group contributions. Small [55] was one of the first to recognize the additive properties of the quantity $(E_i v_i)^{1/2}$, where E_i is the cohesive energy of the group and v_i is the molar volume of the group. Values of this quantity, called F_i with units of $cal^{1/2}$ $cm^{3/2}$, were obtained by regression analysis for various common structural groups in low molecular weight compounds and are reproduced in Table 2.3. This was the first method available for estimating solubility parameters of compounds without recourse to *any physical measurements.*

The values of F_i have been improved and updated by Hoy [62], Konstam

TABLE 2.3

Group Contributions to the Solubility Parameter[a]

Group	F_i (1)	F_i (2)	F_i (3)	Group	F_i (1)	F_i (2)	F_i (3)
—CH$_3$	214	148	205	—CO—	275	263	335
—CH$_2$—	133	131	137	—COO—	310	327	250
—CH<	28	86	68	—C≡N	410	355	480
>C<	−93	32	0	—Cl	250–270	205	230
CH$_2$<	190	127	—	—Br	340	258	300
—CH=	111	122	(109)	—I	425	—	—
—C=	19	(84)	(40)	—CF$_2$—	150	(115)	—
CH≡C—	285	—	—	—CF$_3$	274	(156)	—
—C≡C—	222	—	—	—S—	225	209	225
				—SH	315	—	—
Phenyl	735	683	741	—ONO$_2$	440	—	—
Phenylene	658	705	673	—NO$_2$	440	—	—
Naphthyl	1146	—	—	—PO$_4$	500	—	—
				—C(=O)—O—C(=O)—	—	567	375
Ring, 5-member	105–115	21	—	Urethane	—	—	725
Ring, 6-member	95–105	−23	—	—COOH	—	—	319
Conjugation	20–30	23	—	—CO$_3$—	—	—	375
—H	80–100	—	—	Amide	—	—	600
—O—	70	115	125	—OH	—	226	369
				—F	—	41	80

[a] Adapted from D. W. Van Krevelen, "Properties of Polymers," 2nd ed., p. 134, Elsevier, Amsterdam, 1976. The references referred to for the F_i values are (1) P. A. Small, *J. Appl. Chem.* **3**, 71 (1953); (2) K. L. Hoy, *J. Paint Technol.* **42**, 76 (1970); (3) D. W. Van Krevelen, "Properties of Polymers," 2nd ed., p. 134, Elsevier, Amsterdam, 1976.

and Feairheller [63], and Van Krevelen [64], among others. Several of these sets are included in Table 2.3. It should be pointed out that the best results are obtained by using one consistent F_i data set based on a set of compounds containing a good portion of the structural groups of interest. Accurate densities are also required. The appropriate calculation format is given by

$$\delta = \rho \sum_i F_i / M \qquad (2.55)$$

where the summation is carried over all structural features in the molecule.

Complete estimates of interaction energetics from structure alone can be made by combining the solubility parameter estimate with a scheme

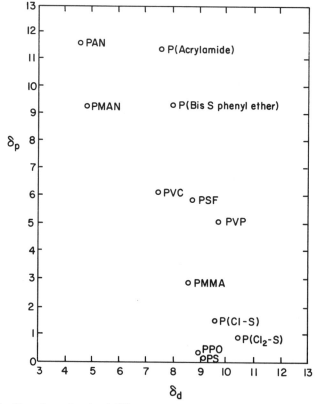

Fig. 2.16. Two-dimensional solubility parameter map showing the location of several polymers. The abscissa, δ_d, is the dispersive contribution to the solubility parameter, while δ_p is the contribution due to polarity. Arcs centered at the origin would represent constant total solubility parameter. [From M. T. Shaw, *J. Appl. Polym. Sci.* **18,** 449 (1974).] (The polymer abbreviations are defined in the appendix.)

2.3. Hildebrand Approach: The Solubility Parameter

such as Gardon's [14] for estimating the importance of polarity. Such has been done by Shaw [65] for several polymer–polymer systems and by Nelson et al. [56] for low molecular weight liquids. A sample of the type of "map" which results from this exercise is shown in Fig. 2.16. It can aid in designing copolymers for increased miscibility. Hydrogen bonding effects may be crudely estimated by assigning donating and accepting group contributions A and B as suggested by Small [55]. Such an approach is far more valid than the single H-bonding parameter used by Hansen [52], Lieberman [66], Crowley et al. [54], and Chen [67, 68]. Nelson et al. [56], attempting to account for the donating and accepting properties of hydrogen bonds, assigned a minus sign to the values of the hydrogen bonding parameters of Crowley et al. As far as is known, no values of the parameters A and B suggested by Small have been published.

A four-parameter characterization (δ_d, δ_p, A, B) of structural groups would make an interesting multiparameter regression analysis problem. A typical objective function is

$$\sum_{i,j} \left(\frac{\Delta H_{ij}}{V} - \phi_i \phi_j [(\delta_{di} - \delta_{dj})^2 + (\delta_{pi} - \delta_{pj})^2 + (A_i - A_j)(B_i - B_j)] \right)^2 \quad (2.56)$$

One wonders, however, if the parameters could have any logical connection with structure once the analysis is complete.

Nonetheless, it is desirable to develop a predictive scheme containing no solution parameters which could predict structures with enhanced probability of exhibiting miscibility. Such a scheme would have definite practical significance.

2.3.4 Predicted Polymer–Polymer Phase Behavior

The spinodals and binodals for a system of two polymer components obeying the free energy relationship given by Eq. (2.39) can easily be calculated [69]. The first step in the process is to calculate the partial molar free energy $\Delta\mu$ from Eq. (2.39).

$$\Delta\mu_i = \frac{\partial \Delta G}{\partial N_i}\bigg|_{P,T,N_j} \quad (2.57)$$

If Eq. (2.39) is rewritten in terms of the size ratio r of components 1 and 2 ($r = V_2/V_1$), and volume fractions are expressed in terms of the number of molecules N_i, then

$$\phi_1 = \frac{N_1}{N_1 + rN_2}, \quad \phi_2 = \frac{rN_2}{N_1 + rN_2} \quad (2.58)$$

This gives for the free energy of mixing

$$\frac{\Delta G^M}{kT} = N_1 \ln \frac{N_1}{N_1 + rN_2} + N_2 \ln \frac{rN_2}{N_1 + rN_2}$$

$$+ \frac{V_1}{kT}(\delta_1 - \delta_2)^2 \frac{rN_1 N_2}{(N_1 + rN_2)} \quad (2.59)$$

Taking the differentials and converting back to volume fractions yields

$$\Delta \mu_1 = kT \left[\ln \phi_1 + \left(1 - \frac{1}{r}\right)\phi_2 + V_1 \frac{(\delta_1 - \delta_2)^2}{kT} \phi_2^2 \right]$$

$$\Delta \mu_2 = kT \left[\ln \phi_2 + (1 - r)\phi_1 + V_2 \frac{(\delta_1 - \delta_2)^2}{kT} \phi_1^2 \right] \quad (2.60)$$

For the system to consist of two phases in equilibrium—the condition described by the binodal—the chemical potential (μ) of a component must be the same in both phases.

$$\Delta(\mu_1)_A = \Delta(\mu_1)_B \quad \text{and} \quad \Delta(\mu_2)_A = \Delta(\mu_2)_B \quad (2.61)$$

A binodal curve is shown schematically in Fig. 2.17. A horizontal tie-line connects the phases A and B, which are in equilibrium at the temperature denoted by the tie-line.

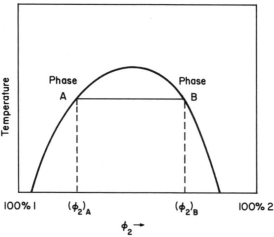

Fig. 2.17. Schematic of the type of phase diagram predicted by the solubility parameter modification of the Flory–Huggins theory. At a given temperature, phases A and B, containing both polymers 1 and 2, are in equilibrium with each other.

2.3. Hildebrand Approach: The Solubility Parameter

Applying the subscripts to the set of equations comprising (2.60) yields the set below where use has been made of $\phi_1 = 1 - \phi_2$.

$$\ln[1 - (\phi_2)_A] + \left(1 - \frac{1}{r}\right)(\phi_2)_A + V_1 \frac{(\delta_1 - \delta_2)^2}{kT}(\phi_2)_A^2$$

$$= \ln[1 - (\phi_2)_B] + \left(1 - \frac{1}{r}\right)(\phi_2)_B + V_1 \frac{(\delta_1 - \delta_2)^2}{kT}(\phi_2)_B^2$$

$$\ln[(\phi_2)_A] + (1 - r)[1 - (\phi_2)_A] + V_2 \frac{(\delta_1 - \delta_2)^2}{kT}[1 - (\phi_2)_A]^2 \quad (2.62)$$

$$= \ln[(\phi_2)_B] + (1 - r)[1 - (\phi_2)_B] + V_2 \frac{(\delta_1 - \delta_2)^2}{kT}[1 - (\phi_2)_B]^2$$

The above two equations contain the two unknowns $(\phi_2)_A$ and $(\phi_2)_B$ and can be solved for various size ratios r, temperatures T, and solubility parameter differences $\delta_1 - \delta_2$. A particularly straightforward case occurs when $r = 1$. Then the spinodal and binodal are found to be perfectly symmetrical; that is, then $(\phi_2)_A = 1 - (\phi_2)_B$ and the above set collapses into one equation.

$$\ln\left[\frac{1}{(\phi_2)_A} - 1\right] = \frac{V_1(\delta_1 - \delta_2)^2}{RT}[1 - 2(\phi_2)_A] \quad (2.63)$$

This yields the expected phase diagram of type B in Fig. 2.1 or Fig. 2.17 of this section. The influence of $(\delta_1 - \delta_2)$ and V_1 is to cause a proportional shift along the reduced temperature axis, as can be seen from Eq. (2.63).

The size ratio r has a skewing effect on the phase diagram. The quantitative effect of r must be determined numerically by the simultaneous solution of the set of equations comprising (2.62).

The spinodals for constant $\delta_1 - \delta_2$ are easily obtained from Eq. (2.60) by applying the criterion

$$\left.\frac{\partial \Delta \mu_1}{\partial \phi_2}\right|_{T,P} = 0 \quad (2.64)$$

which yields

$$0 = \left(1 - \frac{1}{r}\right)(1 - \phi_2) + 2\phi_2(1 - \phi_2)\frac{V_1}{kT}(\delta_1 - \delta_2)^2 - 1 \quad (2.65)$$

This quadratic equation can be solved analytically and supplied with values of r, V, and T to generate the spinodals. Some are included in Fig. 2.18 as dashed lines. Note that the binodal and spinodal are cotangent at $\phi_2 = 0.5$ for the simplified case of equally sized molecules ($r = 1$). The

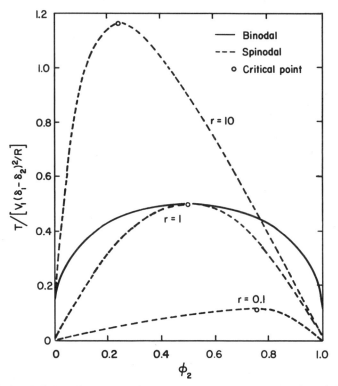

Fig. 2.18. Variation of the shape of the two-phase region and the location of the critical point with the ratio, r, of the size of polymer 2 to polymer 1. The ordinate is a reduced temperature.

point of tangency is called the critical point; spinodal and binodal (or cloud-point curve) are always cotangent at a critical point. The critical point can be determined, given $|\delta_1 - \delta_1|$ and V_1. As suggested above, the point is arrived at by a simultaneous agreement of the binodal and spinodal functions. This occurs when the first and second derivatives of the chemical potential are both zero. All that is required is a zero derivative of the right-hand size of Eq. (2.65).

$$0 = \left(1 - \frac{1}{r}\right) + \frac{2V_1}{kT}(\delta_1 - \delta_2)^2 [2\phi_2 - 1] \tag{2.66}$$

Combined with Eq. (2.65), the solution for ϕ_2 is

$$(\phi_2)_c = 1/(\sqrt{r} + 1) \tag{2.67}$$

2.3. Hildebrand Approach: The Solubility Parameter

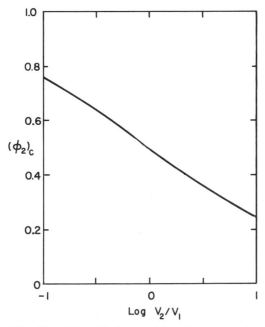

Fig. 2.19. Plot of the critical composition with polymer size ratio.

which has a limit of 0.5 as r approaches 1. The variation of $(\phi_2)_c$ with the size ratio V_2/V_1 is shown in Fig. 2.19. Note that the critical point is moved in the direction of the smaller-sized component.

The remaining calculation is that of the critical temperature given a solubility parameter difference $(\delta_1 - \delta_2)$. This is easily done by substituting Eq. (2.67) into (2.66).

$$T_c = \frac{2V_1(\delta_1 - \delta_2)^2}{k} \frac{r}{(\sqrt{r} + 1)^2} \qquad (2.68)$$

It is also instructive to examine $(\delta_1 - \delta_2)$ given that T_c should be a reasonable temperature, say 400°C, and $r = 1$.

$$(\delta_1 - \delta_2)_c = 40/\sqrt{v_1}, \quad T_c = 400°K, \quad v_1 \text{ in cm}^3/\text{mole}, \quad r = 1 \quad (2.69)$$

If $v_1 = 10^6$ cm^3/mole, or a molecular weight of about 10^6, the value of $(\delta_1 - \delta_2)$ is about 0.04, a very small difference indeed. This points out in a quantitative fashion the difficulty of finding miscible polymer systems.

2.4 FLORY-HUGGINS APPROACH: THE LATTICE THEORY

2.4.1 Background and Assumptions

In principle, there are two distinct equilibrium statistical mechanical methods of formulating a theory of chain molecular liquids and their mixtures. The first seeks to know the process by which the intermolecular forces determine the structure of the system. This is the method of radial distribution functions [70] and it generally attempts to specify the probability of finding polymer segments in particular configurations. Experimental evidence in support of or against this method comes from X-ray and neutron diffraction. The second method proceeds from an assumed structure which usually resembles the regular lattice structure of crystalline solids [71]. This is the so-called "lattice" method and it includes the cell or "free-volume" theories, the hole theories, the tunnel theories, and the "Monte Carlo" as well as the molecular dynamics methods [71]. Of these variations, only the cell and hole theories will be discussed here; the theory of conformal solutions [70] will not be discussed.

The lattice method is considered from two points of view: The first is the purely statistical approach generally referred to as the Flory-Huggins lattice model, or the lattice model for short [72-81]. The other is developed from the statistical mechanical partition function and, according to current literature convention, it is referred to as the equation-of-state approach.

In the description of polymer solutions, Fowler and Rushbrooke [72] were the first to introduce the concept of a regularly arranged set of rigid lattices whose sites are occupied by a solvent molecule or a polymer segment. A polymer chain is conceptionally divided into segments of the same size as the lattice sites, and the size is set equal to that of the solvent molecules. Based on this picture, Chang [73, 74] derived an analytical expression for the entropy of athermal mixing of monomers and dimers. In so doing, he borrowed the most important feature of the theory of strictly regular solutions, namely, that there is no volume change on mixing [7]. Miller [75] extended this formalism to trimers and a further extension was made by Huggins [76, 77], by Staverman and Van Santen [78], and, a little latter, but independently, by Flory [79, 80] for the athermal mixing of monomers and chain molecular liquids. This development has been extended to polymer-polymer mixtures, and the corresponding expression for the entropy of athermal mixing for a binary monodisperse polymer mixture is [81]:

$$\Delta S^M / NR = (\psi_1/r_1) \ln \psi_1 + (\psi_2/r_2) \ln \psi_2 \tag{2.70}$$

where ψ is the segment fraction, and r is the number of segments per chain molecule.

2.4. Flory–Huggins Approach: The Lattice Theory

2.4.2 Non-Athermal Mixing (The Interaction Parameter, χ)

As early as the 1930s, thermodynamic evidence from measurements of solvent activities [82] showed that solutions of polymers in simple liquids are far from ideal and are, therefore, not adequately described by the free energy expression resulting from the purely athermal considerations. Furthermore, such a free energy expression does not account for partial miscibility, a phenomenon already well known even for solvent–polymer mixtures [82]. To overcome these shortcomings, Flory [79, 80] and Huggins [76, 77] added a van Laar-type concentration-independent empirical free energy correction term to the original entropy of athermal mixing to obtain the Gibbs free energy [81]:

$$\Delta G^M/RT = (\psi_1/r_1) \ln \psi_1 + (\psi_2/r_2) \ln \psi_2 + g\psi_1\psi_2 \qquad (2.71)$$

Equation (2.71) differs from the original Flory–Huggins expression, as will be illustrated shortly. This formulation is based on an idea expressed much later by Guggenheim [83] and Maron [84] that the empirical interaction term, which depends on temperature, pressure, and composition, logically belongs with the total free energy rather than with the heat of mixing term as originally assumed by Flory, Huggins, Staverman, and Van Santen [76–80]. Maron's basis was that intermolecular interactions would introduce solution orientation effects which would have noncombinatorial entropy changes associated with them. His idea has been recently generalized by Koningsveld [81], who, like Maron, prefers the free energy of mixing as a starting point. The concept involved in all cases is similar, although not identical, to that originally suggested by Flory and Huggins; consequently, all the various modifications still go under the Flory–Huggins lattice theory.

The chemical potential derived from the original Flory–Huggins theory is

$$\Delta\mu_1/RT = \ln \psi_1 + [1 - (r_1/r_2)]\psi_2 + \chi\psi_2^2 \qquad (2.72)$$

This should be compared to the chemical potential obtained [81] by differentiating Eq. (2.71):

$$\Delta\mu_1/RT = \ln \psi_1 + [1 - (r_1/r_2)]\psi_2 + [g - \psi_1(\partial g/\partial \psi_2)]\psi_2^2 \qquad (2.73)$$

The original Flory–Huggins interaction parameter, χ, was supposed to be independent of concentration, temperature, and molecular weight. Compelling experimental evidence has shown that these assumptions are oversimplifications and χ is now considered dependent on ψ, T, and r. A particular relation suggested by Tompa [85] has been extensively used,

$$\chi = \chi_1 + \chi_2\psi + \chi_3\psi^2 + \cdots \qquad (2.74)$$

and a similar empirical g relation has been used by Koningsveld [81],

$$g = \sum_{k=0}^{n} g_k \psi^k \qquad k = 0, 1, 2, 3, \ldots, n \qquad (2.75)$$

where any coefficient g can be written as a function of temperature:

$$g_k = g_{k,1} + g_{k,2}/T + g_{k,3}T + g_{k,4} \ln T \qquad (2.76)$$

The $g_{k,i}$ in turn depend on measurable physical quantities such as heat of mixing, the specific heat, and the molecular weight. The form of Eq. (2.76) has been validated by more recent investigators [86–89], who have also suggested some molecular interpretations for the $g_{k,i}$. For example, Delmas and Patterson [89] considered the g function in light of the cell theory of Prigogine and discovered that the simultaneous occurrence of the upper and lower critical solution temperature behavior is adequately described when only the second and third terms of Eq. (2.76) are retained. By and large, however, no completely satisfactory molecular interpretations of $g_{k,i}$ exist [90–106] and these coefficients remain, at best, empirical. A relationship between g and χ is obtained from a comparison of Eqs. (2.72)–(2.75),

$$\chi = \sum_{k=0}^{n} (k+1)(g_k - g_{k-1})\psi^k \qquad (2.77)$$

where n denotes the number of terms necessary for adequate representation of experimental data.

The treatment so far has been limited to the case of binary polymer mixtures (made up of two monodisperse polymers), the polydisperse version of Eq. (2.71) is [99, 100]

$$\frac{\Delta G^M}{RT} = \sum_i^n \frac{\psi_{1,i}}{r_{1,i}} \ln \psi_{1,i} + \sum_j^n \frac{\psi_{2,j}}{r_{2,j}} \ln \psi_{2,j} + \Gamma(P, T, \psi_2) \qquad (2.78)$$

where

$$\Gamma = g(P, T, \psi_2)\psi_1 \psi_2 \qquad (2.79)$$

and

$$\psi_1 = \sum_i^n \psi_{1,i}, \qquad \psi_2 = \sum_j^n \psi_{2,j} \qquad (2.80)$$

2.4.3 Polymer–Polymer Phase Behavior

For the sake of describing the thermodynamic stability of a mixture, and therefore its phase behavior, we need to generate (i) the binodal and (ii) the spinodal.

2.4. Flory–Huggins Approach: The Lattice Theory

The binodal curve of the phase boundary for a two-phase binary polymer mixture is given by equating the chemical potential of each component in each phase [107].

$$\Delta\mu_1 = \Delta\mu_1' \tag{2.81}$$

$$\Delta\mu_2 = \Delta\mu_2' \tag{2.82}$$

where the unprimed and primed values, respectively, refer to the equilibrium chemical potentials for the two phases and are given by appropriate substitution of Eq. (2.73).

The curve could be generated, based on Eqs. (2.81) and (2.82), via the procedure of Flory [108, 109] or that of Maron [110]. Alternatively, one can proceed, as did Koningsveld [81], from a direct numerical iterative construction of the double tangent to the ΔG^M curve.

For a system that is strictly binary, the stability limit of the spinodal is also given by the curvature of the free energy curve (107):

$$\left(\frac{\partial^2 \Delta G^M}{\partial \psi_2^2}\right)_{P,T} = \left(\frac{\partial \Delta\mu_2}{\partial \psi_2}\right)_{P,T} = 0 \tag{2.83}$$

Substituting Eq. (2.71) one obtains the spinodal equation [81, 98–100]. For a mixture of two polydisperse polymers (quasi-binary mixture), the spinodal equation reads [81, 98–100]

$$2g - 2(\psi_1 - \psi_2)\frac{\partial g}{\partial \psi_2} - \psi_1\psi_2\frac{\partial^2 g}{\partial \psi_2^2} = \frac{1}{\psi_1 r_{w_1}} + \frac{1}{\psi_2 r_{w_2}} \tag{2.84}$$

where, for a quasi-binary mixture, ψ_1 and ψ_2 are given by Eq. (2.80) and the r_{w_i} are given as

$$r_{w_1} = \sum_{i=1}^{n} \psi_{1,i} r_i / \psi_1, \qquad r_{w_2} = \sum_{j=1}^{n} \psi_{2,j} r_j / \psi_2 \tag{2.85}$$

The critical point or the consolate state is the point where the inflection of the free energy curve coincides with the double tangents defined by Eqs. (2.81) and (2.82). The temperature, pressure, concentration, and molecular weight dependencies of the critical points are obtained by the simultaneous solution of Eq. (2.84) and the following [107]:

$$\left(\frac{\partial^3 \Delta G^M}{\partial \psi_2^3}\right)_{P,T} = 0 \tag{2.86}$$

Substituting Eq. (2.78) into (2.86) yields [81, 98–100]

$$-6\frac{\partial g}{\partial \psi_2^2} + 3(\psi_1 - \psi_2)\frac{\partial^2 g}{\partial \psi_2^2} + \psi_1\psi_2\frac{\partial^3 g}{\partial \psi_2^3} = \frac{r_{z_2}}{r_{w_2}\psi_2} - \frac{r_{z_1}}{r_{w_1}\psi_1} \tag{2.87}$$

where

$$r_{z_1} = \frac{1}{\psi_1 r_{w_1}} \sum_{i=1}^{n} \psi_{1,i} r_i^2, \quad r_{z_2} = \frac{1}{\psi_2 r_{w_2}} \sum_{j=1}^{n} \psi_{2,j} r_j^2 \quad (2.88)$$

Equation (2.87) solved in conjunction with Eq. (2.84) generates the common tangent points for the binodal and the spinodal. For strictly binary mixture, the critical point situates at the extremum of the spinodal but this is not necessarily the case for a quasi-binary polymer mixture.

The foregoing discussion is applicable to any molecular model so long as its free energy can be expressed in the form of the Flory–Huggins combinatorial entropy terms plus the correction function for concentration, temperature, and molecular weight. The approach, developed by Koningsveld [81], is most general and it is used even with the equation-of-state method of describing polymer–polymer miscibility.

2.4.4 The Lattice Theory and Partial Miscibility

The strength and weakness of the lattice model are best elucidated by inspecting the sensitivity of the predicted miscibility curves to variations in chain length, temperature, and composition and by comparing such predictions with experiment. Chain length sensitivity can be tested using a concentration-independent version of the theory, and the other two with a concentration-dependent version. For all cases, g is assumed to be dependent on P and T; consequently, calculation at atmospheric pressure implies that a g-axis can be substituted for a T-axis.

When the interaction parameter is constant, the spinodal is [81, 98–100]

$$2g_0 = \frac{1}{r_{w_1} \psi_1} + \frac{1}{r_{w_2} \psi_2} \quad (2.89)$$

and the critical point concentration is

$$\psi_{2_c} = 1/[1 + (a_1 r_{w_2}/a_2 r_{w_1})^{1/2}] \quad (2.90)$$

where $a_k = r_{zk}/r_{wk}$.

Figure 2.20 represents the spinodal curve calculated [81, 99] with Eqs. (2.89) and (2.90). The dashed line is the locus of critical points for the quasi-binary mixture with varying weight-average chain length but identical a's. Note that the critical point occurs at the maximum of the spinodal when $a_1 = a_2$ and that it shifts toward higher composition of the polymer with the shorter average chain length. For cases where $r_{w_1} = r_{w_2}$, the spinodal curve is symmetrical. In all cases, however, increasing molecular weight results in decreasing miscibility.

2.4. Flory–Huggins Approach: The Lattice Theory

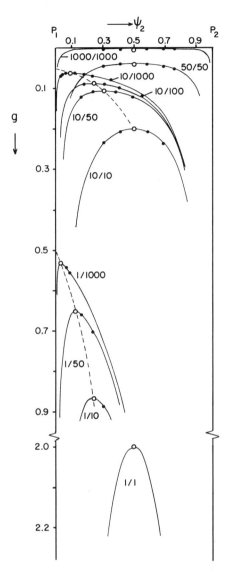

Fig. 2.20. Stability limits for various polymer 1–polymer 2 systems for the indicated r_{w_1}/r_{w_2} values. [Reprinted by permission from R. Koningsveld, L. A. Kleintjens, and H. M. Schoffeleers, *Pure Appl. Chem.* **39**, (1974). Copyright by the International Union of Pure and Applied Chemistry.]

A further illustration of the effect of molecular weight distribution was investigated [81, 99] for two polymers (P_1, P_2) with identical molecular weight averages but three types of distribution. As illustrated in Fig. 2.21, case I is a strictly binary mixture ($r_{w,1} = r_{z,1} = r_{w,2} = r_{z,2} = 50$) of two monodisperse polymers; II, two binary polymers ($r_{w,1} = r_{w,2} = 50$; $r_{z,1} = r_{z,2} = 62.5$); and III, two polydisperse polymers each with an exponential distribution but $r_w = 50$ and $r_n = 62.5$. Note, from Fig. 2.22, that the spinodals are identical, whereas the binodals indicate that system I is the

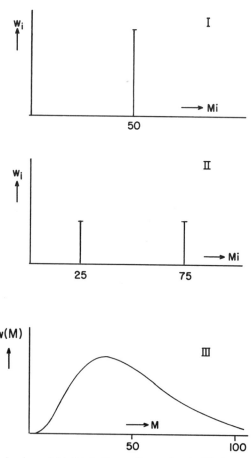

Fig. 2.21. Molecular weight distributions with equal r_w (=50). I, monodisperse; II, binary polymer ($r_z = 62.5$); III, exponential weight distribution ($r_z = 62.5$). [Reprinted by permission from R. Koningsveld, L. A. Kleintjens, and H. M. Schoffeleers, *Pure Appl. Chem.* **39**, 1 (1974). Copyright by the International Union of Pure and Applied Chemistry.]

2.4. Flory–Huggins Approach: The Lattice Theory

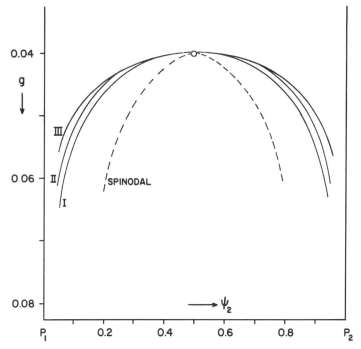

Fig. 2.22. Miscibility gap calculated for a mixture of two polymers (P_1, P_2) with identical molecular weight distributions. The three cases refer to the distributions in Fig. 2.21. [Reprinted by permission from R. Koningsveld, L. A. Kleintjens, and H. M. Schoffeleers, *Pure Appl. Chem.* **39**, (1974). Copyright by the International Union of Pure and Applied Chemistry.]

most miscible while system III is the least favored. The reduced miscibility in III is presumed to be due to the appreciable concentration of the high molecular weight portion of the polymers. It can also be shown that, if $a_1 \neq a_2$, the critical points shifts [81, 99] away from the maximum toward the axis of the constituent with the larger a.

The sensitivity of the predicted phase behavior to temperature, composition, and chain length can be collectively described if the interaction parameter is allowed to depend on ψ and T in the version illustrated by Eqs. (2.75) and (2.76). This aspect of the sensitivity analysis can be carried out first by fitting experimental heat of mixing data to the g function represented by Eqs. (2.75) and (2.76) through

$$\Delta H^M = -RT^2 \psi_1 \psi_2 \, \partial g/\partial T \qquad (2.91)$$

and then by comparing the experimental miscibility envelope with that predicted after the fact. An analysis of this sort has been done by Konings-

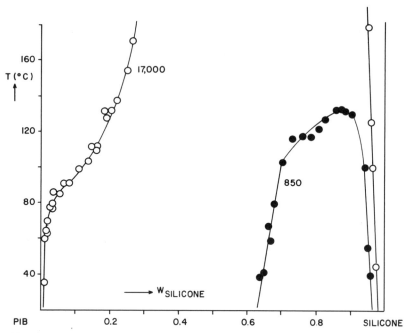

Fig. 2.23. Cloud-point curves for polyisobutene (PIB)–poly(dimethyl siloxane) (silicone) mixtures. [Reprinted by permission from G. Allen, G. Gee, and J. P. Nicholson, *Polymers* **2**, 8 (1961). Copyright by IPC Science and Technology Press Ltd., Surrey, England.]

veld [99, 100] on the polyisobutene–poly(dimethyl siloxane) data of Allen, Gee, and Nicholson [111], whose miscibility gap appears in Fig. 2.23. Figure 2.24 illustrates the rather satisfactory accomplishment of the lattice model modified with Koningsveld's empirical free energy correction [99]. Not only has the modified theory succeeded in outweighing the combinatorial entropy terms by shifting the miscibility gap to the region rich in the constituents with the large average chain length, it also prescribes a possible bimodality. This point is further illustrated by comparing Figs. 2.25 and 2.26 for polystyrene–polyisoprene [99, 112], where the modified theory again describes [99] the general nature of the phase behavior as well as the bimodality, although the model calculation (Fig. 2.26) is based on a system that is only a rough representation of the actual system.

In both of these calculations, the bimodality is attributed, without further molecular interpretations, to a quadratic dependence of g on ψ_2. The temperature dependence of g is absorbed solely by g_0; the

2.4. Flory–Huggins Approach: The Lattice Theory

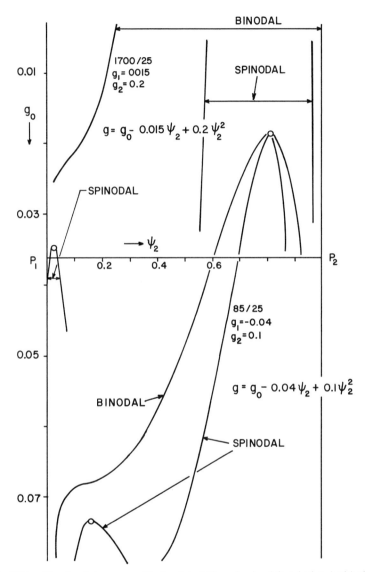

Fig. 2.24. Cloud-point curves, critical points (○), and spinodals calculated with the indicated values of g_1 and g_2 for two sets of r_2/r_1 values, qualitatively representative for the systems in Fig. 2.23. [Reprinted by permission from R. Koningsveld, L. A. Kleintjens, and H. M. Schoffeleers, *Pure Appl. Chem.* **39**, 1, (1974). Copyright by the International Union of Pure and Applied Chemistry.]

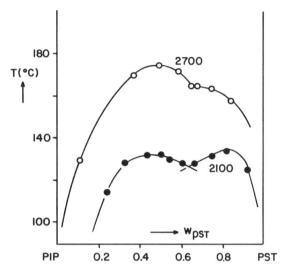

Fig. 2.25. Experimental cloud-point curves of the system polyisoprene (PIP; $M_n = 2700$)–polystyrene (PST; M_n values indicated); w_{pst} = weight fraction polystyrene. [Reprinted by permission from R. Koningsveld, L. A. Kleintjens, and H. M. Schoffeleers, *Pure Appl. Chem.* **39**, 1 (1974). Copyright by the International Union of Pure and Applied Chemistry.]

chain length dependence is reflected in the values of g_1 and g_2 in Fig. 2.24; and the polydispersity dependence in Fig. 2.26 is reflected also by g_0. That is, the theory supplied a semiempirical description of the liquid–liquid phase behavior and the functionality of g must be considered on a case-by-case basis. There are as yet no satisfactory molecular concepts capable of supplying the correct dependence of g on temperature, pressure, composition, molecular weight, and polydispersity [81, 90–106]. Moreover, the theory has provided no molecular interpretation of the bimodality of the miscibility gap or the asymmetry of the critical concentrations. These two phenomena are now well known. Aside from the two experimental examples above, bimodality has been demonstrated for other systems discussed in Section 2.5.4, p. 100.

In an attempt to find a rational basis for these phenomena, Koningsveld [100] recently turned to the new lattice theory of Huggins [113–115]. The tentative explanation is that the shape and location of the miscibility envelope are markedly affected by the chain flexibility dissimilarity between the polymers in the mixture; the less flexible chain molecules hinder the more flexible ones and this is manifested in the unusual shape of the phase envelope [100]. A further possible cause of spinodal bimodality is the nonuniformity of segment densities at both ends of the concentration scale [100].

2.5. Equation-of-State Approach

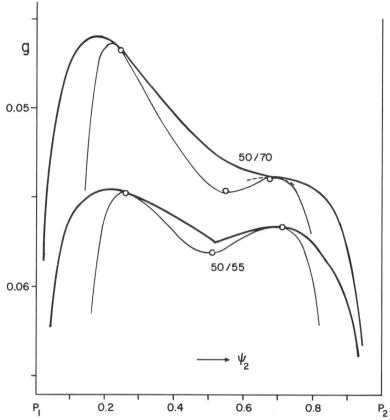

Fig. 2.26. Cloud-point curves (heavy lines), critical points (○), and spinodals (thin lines) calculated with $g = g_o - 0.04\psi_2 + 0.04\psi_2^2$ for two sets of r_2/r_1 values, qualitatively representative for the systems in Fig. 2.25. [Reprinted by permission from R. Koningsveld, L. A. Kleintjens, and H. M. Schoffeleers, *Pure Appl. Chem.* **39**, 1 (1974). Copyright by the International Union of Pure and Applied Chemistry.]

2.5 EQUATION-OF-STATE APPROACH

2.5.1 Background

For thermomechanical systems, an equation of state is a constitutive equation that relates the total stress tensor to the time, the strain, and the temperature history [116]. At equilibrium, time-dependent effects vanish, the total stress tensor reduces to the hydrostatic pressure, and the equation of state now relates the thermodynamic variables of pressure, volume, and temperature. The equation of state could either be deduced empirically from experimental observables or be calculated via statistical mechanics from some assumed knowledge of intermolecular forces. For imperfect crystals and imperfect gases, methods of statistical mechanics have been successful; however, for the liquid state, the story, started around the turn of the twentieth century [117], still continues. The attempts have been to view [118] the constitutive properties of the liquid as similar to the solid at the melting point or to the vapor at the critical point. This permits use of a quasi-lattice model, similar to that used by Einstein [119], in the description of crystals with various levels of modifications to account for communal entropy [120, 121] and density fluctuations [122]. Simple liquids were initially analyzed, but the recent, polymeric models draw immensely from the same concepts; hence, a discussion of simple-liquid theories cannot be bypassed.

We shall first discuss some principles of statistical mechanics, followed by a cursory review of simple-liquid theories. The extension to polymers will then be discussed, followed by a thorough exposition of the Flory equation of state [123] as applied to some aspects of polymer–polymer miscibility. The polymer solution theory of Patterson [124], like that of Flory, derives from the formalism of Prigogine [125] and its essential predictions are formally the same as those of Flory. Hence, it is sufficient for the purposes here to treat only one of the two. The Flory theory [123] has been extensively applied in the literature; therefore, it deserves a proportionate attention here.

2.5.2 Statistical Mechanics of Liquids

Consider a system of N molecules occupying a volume V in phase space [70]. The canonical partition function, Z, is divisible into an internal contribution, Z_{int}, and an external contribution, Z_{ext}.

$$Z = Z_{int} \cdot Z_{ext} \qquad (2.92)$$

The latter represents contributions due to the positions and the momentum of the centers of mass, and the former is due to those motions associated with the internal degrees of freedom, examples of which are rotations and vibration. The external contribution, in this case, is synonymous to the translational part, which is given by the classical integral

$$Z_{tr} = (1/N!)(1/h)^{3N} \int \cdots \int \exp(-H/kT)\, dp_1 \cdots dp_N\, dq_1 \cdots dq_N \qquad (2.93)$$

T is the temperature; k, Boltzmann's constant; $3N$, the total number of translational degrees of freedom; h, Planck's constant; H, the classical Hamiltonian function for the system; dp_i, the differential element of mo-

2.5. Equation-of-State Approach

mentum for molecule i; and dq_i, the differential element of volume for molecule i. After integration over the momenta, Eq. (2.93) becomes

$$Z_{tr} = (2\pi mkT/h^2)^{3N/2} \cdot Q \qquad (2.94)$$

where $Q(V, T, N)$ is the classical canonical configurational integral

$$Q = \frac{1}{N!} \int \cdots \int \exp(-U/kT) \, dq_1 \cdots dq_N \qquad (2.95)$$

U is the potential energy of the system, and m is the mass of one molecule.

In principle, if the potential energy is known, Eq. (2.95) can be solved and an equation of state can be derived from the standard relation [11]:

$$P = kT(\partial \ln Q/\partial V)_{T,N} \qquad (2.96)$$

If a particular structure, such as a cell structure, is assumed, the cell potential and therefore an equation of state can be derived.

a. Simple Liquids. One of the earliest cell theories of liquid, due to Guggenheim [126], equates the cell potential to the potential of the system averaged over all accessible configurations and all molecules. It therefore neglects the changes that would occur in the cell potential as the molecules move around within their respective cells. This smoothed potential model was less acceptable than even the simple van der Waals expression.

More realistic intermolecular forces should incorporate repulsive as well as attractive forces. Some models that did are due to Eyring and Hirschfelder [120], Mie [117], and Eyring and Kincaid [118]. The first introduced variability into an otherwise hard-core cell, while the latter two introduced the soft-core approximation. However, it was not until the classic works of Lennard-Jones and Devonshire (LJD) that a solid foundation was built for the cell model [127, 128].

Aside from the usual assumptions of the lattice model, the LJD model assumes that the molecules can be regarded as moving independently in their cell centers. The consequence of this singular "smearing approximation" is that the potential energy of each cell is identical and that the system's potential is divisible into two independent contributions. One is the lattice energy, which equals the energy of the system when all molecules are at their cell centers (at rest), and the second is due to the motion of the molecules about their cells. For these reasons, the configurational integral is again divisible into two, namely, $Q_{lattice}$ and $Q_{free\ volume}$, where

$$Q = Q_{latt} \cdot Q_{free\ vol} \qquad (2.97)$$

LJD's use of the 6–12 bireciprocal potential field led to a well-defined first approximation to the exact Q [129, 130]. Since the inception of the LJD treatment, a series of successive approximations, mainly having to do with the introduction of holes, has been proposed.

The cell model could not provide adequately for communal entropy [120], density fluctuations [122], thermal expansion, and isothermal compressibility [71]. It is also not in accord with X-ray diffraction results, which indicate [71] that the apparent number of nearest neighbors decreases as liquid is heated. On the other hand, by the introduction of holes, some of the problems could be handled, but others arise [131]. A particularly important question is: How should the free volume depend on the number of empty sites? Both the logarithmic and the linear dependence were variously used. It was also possible to keep the cell size constant [131, 132] while the number of holes changed, or to vary both [133, 134] as the density of the system changed. Henderson's model [135], as in the Eyring theory of significant structure [136], was based on the idea that a molecule acquires gaslike properties as it jumps into a neighboring hole. Both the cell volume and the number of holes were allowed to vary. This theory turned out to be most successful in describing simple liquids. Similar analyses by Simha and Somcynsky [137] and by Sanchez and Lacombe [138] have also yielded a good description of polymeric liquids.

b. Polymer Liquids. Basic to the development of polymeric theories is the idea that the energy of a chain molecule is separable into intra- and intermolecular contributions. Guth and Mark [139, 140] were the first to propose and to utilize this principle in their classic study of the mechanism and statistical mechanics of stretching rubbers. The proposition was justified then [141], as well as more recently [142, 143], by the findings that the configurational energies of polymer chains could be calculated independently of their conformational contributions. This idea has been useful in the description of the polymer crystallization process [144], and recent refinements [145] have introduced Volkenstein's [146] rotational isomeric scheme.

In the description of pressure–volume–temperature (PVT) and related properties of polymers, a similar concept of divisibility of total translational degrees of freedom into "internal" and "external" contributions has facilitated the transition from the theoretical approach for spherical molecular liquids to the case of chain liquids. The formulation of this idea, first introduced by Prigogine [125], dynamically represents N r-mer molecules by a set of $3N$ normal vibrations. There are $3cN$ ($c < 1$) soft modes or degrees of freedom which contribute solely to the configurational partition function. That is, the free volume expression for one segment is similar to that for a spherical molecule, but now there are $3c/r$ external degrees of freedom per molecule rather than the normal 3 translational degrees. In contrast to a

2.5. Equation-of-State Approach

simple molecule, covalent linkages in a chain molecule reduce the external degrees of freedom, the occupied volume, and the coordination number qz. This is so because a number of intermolecular contacts are replaced by intramolecular ones. This apparent loss in the external degrees of freedom is compensated for somewhat by Prigogine's $3cN$ idea, which amounts to stating that some internal degrees of freedom (rotation, vibration, etc.) previously absorbed by Z_{int} of Eq. (2.92) are lumped with the Z_{ext} because the modes are soft enough to be density dependent. Prigogine's empirical c allows for the uncertainty involved in estimating the volume dependent degrees of freedom. Borrowing the same concept involved in deriving Eq. (2.97), the configurational partition function of the Brussels school [10, 37, 38] (Prigogine and co-workers) is

$$Q = \Omega_{comb} \cdot (v^*)^{Nc} (\tilde{v}_f)^{3cN} \exp(-U_0/kT) \quad (2.98)$$

where Ω_{comb} is the combinatorial factor which assumes a value of unity for one component system, v^* is the hard-core volume per segment, tilde represents reduced quantity, $\tilde{v}_f$ is the cell free volume reduced with v^*, and U_0 is the lattice potential energy when all cells are at rest at their cell centers.

Prigogine *et al.* [125, 147, 148] modified the LJD cell model by utilizing different forms of cell potential, and they were able to derive a principle of corresponding states. A similar principle was later developed by Hijmans [149], who essentially proved that statistical mechanics need not be invoked in order to arrive at a corresponding states principle. He showed that only five phenomenological constants would be necessary if the principles of corresponding states were obeyed by polymers.

Furthermore, the equation of state based on Prigogine's square well and harmonic approximations was later tested by Nanda and Simha [150] and by Simha and Havlik [151]. They concluded that the theoretical *PVT* surface, internal pressure, and cohesive energy density do not agree with experiments even though the principle of corresponding states is established.

These failures of the LJD-type cell model were attributed to its inability to yield randomness, a foremost characteristic of the liquid state. Flory [123] sought to remedy this problem by the use of Eyring and Hirschfelder's [120, 129] simple liquid hard-sphere type of repulsive potential and an unspecified soft attraction originally introduced by Hildebrand and Scott [7] and by Frank [152]. The motivation was that, for a hard-sphere liquid, the internal energy in terms of the radial distribution, $g(r)$, is given as [119, 153]

$$\frac{U}{NkT} = \frac{3}{2} + \frac{4\pi}{2kTV} \int_0^\infty u(r)g(r)r^3 \, dr \quad (2.99)$$

Assuming that the radial distribution function is independent of volume, the internal energy U and, consequently, the equilibrium energy U_0 are inversely related to the specific volume. The resultant equation of state, though identical with Eyring–Hirschfelder's theory of simple liquids when allowance is made for the separability of internal and external degrees of freedom for chain molecules, was not committed to a particular structural feature. But then it failed [154] where the cell model failed. In all cases its characteristic scaling parameters have to vary with temperature [155] and pressure [154] in order to achieve adequate representation of experimental data. Despite its inadequacy, the essential simplicity of Flory theory has made it the theory of choice in the description of the thermodynamic aspects of chain-molecular liquid mixtures. This will form the subject of Section 2.5.3.

Other attempts at improving the cell model have included DiBenedetto's introduction [156] of a cylindrically symmetric, square-well potential, as well as other semiempirical modifications due to Bondi [157]. None has particularly excelled. Because of these, Curro has suggested [158] that only a rigorous numerical analysis will do justice to the original LJD–Prigogine cell model in its representation of polymeric systems. His results, cast in the form of the Tait parameters, have not been tested. But then the original model has been found to be insensitive [159] to the form of the bireciprocal potential used and to any averaging technique [160]. Consequently, the approach by Curro may still not suffice.

The introduction of holes into the quasi-lattice model of a polymeric liquid was first undertaken by Hirai and Eyring [161] in their description of bulk viscosity, fluid flow, and diffusion in terms of an activated-state vacancy diffusion mechanism. A PVT representation was derived from their theory but other thermodynamic quantities are not directly obtainable. In the development of thermodynamic equation of state, the general formalism for the introduction of holes was originally presented by Nanda, Simha, and Somcynsky [162]. Later development by Simha and Somcynsky [137] introduced a particular form for the dependence of the statistical mechanical free volume on the vacancy fraction based on the same general assumptions underlying the method of significant structure. The resulting equation of state has been shown, by Quach [163], Olabisi [164], and Wilson [165] all working with Simha, to largely remove the quantitative inadequacies of the cell model. However, the vacancy fraction is defined by a transcendental equation which makes the equation of state implicit. This shortcoming has only recently been corrected by Olabisi and Simha [42, 166], who developed an explicit, reduced equation of state by combining results from the hole theory with a semitheoretical expression. The resulting expression is found to be in excellent agreement with experiment.

Other polymeric hole theories are due to Nose [167], to Ishinabe and

2.5. Equation-of-State Approach

Ishikawa [168], to Prausnitz et al. [169], and to Sanchez and Lacombe [138]. The first assumes a different dependence of the free volume on the number of empty sites, the second is based on DiBenedetto's [156] cylindrically symmetric cell potential, and, in the third, the vacancy fraction was determined from the Monte Carlo calculations on hard sphere. Of all the hole theories, only the "lattice fluid" model of Sanchez and Lacombe [138] has been extended to liquid mixtures.

2.5.3 The Flory Equation-of-State Theory for Polymer Liquids

Following the formalism of Prigogine [125], Flory [123] developed a partition function whose lattice energy, like Eq. (2.99), is inversely related to volume. This potential is formally the same as the van der Waals mean potential.

$$Z_{tr} = \left(\frac{2\pi m_i kT}{h^2}\right)^{3N_i c_i r_i/2} \cdot Q \tag{2.100}$$

$$Q = \Omega_{comb} \cdot (\gamma v_i^*)^{N_i c_i r_i} \cdot (\tilde{v}_i^{1/3} - 1)^{3N_i c_i r_i} \cdot \exp(U_{0i}/kT) \tag{2.101}$$

$\tilde{v}_i$ is the reduced volume per segment; γ, the geometric factor; $3c$ is taken here to be the number of external degrees of freedom for one segment rather than for one chain as originally developed by Prigogine [125]; r_i is the number of segments per chain; and, the lattice energy, U_{0i}, can be expressed [see Eq. (2.121)] as a function of reduced temperature $T = T/T^*$, T^* being the characteristic temperature

$$-U_{0i}/kT = r_i N_i c_i/\tilde{v}_i \tilde{T}_i \tag{2.102}$$

The corresponding equation of state is [8]

$$\tilde{P}_i \tilde{v}_i/\tilde{T}_i = \tilde{v}_i^{1/3}/(\tilde{v}_i^{1/3} - 1) - 1/\tilde{v}_i \tilde{T}_i \tag{2.103}$$

where $\tilde{P}_i = P/P_i^*$; pressure, P, is the total hydrostatic pressure and P^* is the characteristic pressure.

The thermal expansion coefficient, $\alpha = [(\partial V/\partial T)p]/V$, is obtained by differentiating Eq. (2.103) with respect to temperature at constant pressure. In the limit of P going to zero, α is related to temperature and the reduced volume via

$$v/v^* = [(3 + 4\alpha T)/(3 + 3\alpha T)]^3 \tag{2.104}$$

The thermal pressure coefficient $\gamma = (\partial P/\partial T)v$ is similarly obtained in the limit of zero pressure giving the following:

$$P^* = \tilde{v}^2 \gamma T \tag{2.105}$$

The zero pressure limit of Eq. (2.103) gives the characteristic temperature as a function of volume:

$$T^* = T[\tilde{v}^{4/3}/(\tilde{v}^{1/3} - 1)] \qquad (2.106)$$

These equations are formally the same for pure components as for mixtures. In the case of mixtures, Ω_{comb} becomes the Flory–Huggins combinatorial factor, N is the sum of all the N_i, $\bar{r}$ is the number-average chain length, and the total number of volume dependent degrees of freedom is $3\bar{r}Nc$, where

$$c = \sum_{i=1}^{n} \psi_i c_i - \sum_{j=2}^{n} \sum_{i=1}^{j-1} \psi_i \psi_j c_{ij} \qquad (2.107)$$

and c_{ij} is the quadratic correction coefficient to a linear variation in the number of external degrees of freedom. This form of Eq. (2.107) was used by McMaster [3] in his application of Flory theory to polymer–polymer thermodynamics.

For multicomponent systems, Flory assumes that the mixture has only one characteristic hard-core volume, derivable from the relative segment sizes. The definition of the segment is arbitrary; the usual assumption is that a segment corresponds to a monomer unit or a multiple thereof. This is convenient particularly in the treatment of mixtures of molecules differing in size. For a binary mixture, the segment ratio is arbitrarily set equal to the ratio of the pure component hard-core volumes per mole:

$$r_2/r_1 = V_2^*/V_1^* = M_2 v_2^*{}_{spec}/M_1 v_1^*{}_{spec} \qquad (2.108)$$

M_i being the molecular weight of substance i. If V^* is the hard-core volume per mole of the mixtures, then

$$r_1 V^* = V_1^* \quad \text{or} \quad r_2 V^* = V_2^* \qquad (2.109)$$

Since V_1^*, V_2^*, and therefore r_2/r_1 are known, either r_1 or r_2 must be arbitrarily chosen in order to determine V^*. If r_1 is fixed, any r_i is computed from

$$r_i = r_1(V_i^*/V_1^*) \qquad (2.110)$$

In terms of this, the component *segment fraction* is

$$\psi_i = r_i x_i / \sum_{j=1}^{n} r_j x_j = r_i x_i / \bar{r} = N_i r_i / \bar{r} \qquad (2.111)$$

where x_i is the mole fraction.

The lattice energy U_0 is calculated from the assumption that intermolecular interactions occur between the surfaces of adjoining segments. This

2.5. Equation-of-State Approach

assumption has been validated recently by Olabisi and Simha [164], who, working from quite different angles, found that a system's energy at its characteristic states has a $\frac{2}{3}$ dependence on volume. This result, though based on the hole theory and applied to pure polymer systems, should not be limited to a particular molecular model. In the case at hand, U_0 is taken to be related to the intermolecular contact sites through an expression borrowed essentially from the radial distribution function.

$$-U_0 = \sum_{j}^{n} \sum_{i=1}^{j} A_{ij}\eta_{ij}/v \tag{2.112}$$

where η_{ij}/v is the intermolecular energy per i–j contact and A_{ij} represents the number of intermolecular potential interactions between i–j segments and is related to the number of intermolecular contact sites per segment, s_i, through the following:

$$A_{ii} + \sum_{j=1 \, (i \neq j)}^{n} A_{ij} = s_i r_i N_i \tag{2.113}$$

Equations (2.112) and (2.113) can be combined to give

$$-U_0 = \left(\sum_{i}^{n} s_i r_i N_i \eta_{ii} - \sum_{j=2}^{n} \sum_{i=1}^{j-1} A_{ij} \Delta\eta_{ij} \right) \bigg/ 2v \tag{2.114}$$

where $\Delta\eta_{ij} = 2\eta_{ij} - \eta_{ii} - \eta_{jj}$. If it is assumed that the probability (that a species of kind i is next to a given site) equals the segment surface fraction, θ_i,

$$\theta_i = s_i r_i N_i \bigg/ \sum_{j=1}^{n} s_j r_j N_j \tag{2.115}$$

then

$$A_{ij} = s_i r_i N_i \theta_i = s_j r_j N_j \theta_j \tag{2.116}$$

and

$$-U_0 = \frac{s\bar{r}N}{2v} \left(\sum_{i=1}^{n} \theta_i \eta_{ii} - \sum_{j=2}^{n} \sum_{i=1}^{j-1} \theta_i \theta_j \Delta\eta_{ij} \right) \tag{2.117}$$

Also, s, the mixture surface parameter, is

$$s = \sum_{i=1}^{n} s_i r_i N_i / N\bar{r} = \sum_{i=1}^{n} \psi_i s_i \tag{2.118}$$

where $N = \sum_{i}^{n} N_i$.

The various pure component characteristic parameters to be used in converting from the reduced to the actual variables of state are defined as

$$P_i^* = s_i \eta_{ii}/2v_i^{*2} \tag{2.119}$$

$$N_i c_i k T_i^* = P_i^* v_i^* \tag{2.120}$$

By analogy to Eq. (2.119), the exchange energy parameter for unlike interaction can be defined as

$$X_{ij} = s_i \Delta \eta_{ij}/2v_i^{*2} \tag{2.121}$$

and the corresponding P^* for the mixture is

$$P^* = \frac{s}{2v^{*2}} \left(\sum_{i=1}^{n} \theta_i \eta_{ii} - \sum_{j=2}^{n} \sum_{i=1}^{j-1} \theta_i \theta_j \Delta \eta_{ij} \right) \tag{2.122}$$

or

$$P^* = \sum_{i=1}^{n} \psi_i P_i^* - \sum_{j=2}^{n} \sum_{i=1}^{j-1} \psi_i \theta_j X_{ij} \tag{2.123}$$

Because of this, the equilibrium lattice energy for the multicomponent system reduces to

$$-U_0/N\tilde{r} = P^* v^*/\tilde{v} = ckT^*/\tilde{v} \tag{2.124}$$

and the mixture characteristic temperature is directly obtainable from Eq. (2.124):

$$\frac{1}{T^*} = \frac{k}{P^* v^*} \left[\sum_{i=1}^{n} c_i \psi_i - \sum_{j=2}^{n} \sum_{i=1}^{j-1} c_{ij} \psi_i \psi_j \right] \tag{2.125}$$

Given the above mixing rules, we are now in a position to consider the free energy of mixing and therefore the phase behavior.

a. Flory Free Energy Function and the Chemical Potential. The generalized Helmholtz free energy of mixing for a multicomponent system is obtainable from the following.

$$\Delta G^M = -kT \ln \left(Z \bigg/ \prod_{i=1}^{n} Z_i \right) \tag{2.126}$$

where Z_i is given by Eqs. (2.100) and (2.101) and Z is given by Eq. (2.100) (with c_i replaced by c) combined with the following canonical configurational partition function for the mixture:

$$Q = \Omega_{\text{comb}}(\gamma v^*)^{\tilde{r}Nc}(\tilde{v}^{1/3} - 1)^{3\tilde{r}Nc} \exp(-U_{0m}/kT) \tag{2.127}$$

2.5. Equation-of-State Approach

Substitution into Eq. (2.126) yields

$$\frac{\Delta G^M}{kT} = \sum_{i=1}^{n} N_i \ln \psi_i + \sum_{i=1}^{n} 3r_i N_i (c_i - c) \ln[(2\pi m_i kT)^{1/2}/h]$$

$$+ 3\bar{r} N \sum_{j=2}^{n} \sum_{i=1}^{j-1} \psi_i \psi_j c_{ij} \ln[(\gamma v^*)^{1/3}(\tilde{v}^{1/3} - 1)]$$

$$+ 3 \sum_{i=1}^{n} r_i N_i c_i \ln\left[\frac{\tilde{v}_i^{1/3} - 1}{\tilde{v}^{1/3} - 1}\right] + \frac{\bar{r} N v^*}{kT} \left[\sum_{i=1}^{n} \psi_i P_i^* \left(\frac{1}{\tilde{v}_i} - \frac{1}{\tilde{v}}\right)\right]$$

$$+ \frac{\bar{r} N v^*}{kT} \sum_{j=2}^{n} \sum_{i=1}^{j-1} \psi_i \theta_j \left(\frac{X_{ij}}{\tilde{v}} - T\tilde{v}_i Q_{ij}\right) \qquad (2.128)$$

The last term of the last line was added after the fact as an entropic correction factor, which permits a better fit of experiment. The empirical parameter Q_{ij} is supposed to account for the entropy of interaction between unlike segments, and it therefore makes the Flory theory conform to Guggenheim's [126] earlier contention that intermolecular interactions cannot be totally energetic.

The defining equation for the chemical potential of each component in a multicomponent system is

$$\frac{\Delta \mu_k}{kT} = \left.\frac{\partial \Delta G^M}{\partial N_k}\right|_{\substack{T,\tilde{v},N_j \\ j \ne k}} + \left.\frac{\partial \Delta G^M}{\partial \tilde{v}}\right|_{T,N_k,N_j} \cdot \left.\frac{\partial \tilde{v}}{\partial N_k}\right|_{T,V,N_j} \qquad (2.129)$$

In his original work, Flory neglected the second term of Eq. (2.129) but McMaster [3] found that the effect of pressure is incorrectly predicted at any nonzero pressure if the term is neglected. Substitution of Eq. (2.128) into Eq. (2.129) followed by term-by-term differentiation leads to the following equation for any component k of the multicomponent mixture:

$$\frac{\Delta \mu_k}{kT} = \ln \psi_k + \sum_{i=1}^{n} \frac{r_k}{r_i}(\delta_{ki} - \psi_i) + 3r_k(c_k - c)\ln[(2\pi m_k kT)^{1/2}/h]$$

$$- \sum_{i=1}^{n} 3r_k \psi_i \ln[(2\pi m_i kT)^{1/2}/h]$$

$$\times \left[\sum_{l=1}^{n}(\delta_{kl} - \psi_l)c_l - \sum_{j=2}^{n}\sum_{i=1}^{j-1}\{(\delta_{kl} - \psi_l)\psi_j + (\delta_{kj} - \psi_j)\psi_l\}c_{lj}\right]$$

$$+ 3r_k \sum_{j=2}^{n}\sum_{i=1}^{j-1}[\psi_i \psi_j + \psi_j(\delta_{ki} - \psi_i) + \psi_i(\delta_{ki} - \psi_j)]c_{ij}$$

$$\times \ln\{(\gamma v^*)^{1/3}(\tilde{v}^{1/3} - 1)\} + 3r_k c_k \ln\left[\frac{\tilde{v}_k^{1/3} - 1}{\tilde{v}^{1/3} - 1}\right]$$

$$+ \frac{r_k v^*}{kT} \sum_{i=1}^{n} \{\psi_i + (\delta_{ki} - \psi_i)\} P_i^* \left(\frac{1}{\tilde{v}_i} - \frac{1}{\tilde{v}}\right)$$

$$+ \frac{r_k v^*}{kT} \sum_{j=2}^{n} \sum_{i=1}^{j-1} \left[\left\{\psi_i \theta_j + (\delta_{ki} - \psi_i)\theta_j + (\delta_{kj} - \theta_j)\theta_i\left(\frac{S_k}{S_i}\right)\right\}\right.$$

$$\left. + \left(\frac{X_{ij}}{\tilde{v}} - \tilde{v}_i T Q_{ij}\right)\right] - \frac{cr_k E_k(\psi)}{\tilde{v}^{2/3}(\tilde{v}^{1/3} - 1)} + \frac{r_k v^* E_k(\psi)}{kT\tilde{v}^2}$$

$$\times \left\{\sum_{i=1}^{n} \psi_i P_i^* - \sum_{j=2}^{n} \sum_{i=1}^{j-1} \psi_i \theta_j X_{ij}\right\} \qquad (2.130)$$

where

$$E_k(\psi) = \left[\left(\frac{\bar{r}N}{r_k}\right)\frac{\partial \tilde{P}}{\partial N_k} - \frac{1}{\tilde{T}}\left(\tilde{P} + \frac{1}{\tilde{v}^2}\right)\left(\frac{\bar{r}N}{r_k}\right)\frac{\partial \tilde{T}}{\partial N_k}\right] \Big/ \left(\frac{2}{\tilde{v}^2} - \frac{\tilde{T}(\tilde{v}^{1/3} - \frac{2}{3})}{\tilde{v}^{5/3}(\tilde{v}^{1/3} - 1)^2}\right)\right]$$

(2.131)

and δ_{kl} is the Kronecker delta function defined as

$$\delta_{kl} = \begin{cases} +1 & \text{when } k = l \\ 0 & \text{when } k \neq l \end{cases} \qquad (2.132)$$

The partial derivatives in Eq. (2.131) are obtainable via the equation of state for the multicomponent mixture, i.e., Eq. (2.103) with all the variables of state changed to those of the mixtures.

b. The Binodal and the Spinodal Curves. In calculating the binodal curve for a binary mixture, $k = 1,2$ and $n = 2$ are substituted into Eq. (2.130), and Eqs. (2.81) and (2.82) are written in the form

$$\lambda_1 = (\Delta\mu_1 - \Delta\mu_1')/kT = 0 \qquad (2.133)$$

$$\lambda_2 = (\Delta\mu_2 - \Delta\mu_2')/kT = 0 \qquad (2.134)$$

The highly nonlinear algebraic equations in two unknowns ψ_1, ψ_1', or ψ_2, ψ_2' are solved via a nonlinear optimization procedure by finding [63]

$$\min_{\psi}(\lambda_1^2 + \lambda_2^2)/(\psi_1 - \psi_1') \qquad (2.135)$$

To calculate the spinodal curve for the binary polymer mixture, $k = 1,2$ and $n = 2$ are substituted into Eq. (2.128), which is then written in the form of Eq. (2.78). Consequently, Koningsveld's method, embodied in Eq.

2.5. Equation-of-State Approach

(2.84), can be used. After making the necessary substitutions and differentiation, there results the following spinodal equation [3]:

$$\frac{1}{r_1\psi_1} + \frac{1}{r_2\psi_2} + 6\ln\sqrt{m_1/m_2}\,[(c_2 - c_1 + c_{12}(\psi_2 - 2\psi_1)]$$
$$- 6c_{12}\ln\{(2\pi m_2 kT)^{1/2}/h(\gamma v^*)^{1/3}(\tilde{v}^{1/3} - 1)\}$$
$$- \frac{2v^*}{kT}\left(\frac{s_2}{s}\right)^2\left(\frac{s_1}{s}\right)\left(\frac{X_{12}}{\tilde{v}} - T\tilde{v}_1 Q_{12}\right)$$
$$- H(\psi_1, \tilde{v})\left\{\frac{2[c_1 - c_2 + (\psi_1 - \psi_2)c_{12}]}{\tilde{v}^{2/3}(\tilde{v}^{1/3} - 1)}\right.$$
$$\left. - \frac{2v^*}{kT\tilde{v}^2}\left[P_1^* - P_2^* - \frac{s_2}{s}(\psi_2 - \theta_1)X_{12}\right]\right\}$$
$$- \left(\frac{\partial H}{\partial \psi_1} + H\frac{\partial H}{\partial \tilde{v}}\right)\left[\frac{c}{\tilde{v}^{2/3}(\tilde{v}^{1/3} - 1)} - \frac{v^*}{kT\tilde{v}^2}(\psi_1 P_1^* + \psi_2 P_2^* - \psi_1\theta_2 X_{12})\right]$$
$$+ H^2(\psi_1, \tilde{v})\left[\frac{c(\tilde{v}^{1/3} - \frac{2}{3})}{\tilde{v}^{5/3}(\tilde{v}^{1/3} - 1)^2} - \frac{2v^*}{kT\tilde{v}^3}(\psi_1 P_1^* + \psi_2 P_2^* - \psi_1\theta_2 X_{12})\right] = 0$$

(2.136)

where

$$H(\psi_1, \tilde{v}) = E_1(\psi_1, \tilde{v})/\psi_2 \tag{2.137}$$

Equation (2.136) has two solutions or no solution (except for the critical point where it possesses a single solution). McMaster [3] solved for the two equations via a nonlinear optimization procedure by finding

$$\min_{\psi_1, \psi_1'} \left\{\frac{(\partial\mu_1/\partial N_1)(\psi_1)|_{T,V,\mu_2} + (\partial\mu_1/\partial N_1)(\psi_1')|_{T,V,\mu_2}}{(\psi_1 - \psi_1')^2}\right\} \tag{2.138}$$

Finally, Eq. (2.87) is utilized in deriving the following equation for the critical point:

$$18c_{12}\ln\sqrt{m_1/m_2} + \frac{6v^*}{kT}\left(\frac{s_1}{s}\right)\left(\frac{s^2}{s}\right)^2\left(\frac{\psi_1 - \theta_1}{\psi_1\psi_2}\right)\left(\frac{X_{12}}{\tilde{v}} - T\tilde{v}_1 Q_{12}\right)$$
$$+ 3H(\psi_1, \tilde{v})\left[\frac{2c_{12}}{\tilde{v}^{2/3}(\tilde{v}^{1/3} - 1)} - \frac{v^* X_{12}}{kT\tilde{v}^2}\left(\frac{s_1}{s}\right)\left(\frac{s_2}{s}\right)\right]$$
$$- 3H^2(\psi_1, \tilde{v})\left[\frac{[c_1 - c_2 - c_{12}(\psi_2 - \psi_1)][\tilde{v}^{1/3} - \frac{2}{3}]}{\tilde{v}^{2/3}(\tilde{v}^{1/3} - 1)}\right.$$

$$-\frac{v^*}{kT\tilde{v}^2}\left\{P_1^* - P_2^* - \frac{s_2}{s}(\psi_2 - \theta_1)X_{12}\right\}\Bigg]$$

$$+ 3\left(\frac{\partial H}{\partial \psi_1} + H\frac{\partial H}{\partial \tilde{v}}\right)\left[\frac{c_1 - c_2 - c_{12}(\psi_2 - \psi_1)}{\tilde{v}^{2/3}(\tilde{v}^{1/3} - 1)}\right.$$

$$\left. - \frac{v^*}{kT\tilde{v}^2}\left\{P_1^* - P_2^* - \frac{s_2}{s}(\psi_2 - \theta_1)X_{12}\right\}\right]$$

$$- 3H(\psi_1, \tilde{v})\left(\frac{\partial H}{\partial \psi_1} + H\frac{\partial H}{\partial \tilde{v}}\right)\left[\frac{c(\tilde{v}^{1/3} - \frac{2}{3})}{(\tilde{v}^{5/3}(\tilde{v}^{1/3} - 1)^2}\right.$$

$$\left. - \frac{2v^*}{kT\tilde{v}^3}(\psi_1 P_1^* + \psi_2 P_2^* - \psi_1\theta_2 X_{12})\right]$$

$$- H^3(\psi_1, \tilde{v})\left[\frac{c\{\tilde{v}^{1/3}(\tilde{v}^{1/3} - 1) - (\tilde{v}^{1/3} - \frac{2}{3})[5(\tilde{v}^{1/3} - 1) + 2\tilde{v}^{1/3}]\}}{3\tilde{v}^{8/3}(\tilde{v}^{1/3} - 1)^3}\right.$$

$$\left. + \frac{6v^*}{kT\tilde{v}^4}(\psi_1 P_1^* + \psi_2 P_2^* - \psi_1\theta_2 X_{12})\right]$$

$$+ \left[\frac{\partial^2 H}{\partial \psi_1^2} + 2H\frac{\partial^2 H}{\partial \psi_1 \partial \tilde{v}} + H\left(\frac{\partial H}{\partial \tilde{v}}\right)^2 + \frac{\partial H}{\partial \tilde{v}}\frac{\partial H}{\partial \psi_1} + H^2\frac{\partial^2 H}{\partial \tilde{v}^2}\right]\left[\frac{c}{\tilde{v}^{2/3}(\tilde{v}^{1/3} - 1)}\right.$$

$$\left. - \frac{v^*}{kT\tilde{v}^2}(\psi_1 P_1^* + \psi_2 P_2^* - \psi_1\theta_2 X_{12})\right] = \frac{1}{r_2\psi_2^2} - \frac{1}{r_1\psi_1^2} \quad (2.139)$$

The temperature dependence of the critical point composition is obtained by the simultaneous solution of Eqs. (2.136) and (2.139).

For a quasi-binary mixture, the Flory free energy function can again be cast in the form of Eqs. (2.78)–(2.80) with a generalized g given by:

$$g = \frac{1}{\bar{r}_1^\circ N_1^\circ \psi_2^\circ}\left\{\sum_{i=1}^n 3r_i N_i(c_i - c)\ln[(2\pi m_i kT)^{1/2}/h]\right.$$

$$+ 3\bar{r}N\sum_{j=2}^n\sum_{i=1}^{j-1}\psi_i\psi_j c_{ij}\ln[(\gamma v^*)^{1/3}(\tilde{v}^{1/3} - 1)] + 3\sum_{i=1}^n r_i N_i c_i \ln\left[\frac{\tilde{v}_i^{1/3} - 1}{\tilde{v}^{1/3} - 1}\right]$$

$$\left. + \frac{\bar{r}Nv^*}{kT}\left[\sum_{i=1}^n \psi_i P_i^*\left(\frac{1}{\tilde{v}_i} - \frac{1}{\tilde{v}}\right) + \sum_{j=2}^n\sum_{i=1}^{j-1}\left(\frac{\psi_i\theta_j X_{ij}}{\tilde{v}} - T\tilde{v}_i\psi_i\theta_j Q_{ij}\right)\right]\right\}$$

$$(2.140)$$

where

$$N_1^\circ = \sum_{i=1}^m N_i \qquad \bar{r}_i^\circ = \sum_{i=1}^m N_i r_i / N_1^\circ$$

When the molecular weight of the polymer is high, Eq. (2.140) is further

2.5. Equation-of-State Approach

simplified with the assumptions in Table 2.4, the overriding implications being the following:

(1) The pure component properties are independent of chain lengths.
(2) There is always chain length compatibility in the sense that end-group contribution to the free energy of mixing is negligibly small for chemically identical homopolymers with widely varying chain lengths.
(3) In the binary system, the contributions from the two molecularly different segments to the free energy of mixing are reflected in $\bar{c}_{12}$, $\bar{X}_{12}$, and Q_{12}.

Hence, the expression for g becomes

$$g = \frac{1}{\psi_1{}^\circ \psi_2{}^\circ} \{3\psi_1{}^\circ(c_1 - \bar{c})\ln\{(2\pi m_1 kT)^{1/2}/h\}$$

$$+ 3\psi_2{}^\circ(c_2 - \bar{c})\ln\{(2\pi m_2 kT)^{1/2}/h\} + 3\psi_1{}^\circ\psi_2{}^\circ\bar{c}_{12}\ln\{(\gamma v^*)^{1/3}(\tilde{v}^{1/3} - 1)\}$$

$$+ 3c_1\psi_1{}^\circ \ln\left[\frac{\tilde{v}_1^{1/3} - 1}{\tilde{v}^{1/3} - 1}\right] + 3c_2\psi_2{}^\circ \ln\left[\frac{\tilde{v}_2^{1/3} - 1}{\tilde{v}^{1/3} - 1}\right] + \frac{v^*}{kT}\left[P_1{}^*\left(\frac{1}{\tilde{v}_1} - \frac{1}{\tilde{v}}\right)\psi_1{}^\circ\right.$$

$$\left. + P_2{}^*\left(\frac{1}{\tilde{v}_2} - \frac{1}{\tilde{v}}\right)\psi_2{}^\circ + \psi_1{}^\circ\theta_2{}^\circ\left(\frac{\bar{X}_{12}}{\tilde{v}} - T\tilde{v}_1\bar{Q}_{12}\right)\right]\} \quad (2.141)$$

When Eq. (2.141) for g is substituted into Eq. (2.84) we obtain the following general equation for the spinodal curve for a quasi-binary system:

$$\frac{1}{\bar{r}_{w1}\psi_1{}^\circ} + \frac{1}{\bar{r}_{w2}\psi_2{}^\circ} + 6\ln\sqrt{m_1/m_2}\left[(c_2 - c_1) + \bar{c}_{12}(\psi_2{}^\circ - 2\psi_1{}^\circ)\right]$$

$$- 6\bar{c}_{12}\ln\{(2\pi m_2 kT)^{1/2}(\gamma v^*)^{1/3}(\tilde{v}^{1/3} - 1)/h\} - \frac{2v^*}{kT}\left(\frac{s_2}{s}\right)^2\left(\frac{s_1}{s}\right)\left(\frac{\bar{X}_{12}}{\tilde{v}^n} - T\tilde{v}_1\bar{Q}_{12}\right)$$

$$- 2H(\psi_1{}^\circ)\left[\frac{[c_1 - c_2 + (\psi_1{}^\circ - \psi_2{}^\circ)\bar{c}_{12}]}{\tilde{v}^{2/3}(\tilde{v}^{1/3} - 1)} - \frac{v^*}{kT\tilde{v}^2}\right.$$

$$\left. \times \left(P_1{}^* - P_2{}^* - \left(\frac{s_2}{s}\right)(\psi_2{}^\circ - \theta_1{}^\circ)\bar{X}_{12}\right)\right]$$

$$- \left(\frac{\partial H}{\partial \psi_1{}^\circ} + H\frac{\partial H}{\partial \tilde{v}}\right)\left[\frac{\bar{c}}{\tilde{v}^{2/3}(\tilde{v}^{1/3} - 1)} - \frac{v^*}{kT\tilde{v}^2}(P_1{}^*\psi_1{}^\circ + P_2{}^*\psi^\circ - \psi_1{}^\circ\theta_2{}^\circ\bar{X}_{12})\right]$$

$$+ H^2(\psi_1{}^\circ)\left[\frac{\bar{c}(\tilde{v}^{1/3} - \frac{2}{3})}{\tilde{v}^{5/3}(\tilde{v}^{1/3} - 1)^2} - \frac{2v^*}{kT\tilde{v}^3}(P_1{}^*\psi_1{}^\circ + P_2{}^*\psi_2{}^\circ - \psi_1{}^\circ\theta_2{}^\circ\bar{X}_{12})\right] = 0$$

$$(2.142)$$

Similarly, the consolate state is derived from Eqs. (2.87) and (2.88) to yield

$$18\bar{c}_{12} \ln\sqrt{m_1/m_2} + \frac{6v^*}{kT}\left(\frac{s_1}{\bar{s}}\right)\left(\frac{s_2}{\bar{s}}\right)^2\left(\frac{\psi_1° - \theta_1°}{\psi_1°\psi_2°}\right)\left(\frac{\bar{X}_{12}}{\bar{v}} - T\bar{v}_1\bar{Q}_{12}\right)$$

$$+ 3H(\psi_1°)\left[\frac{2\bar{c}_{12}}{\bar{v}^{2/3}(\bar{v}^{1/3} - 1)} - \frac{v^*\bar{X}_{12}}{kT\bar{v}^2}\left(\frac{s_1}{\bar{s}}\right)\left(\frac{s_2}{\bar{s}}\right)\right]$$

$$- 3H^2(\psi_1°)\left[\frac{[c_1 - c_2 - \bar{c}_{12}(\psi_2° - \psi_1°)](\bar{v}^{1/3} - \frac{2}{3})}{\bar{v}^{2/3}(\bar{v}^{1/3} - 1)}\right]$$

$$- \frac{v^*}{kT\bar{v}^2}\left\{P_1^* - P_2^* - \frac{s_2}{\bar{s}}(\psi_2° - \theta_1°)\bar{X}_{12}\right\}\right] - 3H(\psi_1°)\left(\frac{\partial H}{\partial \psi_1°} + H\frac{\partial H}{\partial \bar{v}}\right)$$

$$\times \left[\frac{\bar{c}(\bar{v}^{1/3} - \frac{2}{3})}{\bar{v}^{5/3}(\bar{v}^{1/3} - 1)^2} - \frac{2v^*}{kT\bar{v}^3}(P_1^*\psi_1° + P_2^*\psi_2° - \psi_1°\theta_2°\bar{X}_{12})\right]$$

$$- H^3(\psi_1°)\left[\frac{\bar{c}\{\bar{v}^{1/3}(\bar{v}^{1/3} - 1) - (\bar{v}^{1/3} - \frac{2}{3})[5(\bar{v}^{1/3} - 1) + 2\bar{v}^{1/3}]\}}{3\bar{v}^{8/3}(\bar{v}^{1/3} - 1)^3}\right]$$

$$+ \frac{6v^*}{kT\bar{v}^4}(P_1^*\psi_1° + P_2^*\psi_2° - \psi_1°\theta_2°\bar{X}_{12})\right] + \left(\frac{\partial^2 H}{\partial \psi_1°^2} + 2H\frac{\partial^2 H}{\partial \psi_1°\partial \bar{v}}\right)$$

$$+ H\left(\frac{\partial H}{\partial \bar{v}}\right)^2 + \frac{\partial H}{\partial \bar{v}}\frac{\partial H}{\partial \psi_1°} + H^2\frac{\partial^2 H}{\partial \bar{v}^2}\right)\left[\frac{\bar{c}}{\bar{v}^{2/3}(\bar{v}^{1/3} - 1)}\right]$$

$$- \frac{v^*}{kT\bar{v}^2}(P_1^*\psi_1° + P_2^*\psi_2° - \psi_1°\theta_2°\bar{X}_{12})\right] + \frac{\bar{r}_{z_1}}{\bar{r}_{w_1}\psi_1°^2} - \frac{\bar{r}_{z_2}}{\bar{r}_{w_2}^2\psi_2°^2} = 0 \quad (2.143)$$

Simultaneous satisfaction of Eqs. (2.142) and (2.143) results in the solution for the temperature pressure, concentration, and molecular weight dependencies of the critical point.

It is of interest to see if identical expressions can be recovered for both monodisperse and polydisperse systems if the following substitutions are made:

$$\psi_1 \to \psi_1° \qquad c_{12} \to \bar{c}_{12}$$
$$\psi_2 \to \psi_2° \qquad c \to \bar{c}$$
$$r_1 \to \bar{r}_{w_1} \qquad X_{12} \to \bar{X}_{12}$$
$$r_2 \to \bar{r}_{w_2} \qquad Q_{12} \to \bar{Q}_{12}$$

Indeed, after the substitution, the two spinodal equations are identical but the equations for the consolate state [(2.139) and (2.143)] differ because of chain-length distribution effects. The implication is that, even though the quasi-binary spinodal is identical to the binary spinodal, the exact location of

2.5. Equation-of-State Approach

TABLE 2.4

Assumed Equalities for Polydispersed Polymer Pairs When Molecular Weight Is Very High[a]

Equality	For	Parameter description
$m_i = m_1$	$i = 1, \ldots, m$	Mass/segment
$m_i = m_2$	$i = m + 1, \ldots, n$	Mass/segment
$\tilde{v}_i = \tilde{v}_1$	$i \leq m$	Reduced volume
$\tilde{v}_i = \tilde{v}_2$	$i \geq m + 1$	Reduced volume
$P_i^* = P_1^*$	$i \leq m$	Pressure parameter
$P_i^* = P_2^*$	$i \geq m + 1$	Pressure parameter
$T_i^* = T_1^*$	$i \leq m$	Temperature parameter
$T_i^* = T_2^*$	$i \geq m + 1$	Temperature parameter
$c_{ij} = 0$	$i \leq m; j \leq m$	Degrees of freedom
$= 0$	$i \geq m + 1; j \geq m + 1$	Degrees of freedom
$= \bar{c}_{12}$	$i \leq m; j \geq m + 1$	Degrees of freedom
$X_{ij} = 0$	$i \leq m; j \leq m$	Energy interaction
$= 0$	$i \geq m + 1; j \geq m + 1$	Energy interaction
$= \bar{X}_{12}$	$i \leq m; j \geq m + 1$	Energy interaction
$Q_{ij} = 0$	$i \leq m; j \leq m$	Entropy interaction
$= 0$	$i \geq m + 1; j \geq m + 1$	Entropy interaction
$= \bar{Q}_{12}$	$i \leq m; j \geq m + 1$	Entropy interaction

[a] Reprinted with permission from L. P. McMaster, *Macromolecules* **6**, 760 (1973). Copyright by the American Chemical Society.

the critical point is determined by the ratio of the Z-average to weight-average chain length of each polymer.

2.5.4 The Flory Theory and Partial Miscibility

In order to test the consequences of the equation-of-state theory, McMaster [3] computed a set of binodal and spinodal curves for a representative polymer pair by making perturbations in properties around base case values. The effects of chain length, thermal expansion coefficient, thermal pressure coefficient, exchange energy parameter, and polydispersity were studied.

For polymer–polymer mixtures the theory predicts that miscibility decreases with increasing temperance, i.e., lower critical solution temperature (lcst) behavior is preferred. The lcst behavior is presumed to be due to the significance of the equation-of-state and the noncombinational effects for high polymers. The chain length effect is illustrated by Fig. 2.27. As the molecular weight of component 2 decreases or, as the molecular weight of component 1 increases, the mutual solubility of the two polymers decreases; a difference by a factor of 2 in the average chain length of one of the polymers

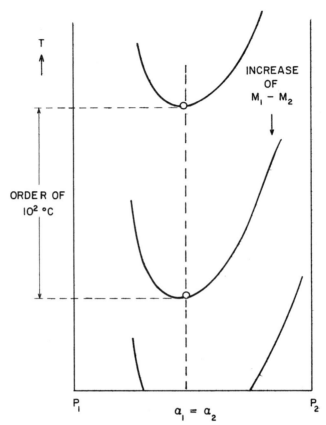

Fig. 2.27. Dependence of the binodal curves (———) and the critical points (○) of two-polymer mixtures on their molecular weight differences. A schematic summary of L. P. McMaster's results [*Macromolecules* **6**, 760 (1973)].

may shift the cloud point more than 100°C. This is expected since the combinatorial entropy term (which is identical to that of the lattice theory) in the free energy of mixing and the free volume decrease monotonically as the molecular weight difference increases.

Similar shifts are brought about by acceptable differences in the thermal expansion coefficients and thermal pressure coefficients of the two polymers (Figs. 2.28 and 2.29). However, while an increase in $(\alpha_1 - \alpha_2)$ only shifts the miscibility curves downward, an increase of $(\gamma_1 - \gamma_2)$ not only shifts the curves downward but also shifts the critical point composition of component 2 to higher values.

Figure 2.30 illustrates the effect of the exchange energy parameter on

2.5. Equation-of-State Approach

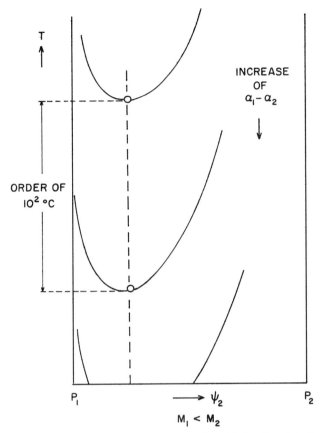

Fig. 2.28. Dependence of the binodal (———) and the critical points (○) of two-polymer mixtures on their thermal expansion coefficient differences. A schematic summary of L. P. McMaster's results [*Macromolecules* **6**, 760 (1973)].

the general phase behavior of a binary monodisperse mixture. As X_{12} increases, miscibility decreases and vice versa; at elevated critical temperatures the curves flatten out. When specific segmental interaction is present, as exemplified by a negative X_{12}, miscibility is all the more favored. Figure 2.30 also shows that lcst and ucst can occur simultaneously within a very narrow range of X_{12}. When the X_{12} values are higher than this range, the two critical points merge, yielding hourglass-shaped binodal and spinodal curves. The temperature of the lcst–ucst merger depends on the difference between the component thermal expansion coefficients; the lower the difference, the higher the temperature at which the two curves meet.

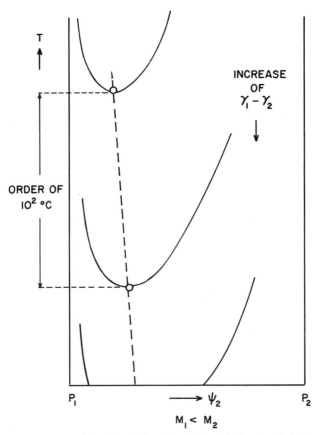

Fig. 2.29. Dependence of the binodal (———) and the critical points (○) of two-polymer mixtures on their thermal pressure coefficient differences. A schematic summary of L. P. McMaster's results [*Macromolecules* **6**, 760 (1973)].

For quasi-binary polymer mixtures, the spinodals remain unchanged as long as the weight-average chain length is substituted in place of the monodisperse chain length for each species. However, the binodal and the location of the consolate state depend on the ratio of the Z-average to weight-average chain length of each component. The consolate state no longer exists at the extremum of the spinodal; instead, it may lie on either ascending branch of the lcst spinodal curve. Figure 2.31 illustrates the effect of polydispersity on the location of the critical point. The first two mixtures have their critical points at the same extremum point in spite of their differences, so long as their polydispersdity ratio equals 1.0. The mixture whose ratio is 0.5 has its critical point to the left ascending branch of the lcst, corresponding to a higher composition of the component with higher

2.5. Equation-of-State Approach

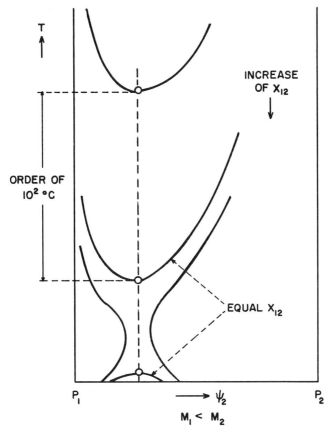

Fig. 2.30. Dependence of the binodal (——) and the critical points (○) of two-polymer mixtures on the exchange energy parameter. A schematic summary of L. P. McMaster's results [*Macromolecules* **6**, 760 (1973)].

polydispersity. With a still lesser ratio, the critical point climbs still higher on the left ascending branch. Conversely, when the polydispersity is greater than 1.0, the critical points move to the right ascending branch of the spinodal, corresponding to a higher composition of the component with the higher polydispersity. Again, the higher the polydispersity ratio, the higher the critical point is located on the right ascending branch.

The foregoing has shown that binary polymer mixtures with acceptable component parameters should, in general, exhibit lcst behavior even though the occurrence of the ucst behavior is not discounted. This prediction seems to be borne out by the following three miscible polymers which demix upon heating: polystyrene–poly(vinyl methyl ether), styrene/acrylonitrile copolymer–poly(methyl methacrylate), and styrene/acrylonitrile copolymer–

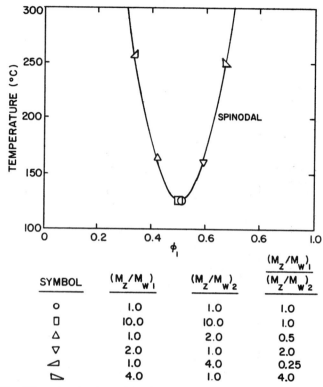

Fig. 2.31. Effect of molecular weight distributions on the spinodal curves and the critical points of two-polymer mixtures. [Reprinted by permission from L. P. McMaster, *Macromolecules* **6**, 760 (1973). Copyright by the American Chemical Society.]

polycaprolacetone. The cloud-point curves for these systems are shown in Fig. 2.32; Table 2.5 gives the molecular weight data [170].

Inasmuch as McMaster's treatment was based on an arbitrary polymer mixture, Olabisi [16] sought to predict the phase behavior of real systems—namely, a polycaprolactone–poly(vinyl chloride) mixture, whose pure component data appear in Table 2.6. He utilized the solvent-probe technique of the inverse chromatography to characterize the nature and magnitude of the specific interactions present in the said mixture and found that the Flory exchange energy parameter is in the range $-10 < X_{12} < -2$ cal/cm^3. The extension of the equation-of-state theory to specific interacting systems was then made by assuming, for a spinodal curve, a constant negative value for X_{12}. A more accurate treatment would incorporate the functional dependence of X_{12} on T, P, ψ, r, and possibly other polymer characteristics. The other parameter of interest, the segmental surface area ratio s_1/s_2, was

2.5. Equation-of-State Approach

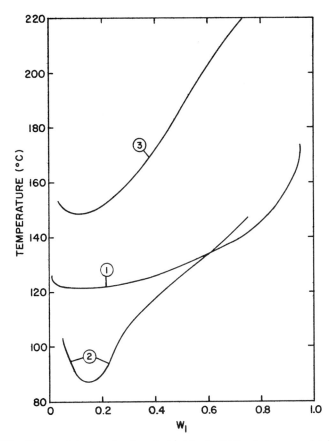

Fig. 2.32. Cloud-point curves for the miscible blends of (1) polystyrene–poly(vinyl methyl ether) (PS–PVME), (2) styrene/acrylonitrile copolymer–polycaprolactone, and (3) styrene/acrylonitrile copolymer–poly(methyl methacrylate). The molecular weights appear in Table 2.5. [Reprinted by permission from L. P. McMaster and O. Olabisi, *Am. Chem. Soc., Div. Org. Coat. Plas. Chem., Pap.* **35**, 322 (1975). Copyright by the American Chemical Society.]

initially estimated from the group contribution format of Bondi [157] to be 2.4. Since Bondi's approach is not entirely satisfactory, Olabisi introduced arbitrary variation within the range $1.0/2.4 \leq s_1/s_2 \leq 2.4$.

The simulated spinodal curves obtained are shown in Fig. 2.33. With s_1/s_2 kept at 2.4 (curves 1–3), variation of X_{12} from -1.0 to -2.0 cal/cm^3 leads to a shift in the minimum temperature from $-67°$ to $65°$C while $\phi_{\min}$ is essentially unchanged at 0.9, i.e., at a high concentration of PCL, whose chains are relatively larger than those of PVC. This phenomenon, referred to as asymmetry of the critical concentration, was discussed earlier in con-

TABLE 2.5

Molecular Weight Properties for Miscible Blends of PS–PVME, SAN Copolymer–PCL, and SAN Copolymer–PMMA[a]

System	Polymer	Commercial code	M_n	M_w	M_z	M_w/M_n	M_z/M_w
1	Polystyrene–	Union Carbide SMD-3500	78,000	237,000	455,000	3.02	1.92
	poly(vinyl methyl ether)	GAF Gantrez M-093	7,700	13,300	21,800	1.73	1.63
	Styrene/acrylonitrile	Union Carbide RMD-4511	88,600	223,000	680,000	2.06	3.05
	copolymer–polycaprolactone	Union Carbide PCL-700	2,200	35,000	61,000	15.6	1.75
3	Styrene/acrylonitrile	Union Carbide RMD-4511	88,600	223,000	680,000	2.06	3.05
	copolymer–poly(methyl methacrylate)	duPont Lucite 140	45,600	92,000	145,000	2.02	1.58

[a] Reprinted with permission from L. P. McMaster and O. Olabisi, *Am. Chem. Soc., Div. Org. Coat. Plast. Chem., Pap.* **35**, 322 (1975). Copyright by the American Chemical Society.

TABLE 2.6

Properties of the PVC–PCL System[a]

Polymer	Commercial code	M_n	M_w	ρ(g/cm^3)	α(°K$^{-1} \times 10^4$)	γ(atm/°K)
Polycaprolactone	Union Carbide PCL-700	15,500	40,500	1.095	7.20	14.0
Poly(vinyl chloride)	Union Carbide QYSA	13,500	30,000	1.406	5.18	22.4
	$-10 < X'_{12} < -2$ cal/cm^3	(Olabisi)				
	$1.0/2.4 < s_1/s_2 < 2.4$	(Value unknown, Bondi's technique gives upper value)				

[a] Reprinted with permission from O. Olabisi, *Macromolecules* **8**, 316 (1975). Copyright by the American Chemical Society.

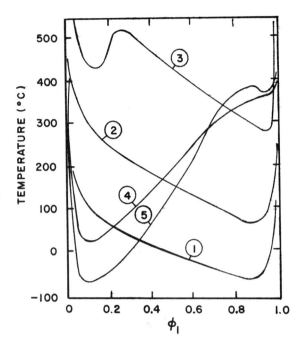

CURVE No.	X_{12} (cal/cm^3)	S_1/S_2
1	1.0	2.4
2	2.0	2.4
3	7.0	2.4
4	2.0	0.71
5	2.0	0.42

Fig. 2.33. Simulated spinodals for the polycaprolactone–poly(vinyl chloride) system whose properties appear in Table 2.6. [Reprinted by permission form O. Olabisi, *Macromolecules* **8**, 316 (1975). Copyright by the American Chemical Society.]

nection with the cloud-point curves in Fig. 2.23. Decreasing X_{12} to -7 cal/cm^3 leads to bimodality of the phase diagram with T_{min} at 280° and 480°C. Bimodality, discussed earlier with respect to Fig. 2.24, has also been demonstrated by Powers [171] for a low molecular weight α-methylstyrene–vinyl toluene copolymer when mixed with a low molecular weight polybutene (Fig. 2.34). Aharoni [172] observed a similar phenomenon with high molecular weight epoxy and copolyester comixed with 1,1′,2,2′-tetrachloroethane. The epoxy is a reaction product of a 50% copolymer of bisphenol A/epichlorohydrin (Union Carbide Phenoxy PKHH grade, $M_w \sim 80{,}000$,

2.5. Equation-of-State Approach

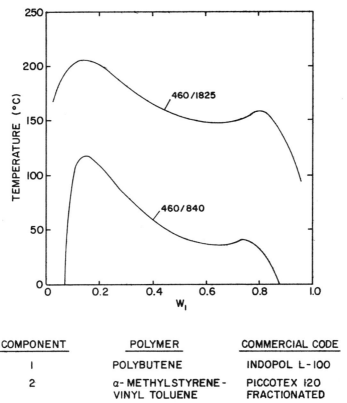

COMPONENT	POLYMER	COMMERCIAL CODE
1	POLYBUTENE	INDOPOL L-100
2	α-METHYLSTYRENE–VINYL TOLUENE	PICCOTEX 120 FRACTIONATED

Fig. 2.34. Cloud-point curves of the miscible blends of polybutene and α-methylstyrene–vinyl toluene copolymer with the indicated r_1/r_2 values. [Reprinted by permission from P. O. Powers, *Polym. Prep., Am. Chem. Soc., Div. Polym. Chem.* **15**, 528 (1974). Copyright by the American Chemical Society.]

$M_n = 23{,}000$), 40% M epoxy intermediate (diglycidyl ether of bisphenol A, Ciba-Geigy Araldite 6099 grade, $M_w \sim 5000$–8000) and 10% low M epoxy (diglycidyl ether of bisphenol A, Ciba-Geigy Araldite 6010 grade, $M_w = 370$–390). The copolyester is a reaction product of terephthalate/isophthalate/sebacate (65/30/5) and ethylene glycol/resorcinol di(β-hydroxyethyl)ether/poly(tetramethylene ether glycol) (70/20/10). A few of Aharoni's results appear in Fig. 2.35. Koningsveld *et al.* [99], who were able to describe all of the above-mentioned behavior with a g defined as a suitably chosen function of temperature, pressure, composition, and chain length on a post-priori basis, concluded that bimodality is not caused by the system polydispersity but by the quadratic dependence of g on ψ_2. The analysis here shows that bimodality is a consequence of the odd combinations of large

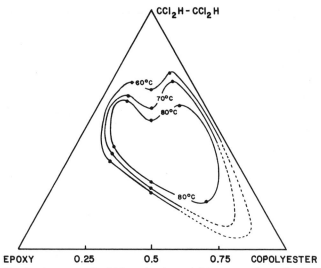

Fig. 2.35. Cloud-point curves for high molecular weight epoxy and copolyester comixed with 1,1′,2,2′-tetrachloroethane. [Reprinted by permission from S. M. Aharoni, *Macromolecules* **11**, 277 (1978). Copyright by the American Chemical Society.]

differences in the pure component properties (which would ordinarily cause immiscibility) with rather large negative exchange energy parameters (which would ordinarily cause miscibility).

All the experimental observations cited are for systems exhibiting ucst behavior; the asymmetry has not been observed and the bimodality has been observed only once for high molecular weight systems exhibiting lcst behavior, namely, styrene/acrylonitrile copolymer mixed with poly(methyl methacrylate) [173]. The reason may lie in the fact that very little effort has been expended so far on trying to observe the phenomena or it may be due to the deleterious effect of polydispersity. All the simulated spinodals in Fig. 2.33 are generated for the polydisperse PCL–PVC systems, but then a quasibinary spinodal curve is invariant for a fixed weight-average chain length. However, the binodal and the location of the consolate state depend on the polydispersity ratio, and the shape of the binodal (which is far more difficult to calculate) would be more reflective of polydispersity.

In Fig. 2.33, note that, when X_{12} is kept constant at -2.0 cal/cm^3 and s_1/s_2 is changed from 2.4 to 0.712, a complete reversion of the spinodal is produced. As the ratio is decreased further to 1/2.4, the bimodality which starts from the previous curve now becomes obvious. It seems safe to conclude that, according to the equation-of-state theory, asymmetry is due to the relative segmental surface area of the pure components; this suggests that this parameter is important, which has been previously unsuspected.

2.5. Equation-of-State Approach

The question as to the goodness of fit of the theory cannot, however, be answered since direct comparison of the predictions with experiment is not possible because of the thermal instability of PVC at high temperature. Unfortunately, binodals have not been computed for those low molecular weight mixtures whose cloud-point curves are available. Such computations might also be useful in identifying whether the critical points are stable or metastable.

2.5.5 Alternative Theories

The two more popular theories have been discussed, and both, in their own way, can give quantitative descriptions of the phase behavior of binary polymer mixtures. The conventional Flory–Huggins lattice theory describes all the observed behavior provided the functionality of $g(T, \psi_2, P)$ is prescribed by appropriate experimental data. That is, g cannot be derived from lattice considerations alone and the theory provides neither the understanding of the origin of the observed behavior nor does it have any predictive capability. The Flory equation-of-state theory defines $g(T, \psi_2, P)$ in terms of a minimum set of fundamental parameters. That is, it attempts to describe the disruption in the local liquid structure that results from mixing two fluids having different packing densities or "free volumes." Notable consequences of this are embodied in the following predictions for any mixture:

(a) There is a nonvanishing "free volume" contribution to the excess enthalpy even when each component has identical reduced volume (provided, of course that there is a volume change on mixing).
(b) There is a nonvanishing "free volume" contribution to the excess enthalpy even when excess volume vanishes (provided the reduced volume of each component is not identical).
(c) The residual entropy is nonvanishing when the excess volume of mixing vanishes.

These assertions are completely at variance with the regular solution theory and, consequently, with the Flory–Huggins theory, which originates with and conserves those important assumptions of the theory of strictly regular solutions. Obviously, the empirical modifications introduced by Koningsveld would describe these features of mixtures on a post-pirori basis; however, such descriptions provide no fundamental insight into the problem.

The ability of the equation-of-state approach to qualitatively predict simultaneous occurrences of upper and lower critical solution temperature behavior is due to its additional allowance for the "free volume" effects. At low temperature, its intermolecular energy dissimilarities cause the occurrence of the ucst behavior; at high temperatures, the "free volume" dissimilarities result in the lcst behavior.

Presently, the theory has not been tested for its quantitative accuracy with respect to the polymer–polymer miscibility, and it is not known where quantitative improvement should be sought. However, a particular shortcoming is now becoming apparent. Recent measurements [100, 173] of polymer–polymer spinodals by "pulse-induced critical" scattering and by conventional light-scattering reveal that the shape and location of the spinodal do indeed depend on polydispersity. Such an occurrence is not accommodated by the Flory equation-of-state theory; neither is it accommodated by Koningsveld's modification of the lattice theory. Only the new lattice theory of Huggins [113–115] has provided [100] a reasonable, albeit inconclusive, basis for these phenomena; chain flexibility dissimilarity between the polymers in the mixture plays a major role in defining the shape and location of the miscibility envelope. However, current thinking [173] ascribes the influence of polydispersity on the shape and location of the spinodal and the binodal (or the cloud-point curve) to nonuniformity of segment densities at both ends of the concentration scale. In other words, the radius of gyration is one of the major factors governing polymer–polymer miscibility.

The recent lattice fluid model of Sanchez and Lacombe [138] awaits a thorough exposition.

2.6 THERMODYNAMICS OF BLOCK COPOLYMER SYSTEMS

Although the field of block copolymers is relatively new, various theoretical treatments of phase separation in block copolymers have appeared; some aspects of the theories have been substantiated by experimental observations. The presence of covalent bonds in copolymers places some constraints on the number of possible arrangements, resulting in an appreciable loss in configurational entropy of mixing. Because of the entropy loss, the critical molecular weight required for microphase separation in copolymers is predicted to be considerably higher than in the analogous homopolymer mixtures. Recent experimental observations support these theoretical predictions. The theories to be discussed below relate to block copolymers, but the general conclusions apply to graft, comb, and radial-type block copolymers.

The available theories of the phase separation phenomena are quite rigorous, but they are inadequate, as are many treatments of simpler polymer mixtures. They rely on the interaction parameter, χ_{AB}, which is at best a

2.6. Thermodynamics of Block Copolymer Systems

rough approximation, poorly defined for polymer–polymer interactions. χ_{AB} can be accurately estimated for polymer systems exhibiting purely dispersive interactions, but it will invariably be erroneous when specific interactions exist. However, several experimental techniques do exist where values of χ_{AB} are estimated in cases involving specific interactions [174–178].

Of the various theoretical treatments of phase separation in block copolymers, the methods of Krause [174, 175] and of Meier [176–178] will be summarized here. Krause's approach is concerned only with thermodynamic phase separation and not with the morphology of the domain structure. The treatment is not limited to a particular type of block copolymer (i.e., A–B); it is applicable to $(AB)_n$ systems, as well as to mixtures of one or more of the copolymers with the constituent homopolymers. Phase separation resulting from a crystallizable constituent is also covered.

The following assumptions are basic to the development of Krause's method:

(1) The block copolymers are monodisperse.
(2) The same number of blocks and sequence distribution exists within each molecule.
(3) Complete separation between the constituent phases exists.
(4) Any homopolymer present is monodisperse and is miscible only with its analogous phase.
(5) The lattice model is applicable.

With these assumptions, the enthalpy change for amorphous block copolymer constituents upon phase separation is

$$\Delta H = -(kTV/V_r)V_A^M V_B^M \chi_{AB}(1 - 2/z) \quad (2.144)$$

where V is the total volume of the system, T is the absolute temperature, V_A and V_B are the volume fraction of monomer repeat units for constituents A and B, z is the coordination number of the lattice, V_r is the volume of the lattice site, and superscript M refers to the total mixture.

The total entropy change on phase separation is given by

$$\frac{\Delta S}{k} = N_c[V_A^C \ln V_A^M + V_B^C \ln V_B^M] + N_{HA} \ln V_A^M$$
$$- 2N_c(m - 1)(\Delta S_{dis}/R) + N_c \ln(m - 1) \quad (2.145)$$

where N_{HA} represents the number of homopolymer molecules, m is the number of blocks in the block copolymer, $\Delta S_{dis}/R$ is the disorientational gain on fusion per segment of a molecule, and superscript C refers to the block copolymer.

At equilibrium it is assumed* that $\Delta G = 0$ and $\Delta H = T\Delta S$; hence, the critical interaction parameter $(\chi_{AB})_{cr}$ for phase separation is

$$(\chi_{AB})_{cr} = \frac{zV_r}{(z-2)V_B V_A^M} n_B^C \left[-V_A^C \ln V_A^M - V_B^C \ln V_B^M \right.$$

$$\left. - \frac{N_{HA}}{N_c} \ln(V_A^M) + 2(m-1)\frac{\Delta S_{dis}}{R} - \ln(m-1) \right] \quad (2.146)$$

Values of $z = 8$ and $\Delta S_{dis}/R = 1.0$ were assumed by Krause.

The predictions of the theory have been compared with the experimental data of polystyrene–poly(α-methylstyrene) AB block copolymer by Robeson et al. [179], who investigated AB block copolymers with molecular weights of the individual blocks of up to 210,000 M_w ($M_w/M_n \simeq 1.1$). Although some broadening of the glass transition occurred with molecular weight increase, single-phase behavior was clearly indicated by the single mechanical loss peak, as illustrated in Fig. 2.36. On the other hand, Dunn and Krause [180] observed two-phase behavior for the AB block copolymer at a total copolymer molecular weight higher than 500,000 M_w. At a value of $(\chi_{AB})_{cr} = 0.002$ for polystyrene and poly(α-methylstyrene) blends [181], phase separation is predicted to occur at a block molecular weight greater than 400,000 for the AB block copolymer with equivalent volume fractions of the constituent blocks. Considering the error involved with the $(\chi_{AB})_{cr}$ estimation, this agreement of theory and experimental data is reasonable.

Blends of polystyrene and poly(α-methylstyrene) homopolymer constituents with 420,000 M_w polystyrene–poly(α-methylstyrene) AB block copolymer provide an interesting observation. Polystyrene, when mixed with the block copolymer, did phase-separate, whereas the mixture of poly(α-methylstyrene) with the block copolymer exhibited single-phase behavior. The theory erroneously predicts single-phase behavior for both blends. Finally, recent calculations [181] show that a value of χ_{AB} between 0.0030 and 0.0036 will satisfactorily represent the experimental data.

In the analysis by Meier [176–178] for establishing the criteria for phase separation in AB block copolymers, three contributions are required to account for the entropy difference between the block copolymer and that of a simple mixture of the components. The restriction of the position of the A–B covalent bonds between the domains results in an entropy decrease termed the placement entropy ΔS_p. Based on the lattice model, one has

$$\Delta S_p = -kN_{AB} \ln[3\sigma_A \Delta R/(\sigma_A + \sigma_B R)] \quad (2.147)$$

---

* This equilibrium condition is only valid for a one-component system. Hence, it is valid for the treatment of phase separation in block copolymers but not for mixtures of copolymers with homopolymers. In multicomponent mixtures phase separation can occur even if $\Delta G < 0$ [105]; see also Fig. 2.4 and the accompanying discussion.

2.6. Thermodynamics of Block Copolymer Systems

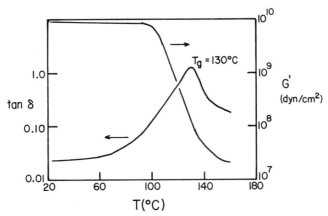

Fig. 2.36. Mechanical loss and shear modulus data on polystyrene–poly(α-methylstyrene) AB block copolymer. $M_w = 420,000$; $M_n = 390,000$; 50 mole% each constituent. [Reprinted by permission from L. M. Robeson, M. Matzner, L. J. Fetters, and J. E. McGrath, in "Recent Advances in Polymer Blends, Grafts, and Blocks" (L. H. Sperling, ed.), p. 281. Copyright 1974 by Plenum Publishing Corp.]

where σ_A and σ_B represent the number of A and B segments in the N_{AB} copolymer molecules, R is the domain radius, and ΔR is the region occupied by the A–B covalent bonds.

A reduction in entropy, due to the requirement that A segments remain in domain space of A and that B segments remain in B domain, arises and it is termed the "restricted volume" entropy difference, ΔS_v [182]:

$$\Delta S_v = N_{AB}k[\ln P(\sigma_A; r', r < R) + \ln P(\sigma_B; r', r > R)] \quad (2.148)$$

where $P(\sigma_B; r', r > R)$ is the probability that all σ_B chain elements are outside the A domain and $P(\sigma_A; r', r < R)$ is the probability that all σ_A chain elements are inside the domain when the chain origin is at r'. The domain radius is given by $R = (\frac{4}{3})(\sigma_A l^2)^{1/2}$. The following represent the functional relationship of the probabilities:

$$P(\sigma_A; r', r < R) = 2 \sum_{i=1}^{\infty} (-1)^{i+1} j_0(i\pi r'/R) \exp\{-i^2\pi^2\sigma_A l^2/6R^2\} \quad (2.149)$$

$$P(\sigma_B; r', r > R) = 1.0 - (R/r') \operatorname{Erfc}\left(\frac{3}{2\sigma_B l^2}\right)^{1/2}(r' - R) \quad (2.150)$$

A third entropy difference (also a decrease relative to the simple mixture) arises for the perturbation of chain dimensions in the domain system from the random-flight values. This is termed the "elasticity entropy," ΔS_{el}:

$$\Delta S_{el} = -(\tfrac{3}{2})N_{AB}k(\alpha^2 - 1 - 2\ln \alpha) \quad (2.151)$$

where α is the ratio of perturbed to unperturbed end-to-end chain distances.

The enthalpy change is considered equivalent to the negative heat of mixing of simple mixtures of the constituent blocks:

$$\Delta H^M = N_{AB} k T \chi_{AB} \phi_A \qquad (2.152)$$

where $\phi_A = \sigma_A/(\sigma_A + \sigma_B)$.

Finally, the surface free energy contribution, G_s, is given as

$$G_s = \frac{9}{4} \frac{N_{AB} M_A^{1/2} \gamma}{\bar{A} \rho \alpha K} \qquad (2.153)$$

G_s represents the interaction of A and B segments at the domain interface. $\bar{A}$ is Avogadro's number and ρ is the density.

Summing all the contributions, the free energy change for phase separation of a miscible mixture of the AB block copolymer molecules is

$$\Delta G = N_{AB} k T \left\{ \ln \frac{(\sigma_A + \sigma_B) R}{3 \sigma_A \Delta R} - \ln P(\sigma_A; r', r < R) - \ln P(\sigma_B; r', r > R) \right.$$

$$\left. + \tfrac{3}{2}(\alpha^2 - 1 - 2 \ln \alpha) + \frac{9 M_A^{1/2} \gamma}{4 \alpha K \bar{A} \rho k T} - \chi_{AB}\left(\frac{\sigma_A}{\sigma_A + \sigma_B}\right) \right\} \qquad (2.154)$$

The minimum molecular weights required for domain formation for the AB block copolymer versus phase separation of simple polymer mixtures were computed by Meier. For a simple mixture of homopolymers,

$$\Delta G^M = kT(n_A \ln \phi_A + n_B \ln \phi_B + \chi_{AB} n_B \phi_A) \qquad (2.155)$$

where n_A and n_B are the number of A and B molecules ($n_A = n_B$ for this analysis). At equilibrium phase separation, he assumed* that $\Delta G^M = 0$;

$$\chi_{AB} \sigma_A/(\sigma_A + \sigma_B) = \ln\left\{\frac{(\sigma_A + \sigma_B)^2}{\sigma_A \sigma_B}\right\} \qquad (2.156)$$

For $\sigma_B \gg \sigma_A$, the ratio of σ_A^d for copolymer domain formation and σ_A^m for homopolymer phase separation is obtained by combining Eqs. (2.154) and (2.155) with the substitution of α_m (the equilibrium chain expansion parameter: $\alpha_m^3 - \alpha_m = 3 M_A^{1/2} \gamma / 4 K k \bar{\rho} \bar{A} T$) for α.

$$\frac{\sigma_A^d}{\sigma_A^m} = \frac{\ln[(\sigma_B/3\sigma_A^d)(R/\Delta R)] - \ln P(\sigma_A; r', r < R) - \ln P(\sigma_B; r', r > R)}{\ln(\sigma_B/\sigma_A^m)}$$

$$+ \frac{(\tfrac{9}{2})(\alpha_m^2 - 1) - 3 \ln \alpha_m}{\ln(\sigma_B/\sigma_A^m)} \qquad (2.157)$$

* This equilibrium condition is only valid for a one-component system. Hence, it would be valid if used for Eq. (2.154) but not for Eq. (2.155). In multicomponent mixtures, phase separation can occur even if $\Delta G < 0$ [105]; see also Fig. 2.4 and the accompanying discussion.

2.6. Thermodynamics of Block Copolymer Systems

Using appropriate estimations, Meier determined that the ratio of critical molecular weights σ_A^d/σ_A^m will be 2.5–5. Thus, phase separation in a block copolymer requires a significantly higher molecular weight than in the analogous homopolymer mixtures.

Meier [178] extended his treatment of domain formation in AB block copolymers so as to estimate the most stable configuration as a function of block copolymer volume fraction. Spherical domains are preferred below volume fractions of 0.20; cylindrical domains are preferred between 0.20 and 0.30 with lamellar structure existing above 0.30. Experimental evidence confirms the theoretical predictions illustrated in Fig. 2.37.

The statistical thermodynamics of ABA block copolymer domain formation have been developed by Leary and Williams [183, 184]. A major difference between their treatment and Meier's is their incorporation of a mixed region at the interface which is assumed to exhibit single-phase behavior. The entropy of phase separation was considered to arise from three individual contributions, namely,

$$\Delta S_{\text{tot}} = \Delta S_1 + \Delta S_A + \Delta S_B \tag{2.158}$$

where ΔS_1 is the entropy change due to restricting an A–B junction of the block copolymer to a position in the mixed region, ΔS_A is due to the requirement that A blocks remain in A domains or the mixed region, and ΔS_B results from locating the B blocks outside of the A domain with both block ends in the mixed region. The model actually predicts the specific microstructure which is thermodynamically favored.

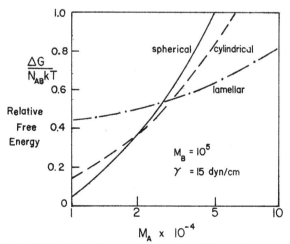

Fig. 2.37. Predicted relative free energies for three different domain morphologies of an AB block copolymer as a function of molecular weight (B block molecular weight held constant). [Reprinted by permission from D. J. Meier, *Polym. Prepr., Am. Chem. Soc., Div. Polym. Chem.* **11,** 400 (1970). Copyright by the American Chemical Society.]

Other theoretical treatments of phase separation in block copolymers include that due to Pouchly *et al.* [185, 186], to Bianchi *et al.* [187, 188], to Marker [189], and to Helfand [190, 191]. Further refinements of block copolymer phase behavior were recently published by Meier [192, 193].

On the experimental scene, electron microscopy studies for the polysty-

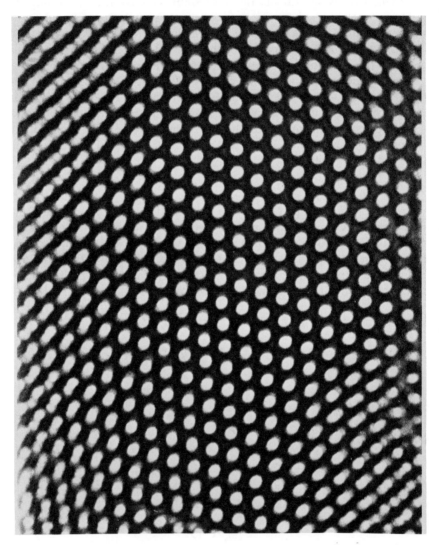

Fig. 2.38. Morphology of a radial block copolymer (styrene-*b*-butadiene). M_w of star = 8.95×10^5; domain diameter = 280 Å; cubic unit cell distance = 457 Å; 27 wt% polystyrene. [Photomicrograph kindly provided by Dr. L. J. Fetters of the University of Akron, Akron, Ohio.]

rene–diene block copolymers have been extensive [194–197]. These studies have shown the ordered structures of these phase-separated block copolymers with domain configurations similar to those predicted by theory. An example of this ordered morphology is shown in Fig. 2.38 for a radial block copolymer of polystyrene and polybutadiene.

While the thermodynamics of phase separation for block copolymers has been extensively studied, a similar analysis for interpenetrating networks has not been forthcoming, presumably because of the complexity of such mixtures. Donatelli *et al.* [198] presented a semiempirical approach for determining the domain size of a second component as a function of cross-link density, interfacial energy, and composition. Assuming that the density of both constituents equals $1.0 \, \text{g/cm}^3$ [i.e., the weight fraction (w_i) and volume fraction (V_i) are equivalent], the domain size for the second component is

$$D_2 = \frac{2\gamma w_2}{RT v_i [(1 - w_2)^{2/3} + w_2/v_i M_2 - \frac{1}{2}]} \qquad (2.159)$$

where v_i is the cross-link density of component 1, γ is the interfacial energy, and M_2 is the molecular weight of component 2.

REFERENCES

1. P. H. Geil, *Ind. Eng. Chem., Prod. Res. Dev.* **14**, 59 (1975).
2. A. J. Yu, *Adv. Chem. Ser.* **99**, 2 (1971).
3. L. P. McMaster, *Macromolecules* **6**, 760 (1973).
4. S. H. Maron and C. F. Prutton, "Principles of Physical Chemistry." Macmillan, New York, 1958.
5. B. A. Wolf and G. Blaum, *Am. Chem. Soc., Div. Org. Coat. Plast. Chem., Pap.* **37**, 11 (1977).
6. D. D. Patterson, *J. Paint Technol.* **45**, 37 (1973).
7. J. H. Hildebrand and R. L. Scott, "The Solubility of Non-Electrolytes." Dover, New York, 1964.
8. L. Bohn, *Rubber Chem. Technol.* **41**, 495 (1968).
9. P. J. Flory, "Principles of Polymer Chemistry." Cornell Univ. Press, Ithaca, New York, 1953.
10. R. L. Scott, *J. Chem. Phys.* **17**, 279 (1949).
11. C. H. Hinshelwood, "The Structure of Physical Chemistry," Chapter XIII. Oxford Univ. Press, London and New York, 1951.
12. A. D. Buckingham, *Adv. Chem. Phys.* **12**, 107 (1967).
13. A. L. McClellan, "Tables of Experimental Dipole Moments." Freeman, San Francisco, California, 1963.
14. J. L. Gardon, *J. Paint Technol.* **38**, 43 (1966).
15. G. C. Pimental and A. L. McClellan, "The Hydrogen Bond." Freeman, San Francisco, California, 1960.
16. O. Olabisi, *Macromolecules* **8**, 316 (1975).
17. N. S. Schneider and C. S. Paik Sung, *Polym. Eng. Sci.* **17**, 73 (1977).
18. K. L. Smith, A. E. Winslow, and D. E. Peterson, *Ind. Eng. Chem.* **51**, 1361 (1959).

19. G. R. Williamson and B. Wright, *J. Polym. Sci., Part A* **3**, 3885 (1965).
20. Y. Osada and H. Sato, *Polym. Lett.* **14**, 129 (1976).
21. K. L. Smith, A. E. Winslow, and D. E. Peterson, *Ind. Eng. Chem.* **51**, 1361 (1959).
22. L. A. Bimendina, V. V. Roganov, and Ye. A. Bekturov, *Polym. Sci. USSR (Engl. Transl.)* **16**, 3274 (1974).
23. H. Sato and A. Nakajima, *Polym. J.* **7**, 241 (1975).
24. A. S. Hoffman, R. W. Lewis, and A. S. Michaels, *Am. Chem. Soc., Div. Org. Coat. Plast. Chem., Pap.* **29**, 236 (1969).
25. T. Sulzberg and R. J. Cotter, *J. Polym. Sci., Part A 1* **8**, 2747 (1970).
26. W. Gibbs, "Collected Works," Vol. I, Longmans, Green, and Co., New York, 1928.
27. W. Ostwald, as quoted by M. Volmer, "Kinetik der Phasenbildung." Edwards, Ann Arbor, Michigan, 1945.
28. M. Volmer, "Kinetik der Phasenbildung." Edwards, Ann Arbor, Michigan, 1945.
29. R. Becker and W. Doering, *Ann. Phys. (Leipzig)* [5] **24**, 719 (1935).
30. J. Frenkel, "Kinetic Theory of Liquids," Chapter VII. Oxford Univ. Press, London and New York, 1946.
31. J. W. Cahn, *Trans. Metall. Soc. AIME* **242**, 166 (1968).
32. V. K. La Mer, *Ind. Eng. Chem.* **44**, 1270 (1952).
33. L. P. McMaster, *Adv. Chem. Ser.* **142**, 43 (1975).
34. I. M. Lifshitz and V. V. Slyozov, *J. Phys. Chem. Solids, Lett. Sect.* **19**, 35 (1961).
35. T. Nishi, T. T. Wang, and T. K. Kwei, *Macromolecules* **8**, 227 (1975).
36. J. S. Langer, in "Fluctuations, Instabilities, and Phase Transitions" (T. Riste, ed.), p. 19. Plenum, New York, 1975.
37. J. W. Cahn, *J. Chem. Phys.* **42**, 93 (1965).
38. P. J. Debye, *Chem. Phys.* **31**, 680 (1959).
39. J. W. Cahn and J. E. Hilliard, *Chem. Phys.* **28**, 258 (1958); **31**, 688 (1959).
40. J. J. van Aarsten, *Eur. Polym. J.* **6**, 919 (1970).
41. J. E. Hilliard, "Phase Transformations," Chapter 12. Am. Soc. Met., Metals Park, Ohio, 1970.
42. O. Olabisi and R. Simha, *J. Appl. Polym. Sci.* **21**, 149 (1977).
43. J. E. Guillet, British Patent 1,331,429 (1973).
44. R. D. Newman and J. M. Prausnitz, *J. Phys. Chem.* **76**, 1492 (1972).
45. A. J. Manning and F. Rodriguez, *J. Appl. Polym. Sci.* **17**, 1651 (1973).
46. W. R. Song and D. W. Brownawell, *Polym. Eng. Sci.* **10**, 222 (1970).
47. B. A. Wolf, *Makromol. Chem.* **178**, 1869 (1977).
48. R. J. Kern, *J. Polym. Sci.* **21**, 19 (1956).
49. J. W. Schurer, A. de Boar, and G. Challa, *Polymer* **16**, 201 (1975).
50. A. E. van Arkel and S. E. Vles, *Trans. Faraday Soc.* **42B**, 81 (1946).
51. C. M. Hansen, *J. Paint Technol.* **39**, 104 (1967).
52. C. M. Hansen, *Ind. Eng. Chem., Prod. Res. Dev.* **8**, 2 (1969).
53. R. F. Blanks and J. M. Prausnitz, *Ind. Eng. Chem., Fundam.* **3**, 1 (1964).
54. J. D. Crowley, G. S. Teague, Jr., and J. W. Lowe, Jr., *J. Paint Technol.* **38**, 269 (1966).
55. P. A. Small, *J. Appl. Chem.* **3**, 71 (1953).
56. R. C. Nelson, R. W. Hemwell, and G. D. Edwards, *J. Paint Technol.* **42**, 636 (1970).
57. E. B. Bagley, T. P. Nelson, and J. M. Scigliano, *J. Paint Technol.* **43**, 35 (1971).
58. D. M. Koenhen and C. A. Smolders, *J. Appl. Polym. Sci.* **19**, 1163 (1975).
59. R. Bonn and J. J. van Aartsen, *Eur. Polym. J.* **8**, 1055 (1972).
60. K. S. Siow and D. Patterson, *Macromolecules* **4**, 26 (1971).
61. J. L. Gardon, *J. Phys. Chem.* **67**, 1935 (1963).
62. K. L. Hoy, *J. Paint Technol.* **42**, 76 (1970).

References

63. A. H. Konstam and W. R. Feairheller, Jr., *AIChE J.* **16**, 837 (1960).
64. D. W. Van Krevelen, "Properties of Polymers, Correlations with Chemical Structure," p. 85. Elsevier, Amsterdam, 1972.
65. M. T. Shaw, *J. Appl. Polym. Sci.* **18**, 449 (1974).
66. E. P. Lieberman, *Off. Dig., Fed. Soc. Paint Technol.* **34**, 30 (1962).
67. S. A. Chen, *J. Appl. Polym. Sci.* **15**, 1247 (1971).
68. S. A. Chen, *J. Appl. Polym. Sci.* **16**, 1603 (1972).
69. R. L. Scott, *J. Chem. Phys.* **17**, 279 (1949).
70. T. L. Hill, "An Introduction to Statistical Thermodynamics." Addison-Wesley, Reading, Massachusetts, 1960.
71. J. A. Barker, "Lattice Theories of the Liquid State." Macmillan, New York, 1963.
72. R. H. Fowler and G. S. Rushbrooke, *Trans. Faraday Soc.* **33**, 1272 (1937).
73. T. S. Chang, *Proc. R. Soc. London, Ser. A* **169**, 512 (1939).
74. T. S. Chang, *Proc. Cambridge Philos. Soc.* **35**, 265 (1939).
75. A. R. Miller, *Proc. Cambridge Philos. Soc.* **38**, 109 (1942).
76. M. L. Huggins, *J. Chem. Phys.* **9**, 440 (1941).
77. M. L. Huggins, *J. Phys. Chem.* **46**, 151 (1942).
78. A. J. Staverman and J. H. Van Santen, *Recl. Trav. Chim. Pays Bas.* **60**, 76 (1941).
79. P. J. Flory, *J. Chem. Phys.* **9**, 660 (1941).
80. P. J. Flory, *J. Chem. Phys.* **10**, 51 (1942).
81. R. Koningsveld, Ph.D. Thesis, University of Leiden (1967).
82. A. Dobry, *J. Chim. Phys.* **35**, 387 (1939).
83. E. A. Guggenheim, *Trans. Faraday Soc.* **44**, 1007 (1948).
84. S. H. Maron, *J. Polym. Sci.* **38**, 329 (1959).
85. H. Tompa, "Polymer Solutions." Butterworth, London, 1956.
86. R. Koyama, *J. Polym. Sci.* **35**, 247 (1949).
87. G. Rehage, *Kunststoffe* **53**, 605 (1963).
88. A. J. Staverman, *in* "Handbuch der Physik" (S. Flügge, ed.), Vol. 13, p. 456. Springer-Verlag, Berlin and New York, 1962.
89. G. Delmas and D. Patterson, *J. Polym. Sci.* **57**, 79 (1962).
90. R. Koningsveld and A. J. Staverman, *J. Polym. Sci., Part C-1* **6**, 1775 (1967).
91. R. Koningsveld and A. J. Staverman, *J. Polym. Sci., Part A-2* **6**, 325 (1968).
92. M. Gordon, H. A. G. Chermin, and R. Koningsveld, *Macromolecules* **2**, 207 (1969).
93. R. Koningsveld, L. A. Kleintjens, and A. R. Shultz, *J. Polym. Sci., Part A-2* **8**, 1261 (1970).
94. R. Koningsveld, H. A. G. Chermin, and M. Gordon, *Proc. R. Soc. London, Ser. A* **319**, 331 (1970).
95. R. Koningsveld and L. A. Kleintjens, *Macromolecules* **4**, 637 (1971).
96. J. W. Kennedy, M. Gordon, and R. Koningsveld, *J. Polym. Sci., Part C* **39**, 43 (1972).
97. R. Koningsveld, *Chem. Zvesti* **26**, 263 (1972).
98. R. Koningsveld and L. A. Kleintjens, *Pure Appl. Chem., Macromol. Suppl.* **8**, 197 (1973).
99. R. Koningsveld, L. A. Kleintjens, and H. M. Schoffeleers, *Pure Appl. Chem.* **39**, 1 (1974).
100. R. Koningsveld and L. A. Kleintjens, *Br. Polym. J.* **9**, 212 (1977).
101. R. Koningsveld, W. H. Stockmayer, J. W. Kennedy, and R. Kleintjens, *Macromolecules* **1**, 73 (1974).
102. R. Koningsveld, *Discuss. Faraday Soc.* **49**, 144 (1970).
103. R. Koningsveld and A. J. Staverman, *J. Polym. Sci., Part A-2* **6**, 349 (1968).
104. K. W. Derham, J. Goldsborough, M. Gordon, R. Koningsveld, and L. A. Kleintjens, *Makromol. Chem., Suppl.* **1**, 28 (1975).
105. R. Koningsveld and L. A. Kleintjens, *J. Polym. Sci., Polym. Symp.* **61**, 221 (1977).
106. L. A. Kleintjens, R. Koningsveld, and W. H. Stockmayer, *Br. Polym. J.* **8**, 144 (1976).

107. J. W. Gibbs, "The Scientific Papers of J. Willard Gibbs," Vol. I. Dover, New York, 1971.
108. P. J. Flory, *J. Chem. Phys.* **10**, 51 (1942).
109. P. J. Flory, *J. Chem. Phys.* **12**, 425 (1944).
110. S. H. Maron and N. Nakajima, *J. Polym. Sci.* **54**, 587 (1964).
111. G. Allen, G. Gee, and J. P. Nicholson, *Polymer* **2**, 8 (1961).
112. D. McIntyre, N. Rounds, and E. Campus-Lopez, *Polym. Prepr., Am. Chem. Soc., Div. Polym. Chem.* **10**, 531 (1969).
113. M. L. Huggins, *J. Phys. Chem.* **74**, 371 (1970).
114. M. L. Huggins, *J. Phys. Chem.* **75**, 1255 (1971).
115. M. L. Huggins, in "Macromolecular Science" (C. E. H. Bawn, ed.), p. 132. Am. Chem. Soc., Washington, D.C., 1974.
116. C. Truesdell, "The Elements of Continuum Mechanics." Springer-Verlag, Berlin and New York, 1965.
117. von Gustav Mie, *Ann. Phys. (Leipzig)* [4] **11**, 657 (1903).
118. H. Eyring and J. F. Kincaid, *J. Phys. Chem.* **43**, 37 (1939).
119. E. Schrodinger, "Statistical Thermodynamics." Cambridge Univ. Press, London and New York, 1949.
120. H. Eyring and J. O. Hirschfelder, *J. Phys. Chem.* **41**, 249 (1937).
121. H. Eyring, *J. Chem. Phys.* **4**, 283 (1936).
122. O. K. Rice, *J. Chem. Phys.* **6**, 476 (1938).
123. P. J. Flory, *J. Am. Chem. Soc.* **86**, 1833 (1965).
124. D. Patterson, *Macromolecules* **2**, 672 (1969).
125. I. Prigogine, "The Molecular Theory of Solutions." North-Holland Pub., Amsterdam, 1957.
126. E. A. Guggenheim, *Proc. R. Soc. London, Ser. A* **135**, 181 (1932).
127. J. E. Lennard-Jones and A. F. Devonshire, *Prov. R. Soc. London, Ser. A* **163**, 53 (1937).
128. J. E. Lennard-Jones and A. F. Devonshire, *Proc. R. Soc. London, Ser. A* **165**, 1 (1938).
129. J. O. Hirschfelder, C. F. Curtiss, and R. B. Bird, "Molecular Theory of Gases and Liquids." Chapman & Hall, London, 1954.
130. J. G. Kirkwood, *J. Chem. Phys.* **18**, 380 (1950).
131. J. S. Rowlinson and C. F. Curtiss, *J. Chem. Phys.* **19**, 1519 (1951).
132. F. Cernuschi and H. Eyring, *J. Chem. Phys.* **7**, 549 (1939).
133. R. O. Davis and J. O. Jones, *Adv. Phys.* **2**, 370 (1953).
134. H. M. Peek and T. L. Hill, *J. Chem. Phys.* **18**, 1252 (1950).
135. D. Henderson, *J. Chem. Phys.* **37**, 631 (1962).
136. H. Eyring and T. Ree, *Proc. Natl. Acad. Sci. U.S.A.* **47**, 526 (1961).
137. R. Simha and T. Somcynsky, *Macromolecules* **2**, 342 (1969).
138. I. C. Sanchez, in "Polymer Blends, Vol. I" (D. P. Paul and S. Newman, eds.) Chapter 3, p. 115. Academic Press, New York, 1978.
139. E. Guth and H. Mark, *Monatsh. Chem.* **65**, 93 (1934).
140. E. Guth, H. M. James, and H. Mark, *Adv. Colloid Sci.* **2**, 253 (1946).
141. A. Muller, *Proc. R. Soc. London, Ser A* **154**, 624 (1936); **178**, 227 (1941).
142. A. Odajima and T. Maeda, *J. Polym. Sci., Part C* **15**, 55 (1966).
143. B. Wunderlich, *J. Chem. Phys.* **37**, 1207 (1962).
144. T. Alfrey and H. Mark, *J. Phys. Chem.* **46**, 112 (1942).
145. T. M. Birshtein and O. B. Ptitsyn, "Conformations of Macromolecules." Wiley (Interscience), New York, 1966.
146. M. V. Volkenstein, "Configurational Statistics of Polymeric Chains." Wiley (Interscience), New York, 1963.

References

147. I. Prigogine, N. Trappeniers, and V. Mathot, *Discuss. Faraday Soc.* **15**, 93 (1953).
148. I. Prigogine, A. Bellamans, and C. Naar-Colin, *J. Chem. Phys.* **26**, 75 (1957).
149. J. Hijmans, *Physica (Utrecht)* **27**, 433 (1961).
150. V. S. Nanda and R. Simha, *J. Phys. Chem.* **68**, 3158 (1964).
151. R. Simha and A. J. Havlik, *J. Am. Chem. Soc.* **86**, 197 (1964).
152. H. S. Frank, *J. Chem. Phys.* **13**, 478 (1945).
153. T. Hill, "Statistical Mechanics." McGraw-Hill, New York, 1956.
154. R. A. Orwoll and P. J. Flory, *J. Am. Chem. Soc.* **89**, 6814 (1967).
155. H. Shih and P. J. Flory, *Macromolecules* **5**, 758 (1972).
156. A. T. DiBenedetto, *J. Polym. Sci., Part A* **1**, 3459 (1963).
157. A. Bondi, *J. Phys. Chem.* **70**, 530 (1966); **68**, 441 (1964).
158. J. G. Curro, *J. Chem. Phys.* **56**, 5739 (1972).
159. I. Prigogine and G. Garikian, *J. Chim. Phys.* **45**, 273 (1948).
160. R. H. Wentorf, R. J. Buehler, J. O. Hirschfelder, and C. F. Curtiss, *J. Chem. Phys.* **18**, 1484 (1950).
161. N. Hirai and H. Eyring, *J. Polym. Sci.* **37**, 51 (1959).
162. V. S. Nanda, R. Simha, and T. Somcynsky, *J. Polym. Sci., Part C* **12**, 277 (1966).
163. A. Quach and R. Simha, *J. Appl. Phys.* **42**, 4392 (1971).
164. O. Olabisi and R. Simha, *Macromolecules* **8**, 211 (1975).
165. R. Simha and P. Wilson, *Macromolecules* **6**, 908 (1973).
166. O. Olabisi, Ph.D. Thesis, Case Western Reserve University, Cleveland, Ohio (1973).
167. T. Nose, *Polym. J.* **2**, 445 (1971).
168. T. Ishinabe and K. Ishikawa, *J. Appl. Phys. Jpn.* **7**, 462 (1968).
169. D. C. Bonner, A. Bellemans, and J. M. Prausnitz, *J. Polym. Sci., Part C* **39**, 1 (1972).
170. L. P. McMaster and O. Olabisi, *Am. Chem. Soc., Div. Org. Coat. Plast. Chem., Pap* **35**, 322 (1975).
171. P. O. Powers, *Am. Chem. Soc., Polym. Prepr., Div. Polym. Chem.* **15**, 528 (1974).
172. S. M. Aharoni, *Macromolecules* **11**, 277 (1978).
173. R. Koningsveld, *Bull. Soc. Chim. Beograd*, **44**, 5 (1979).
174. S. Krause, *J. Polym. Sci., Part A-2* **7**, 249 (1968).
175. S. Krause, *Macromolecules* **3**, 84 (1970).
176. D. J. Meier, *J. Polym. Sci., Part C* **26**, 81 (1969).
177. D. J. Meier, *in* "Sagamore Conference on Block and Graft Copolymers.. (J. J. Burke and V. Weiss, eds.), p. 105. Syracuse Univ. Press, Syracuse, New York, 1973.
178. D. J. Meier, *Polym. Prepr., Am. Chem. Soc., Div. Polym. Chem.* **11**, 400 (1970).
179. L. M. Robeson, M. Matzner, L. J. Fetters, and J. E. McGrath, *in* "Recent Advances in Polymer Blends, Blocks, and Grafts" (L. H. Sperling, ed.), p. 281. Plenum, New York, 1974.
180. D. J. Dunn and S. Krause, *J. Polym. Sci., Polym. Lett.* **12**, 591 (1974).
181. S. Krause, D. J. Dunn, A. Seyed-Mozzaffari, and A. M. Biswas, *Macromolecules* **10**, 785 (1977).
182. D. J. Meier, *J. Phys. Chem.* **71**, 1861 (1967).
183. D. F. Leary aüd M. C. Williams, *J. Polym. Sci., Polym. Phys. Ed.* **11**, 345 (1973).
184. D. F. Leary and M. C. Williams, *J. Polym. Sci., Polym. Phys. Ed.* **12**, 265 (1974).
185. J. Pouchly, A. Zivny, and A. Sikora, *J. Polym. Sci., Part C* **39**, 133 (1972).
186. J. Pouchly, A. Zivny, and A. Sikora, *J. Polym. Sci., Part A-2* **10**, 151 (1972).
187. U. Bianchi, E. Pedemonte, and A. Turturro, *J. Polym. Sci., Part B* **7**, 785 (1969).
188. U. Bianchi, E. Pedemonte, and A. Turturro, *Polymer* **11**, 268 (1970).
189. L. Marker, *Polym. Prepr., Am. Chem. Soc., Div. Polym. Chem.* **10**, 524 (1969).

190. E. Helfand, *J. Chem. Phys.* **62**, 999 (1975).
191. E. Helfand, *Macromolecules* **8**, 552 (1975).
192. D. J. Meier, *Am. Chem. Soc., Div. Org. Coat. Plast., Chem. Pap.* **37**, 246 (1977).
193. D. J. Meier, *Polym. Prepr., Am. Chem. Soc., Div. Polym. Chem.* **18**, 340 (1977).
194. M. Matsuo, S. Sagae, and H. Asai, *Polymer* **10**, 79 (1969).
195. T. Inoue, T. Seon, T. Hashimoto, and H. Kawai, *Macromolecules* **3**, 87 (1970).
196. T. Miyamoto, K. Kodama, and K. Shibayama, *J. Polym. Sci., Part A-2* **8**, 2095 (1970).
197. T. Uchida, T. Soen, T. Inoue, and H. Kawai, *J. Polym. Sci., Part A-2* **10**, 101 (1972).
198. A. A. Donatelli, L. H. Sperling, and D. A. Thomas, *J. Appl. Polym. Sci.* **21**, 1189 (1977).

Chapter 3

Methods for Determining Polymer–Polymer Miscibility

3.1 CRITERIA FOR ESTABLISHING MISCIBILITY

Small-angle neutron scattering in one-component amorphous polymers has established that the polymer chain in the bulk state is essentially randomly placed [1, 2]. This conclusion supports the vast body of other evidence for random statistics. An example from this evidence is the high dependence of viscosity on molecular weight above a critical molecular weight, implying the presence of highly entangled chains. The appropriate picture for one-component, amorphous polymers is that of A rather than B or C in Fig. 3.1.

Polymers dissolved in solvents usually are "expanded" by the interaction of the solvent with the chain segments. Expanded means simply that the average end-to-end distance is increased over that of the bulk. A polymer dissolved in a polymer solvent might be expanded by favorable interactions, in which case a structure represented by Fig. 3.1C would result. In the extreme, a polymer–polymer adduct such as DNA might form. The situation in Fig. 3.1B implies some segregation on a segmental scale, but a random dispersing of molecular *centers*. Methods with high resolution, such as X-ray scattering, small-angle neutron scattering, nmr relaxation, and electron microscopy, suggest that many miscible systems fall between A and B or A and C; i.e., the components are not as randomly mixed as the molecules in a single-component system.

As pointed out by Yu [3], the homogeneity of the polymer–polymer solution, because of its high viscosity, will depend a great deal on the methods of preparation and the time and temperature (energy) to which the mixture is subjected. He felt that concentration equilibrium might be approachable

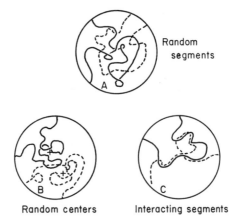

Fig. 3.1. Variations in the placement of two different polymer molecules in a miscible system.

only as an asymptote. More recent evidence shows that, with reasonable care, thermodynamic equilibrium can be bracketed fairly easily. By taking advantage of spinodal decomposition, a one-phase mixture can be transformed into a two-phase mixture regardless of the diffusional barriers. Returning to the one-phase region involves a longer wait [4] or gentle mixing in the melt. Preparation of mixed polymer systems with the aid of solvents can lead to spurious results [5, 6]. Shown schematically in Fig. 3.2 is an extreme case, demonstrated by Robard et al. [6], for the system PS–PVME–CH_3Cl. Removal of solvent during the preparation of polymer mixtures should be accompanied by mixing, such as with a two-roll mill, to aid in the equilibration of the polymer phases, or by annealing at a suitable temperature [4].

It must be emphasized that any experiments on polymer mixtures performed at temperatures other than the temperature of equilibration will be subject to unknown effects due to the slow process of reequilibration at the

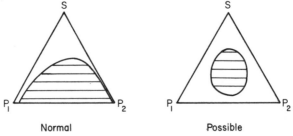

Fig. 3.2. Schematic representation of ternary phase behavior of a system containing two polymers and a solvent.

test temperature. For example, the presence of two T_g's (or two phases by microscopy) for a glassy sample quenched to room temperature from the melt does not mean that the components were immiscible in the melt. Of course, the reverse situation is true as well.

3.2 GLASS TRANSITION TEMPERATURE

Polymers, as with many common liquids, exhibit certain characteristics similar to a second-order transition, if indeed one exists, upon sufficient supercooling below their crystalline melting point. The temperature and pressure derivatives of the thermodynamic quantities of energy, E, enthalpy, H, entropy, S, and volume, V, exhibit a discontinuity at this transition, whereas E, H, S, and V, all first-order derivatives of the free energy, are continuous quantities through this transition. The viscous liquid (or flexible, rubbery material in the case of high molecular weight polymers) is transformed into a hard, glassy material upon passing through this transition. This glass transition is characteristic of the polysilicates more commonly referred to as glasses. However, polyphosphate glasses, organic liquids, and organic polymers also exhibit the features of this glass transition. A glass transition temperature has even been assigned to water (via extrapolation techniques) with a value of 128°K [7]. Highly crystalline materials (e.g., metals) may indeed exhibit glass transitions but the experimental difficulty of supercooling to a glassy state prior to crystallization limits investigation of the glassy state. Crystallizable polymers, however, do not achieve total crystallinity, even if supercooling to a truly amorphous state is not possible. The amorphous structure left between the crystalline regions allows determination of the glass transition.

The nature of the glassy state of liquids and polymers has been the subject of many investigations; thus, various theories and interpretations have been forwarded. Due to the similarity of the glass transition to a second-order thermodynamic transition, experimental investigations of the second derivatives of the free energy, G, for polymers have been compared to those of true second-order transitions. Rehage and Borchard [8] have contrasted glass transition behavior with true second-order thermodynamic transitions (e.g., rotational transitions as well as the liquid helium transition at 2.2°K). For second-order transitions, the second derivatives of G (specific heat at constant pressure, c_p; isothermal compressibility, k; and the coefficient of thermal expansion, α) have lower values above the T_g than below. This is in contrast to the behavior observed with the glass transition. The glass transition temperature shifts to higher temperatures with increasing cooling rates, also in general contrast to true second-order transitions. Thus, it was concluded that the glass transition cannot be considered to be a true second-order thermodynamic transition.

Semiempirical free-volume treatments of viscosity relationships for liquids and polymers have been proposed to account for the rapid change in viscosity with temperature [9–11]. From the same basic approach, Williams, Landel, and Ferry [12] proposed a universal relationship

$$\log a_T = -17.44(T - T_g)/[51.6 + (T - T_g)] \quad (3.1)$$

where a_T represents the temperature variation of the segmental friction coefficient for mechanical relaxations. This empirical relationship has been applied successfully to describe the relaxation or viscosity variation of polymers in the temperature range of $T_g < T < (T_g + 100°C)$.

Elegant theoretical treatments of the glass transition temperature have been proposed by Gibbs and DiMarzio [13] and Nose [14] using lattice models that allow vacant sites. The basic difference between these two approaches is the assumption of a true second-order transition for the Gibbs–DiMarzio treatment, unlike that of Nose's.

In summary, the glass transition temperature has been viewed as a second-order transition, an isoviscous state, an isoconfigurational state, and an iso-free-volume state. For a detailed discussion of the nature of the glassy state, the cited references [15–17] will provide an excellent background.

The most commonly used method for establishing miscibility in polymer–polymer blends or partial phase mixing in such blends is through determination of the glass transition (or transitions) in the blend versus those of the unblended constituents. A miscible polymer blend will exhibit a single glass transition between the T_g's of the components with a sharpness of the transition similar to that of the components. In cases of borderline miscibility, broadening of the transition will occur. With cases of limited miscibility, two separate transitions between those of the constituents may result, depicting a component 1-rich phase and a component 2-rich phase. In cases where strong specific interactions occur, the T_g may go through a maximum as a function of concentration. The basic limitation of the utility of glass transition determinations in ascertaining polymer–polymer miscibility exists with blends composed of components which have equal or similar ($<20°C$ difference) T_g's, whereby resolution by the techniques to be discussed of two T_g's is not possible.

In the analysis of polymer–polymer miscibility via the utilization of macroscopic techniques to observe the glass transition, certain questions have been posed for which unambiguous answers do not presently exist. The basic question revolves around the level of molecular mixing required to yield single glass transition temperatures for miscible polymer mixtures. The level of molecular mixing to yield a single T_g in polymer mixtures is not

3.2. Glass Transition Temperature

clearly defined presently, and experimental investigations recently reported and cited in other sections of this treatise have been directed toward this specific question.

The question rephrased is what size of a "domain" or "phase" of composition different than that of the bulk mixture is required to yield distinct macroscopic property characteristics (i.e., T_g)? In some blends, microscopic evidence of phase structure has been observed where single-T_g behavior was determined. This, of course, posed the question of domain size required for unique T_g behavior of the individual domains. Recent studies of the physical structure of the amorphous state may provide a clue to this anomaly [18, 19]. With amorphous homopolymers, electron microscopy has shown that domains of local order may exist in the amorphous state [20]. Small-angle neutron scattering experiments demonstrate that local order does not exist in the amorphous state [1, 2]; the polymeric chains are in a random conformation. While much of the effort is presently directed toward resolving the differences found by these two experimental techniques for unblended homopolymers, the results will have direct bearing on resolving the question of the level of molecular mixing in polymer blends as derived from macroscopic glass transition determinations.

As the glass transition value is inherent in the property characteristics (e.g., viscosity, crystallization kinetics, thermomechanical properties) of a material, the existence of a single and sharp, single and broad, shifted, or individual transition for a blend reveals the macroscopic property characteristics of the blend. Thus, while there may exist debate concerning the level of molecular mixing, the glass transition behavior of the blend will remain an extremely important characteristic. For the present, these features of the T_g behavior will be assumed to assess qualitatively the level of miscibility. Hopefully, when the above questions on the effect of domain size and level of molecular mixing on transition behavior are answered, a quantitative assessment of the level of miscibility will be possible. An interesting review of the above question has been presented by Kaplan [21], in which he has assigned a value of 150 Å as the domain size required to contain a "universal" segmental length associated with the glass transition. Further investigations are necessary to determine if indeed this value is "universal."

3.2.1 Mechanical Methods

Mechanical methods for determination of the transitional behavior of polymers and polymer blends have been cited more frequently than the other techniques to be discussed. The elastic and viscoelastic properties of polymers derived by subjecting polymers to small-amplitude cyclic deformation can yield important information concerning transitions occurring on the molecular scale. Data obtained over a broad temperature range can be used to ascertain the molecular response of a polymer in blends with other polymers. In a highly phase-separated polymer blend, the transitional behavior of the individual components will be unchanged. Likewise, in a miscible blend, a single and unique transition corresponding to the glass transition will appear. Dynamic mechanical testing can be accomplished using various experimental arrangements. In this discussion, both free and forced vibrational techniques will be covered. Free-vibration dynamic mechanical testing devices include the torsion pendulum, freely vibrating reed, and the torsional braid analyzer. Forced vibration techniques employ the viscoelastometer or a forced vibrating reed.

The torsion pendulum consists of an inertial source (disk or rod) connected to a polymer specimen (e.g., a rectangle with length $\gg$ width $\gg$ thickness) which is firmly fixed at the other end. The inertial source is angularly displaced and released, allowing the specimen to vibrate freely. The resultant damped sinusoidal wave is then determined using a suitable recording device such as a rotary variable differential transformer, linear variable differential transformer, or a mirror system. The damped sinusoidal wave can be used to calculate the shear modulus, G', the loss modulus, G'', and mechanical loss, tan δ, defined as G''/G'. Tan δ can also be calculated more directly as

$$\tan \delta = \ln (A/B)/N\pi \qquad (3.2a)$$

where A and B are the magnitudes of individual cycles (A > B) and N is the number of cycles between A and B. The shear modulus, G', is calculated from

$$G' = 4\pi^2 I f^2/k \qquad (3.2b)$$

where k is a geometrical shape factor determined from sample dimensions, I is the inertial force, and f is the frequency of the sinusoidal wave (cycles/sec).

Generalized data (tan δ, G', G'') for polymer–polymer blends versus temperature are illustrated in Fig. 3.3 for behavior expected of two-phase blends. In Fig. 3.4, the generalized data expected of miscible, one-phase blends are depicted. Actual experimental data for the miscible polymer blend poly(vinyl chloride)–(ethylene/ethyl acrylate/carbon monoxide) terpolymer are illustrated in Fig. 3.5, which clearly shows an intermediate glass transition temperature for the blend.

The torsional braid analyzer [22], a variation of the torsion pendulum, has the advantage of being capable of handling very brittle materials as well as very fluid materials. A fiberglass braid or other suitable support is impregnated with the material to be tested. One end is firmly fixed while the other end is attached to an inertial system. Absolute values of the shear modulus, G', and the loss modulus, G'', cannot be determined using this

3.2. Glass Transition Temperature

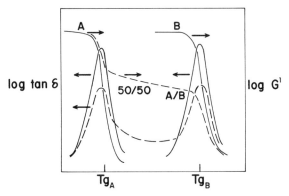

Fig. 3.3. Generalized behavior of the dynamic mechanical properties of a two-phase blend. ——, pure components; ———, mixture.

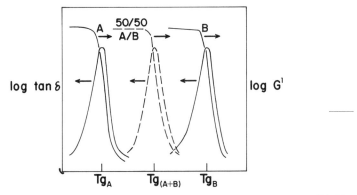

Fig. 3.4. Generalized behavior of the dynamic mechanical properties of a miscible blend. ——, pure components; ———, mixture.

technique. However, the mechanical loss and a relative modulus (cycles/sec)² can be used to ascertain transitions occurring in the experimental specimens.

The vibrating reed arrangement [23] can be used to measure polymer transitions via determination of the tensile modulus and mechanical loss in either free or forced vibration. The experimental apparatus for forced vibration consists of a polymer strip rigidly fixed at one end and forced to vibrate transversely via an electromagnetic vibrator driven by a variable frequency source. At the resonance frequency, a maximum in the deflection of the free end is observed. The tensile modulus, E, is calculated from [23]

$$E = 38.24 dL^4 f_r^2 / D^2 \tag{3.3}$$

where d = density, L = length of polymer strip, D = thickness, and f_r =

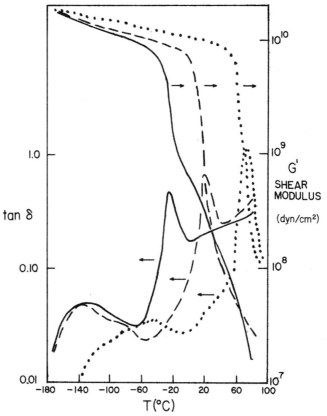

Fig. 3.5. Mechanical loss and shear modulus versus temperature data for: ethylene/ethylacrylate/carbon monoxide (E/EA/CO) (71.8/10.5/17.7) terpolymer, ———; 50/50 blend of terpolymer with poly(vinyl chloride), ----; and poly(vinyl chloride), ····. [From L. M. Robeson and J. E. McGrath, *Polym. Eng. Sci.* **17**, 300 (1977).]

resonance frequency (cycles/sec). The mechanical loss is given by

$$\tan \delta = E''/E' = F/A_0^3 \tag{3.4}$$

where

$$F = [-5.478 + 2(7.502 + 6.15M^2)^{1/2}]/1.689M^2 \tag{3.5}$$

and M is the ratio of amplitude of the free end to the clamped ends of the plastic strip. $A_0 = 1.875$.

Dynamic mechanical testing of materials subjected to a cyclic tensile strain (forced vibration) is another method commonly employed to measure polymeric transitions. The instrument, commonly referred to as a viscoelastometer [24, 25], operates on the principle that an applied sinusoidal tensile

3.2. Glass Transition Temperature

strain applied to the specimen generates a sinusoidal stress with a phase angle δ. The horizontal specimen is attached at one end to a driver unit providing oscillatory motion while the other end is connected to a load transducer. Outputs of the stress and strain transducers are converted to provide direct tan δ readings. The absolute value of the complex tensile modulus E^* ($E^* = E' + iE''$) is given by

$$|E^*| = Fl/\Delta l\, A \qquad (3.6)$$

where F = tensile force, A = cross-sectional area, l = length of specimen, and Δl = amplitude of elongation. Then E' and E'' can be calculated with the following relationships.

$$E' = E^* \cos \delta \qquad (3.7)$$

$$E'' = E' \tan \delta \qquad (3.8)$$

An example of the use of viscoelastometer data to determine polymer–polymer miscibility is given in Fig. 3.6 for the blend of poly(vinylidene fluoride) and poly(methyl methacrylate) [26].

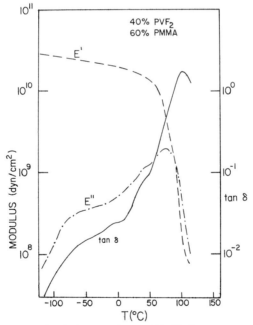

Fig. 3.6. Dynamic mechanical properties at 110 Hz for an annealed poly(vinylidene fluoride)–poly(methyl methacrylate) (40/60) blend using a Rheovibron Viscoelastometer. [Reprinted with permission from D. R. Paul and J. O. Altamirano, *Adv. Chem. Ser.* **142**, 371 (1975). Copyright by the American Chemical Society.]

Another method of mechanically determining the glass transition of polymers is by simultaneously measuring the modulus and resilience. This method involves the measurement of the stress–strain curve while elongating the specimen to 1% (or lower) strain and then reversing the direction of strain back to 0% strain at the same testing rate (e.g., 0.1 in./min per inch of test specimen length). The strain required to reach zero stress on the return curve is then divided by the strain reached before reversal (i.e., 1%) to yield a value times 100 termed percent resilience. Generalized data illustrated in Fig. 3.7 provide the basis for defining percent resilience:

$$\text{percent resilience} = (AB/OA) \times 10^2 \tag{3.9}$$

The modulus can be determined from the slope of the stress–strain curve. The modulus and percent resilience data plotted against temperature can be used to determine one-phase versus two-phase behavior, as illustrated by the generalized curves depicted in Figs. 3.8 and 3.9.

The modulus–temperature data obtained via this technique are as accurate as those obtained by the previously described mechanical methods; however,

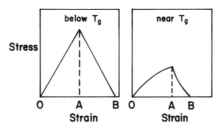

Fig. 3.7. Generalized stress–strain data utilized for resilience determination.

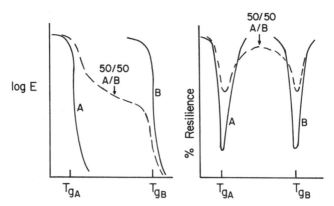

Fig. 3.8. Generalized modulus and resilience versus temperature data for a two-phase polymer blend.

3.2. Glass Transition Temperature

resilience is not as sensitive as mechanical loss. Nevertheless, this type of data has been used by several investigators to characterize polymer blends [27, 28], as illustrated in Fig. 3.10 for blends of tetramethyl bisphenol A polycarbonate and polystyrene, which were found to exhibit a high level of miscibility [28].

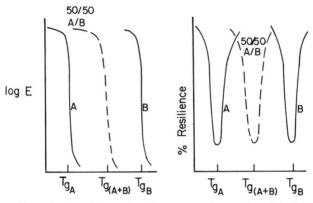

Fig. 3.9. Generalized modulus and resilience versus temperature data for a single-phase polymer blend.

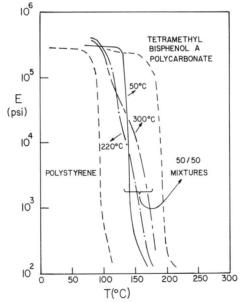

Fig. 3.10. Modulus–temperature data for a 50/50 blend of polystyrene with tetramethyl bisphenol A polycarbonate (mixtures prepared at temperatures indicated). [From M. T. Shaw, *J. Appl. Polym. Sci.* **18**, 449 (1974).]

3.2.2 Dielectric Methods

The electrical properties of polymers are analogous to mechanical properties in that the dielectric constant, ε', is similar to compliance, the dielectric loss factor, ε'', is similar to mechanical loss, and the dielectric strength is analogous to tensile strength. The dielectric loss factor and the dissipation factor, tan δ ($\varepsilon''/\varepsilon'$), are of primary interest in this discussion as they are commonly used to ascertain polymeric transitions such as the glass transition. The experimental advantage of obtaining transition data from electrical measurements over dynamic mechanical testing is in the ease of changing frequency. The major disadvantage is the difficulty in determining the transitions of nonpolar polymers. Generally nonpolar polymers will require slight modification, such as oxidation, to provide sufficient polarity to resolve adequately secondary loss transitions as well as glass transitions in blends.

For polar polymers, if one represents the dipole by a single relaxation time, τ, then the constituents of the complex dielectric constant, ε^*, are defined as

$$\varepsilon^* = \varepsilon' - i\varepsilon'' \tag{3.10}$$

$$\varepsilon' = \varepsilon_\infty + (\varepsilon_0 - \varepsilon_\infty)/(1 + \omega^2\tau^2) \tag{3.11}$$

$$\varepsilon'' = (\varepsilon_0 - \varepsilon_\infty)\omega\tau/(1 + \omega^2\tau^2) \tag{3.12}$$

where ε_0 and ε_∞ are the limits of ε' at zero frequency and infinite frequency, respectively. The loss factor goes through a maximum when $\omega\tau = 1$.

The dielectric constant increases as molecular motion in a polymer increases; thus, large secondary relaxations and the glass transition will yield increasing values. The generalized behavior for the dielectric constant and the dielectric loss factor yields the schematics for miscible or phase-separated polymer blends illustrated in Fig. 3.11. An experimental example of the dielectric method for establishing the miscibility of polymer blends is illustrated in Fig. 3.12 for poly(2,6-dimethyl-1,4-phenylene oxide)–polystyrene blends [29].

A technique in which the change of dielectric loss is measured under a definite temperature program is termed the thermodielectric loss measurement. This recent technique has been used for estimating the level of polymer–polymer miscibility [30]. As the dielectric loss of a sample is dissipated in the form of heat, a differential thermal analyzer has been utilized to measure ε'' in this approach. This new technique is claimed to be more sensitive for measuring the degree of miscibility than other methods. A brief outline of the theory and data analysis will be presented.

3.2. Glass Transition Temperature

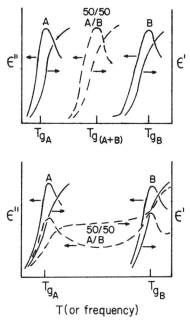

Fig. 3.11. Generalized dielectric loss factor, ε'', and dielectric constant, ε', versus temperature (or frequency) data for single-phase and two-phase polymer blends.

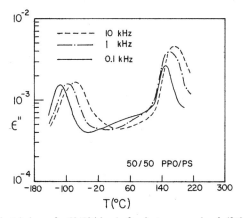

Fig. 3.12. Dielectric loss of a 50/50 blend of polystyrene and poly(2,6-dimethyl-1,4-phenylene oxide) as functions of temperature and frequency. [Reprinted with permission from W. J. MacKnight, J. Stoelting, and F. E. Karasz, *Adv. Chem. Ser.* **99**, 26 (1971). Copyright by the American Chemical Society.]

The apparatus consists of an arrangement to measure the dielectric loss of a sample in one chamber of a differential thermal analyzer. A reference sample is used such that the dielectric loss is determined from the temperature difference between the reference and the sample in the electric field. The dielectric loss is determined from

$$\varepsilon'' = 4Q/E_0^2 f \qquad (3.13)$$

where Q is the heat generated per unit volume per second, f is the cyclic frequency, and E_0 is the electric field intensity. The temperature difference, ΔT, between the sample in the electric field and the reference position is assumed proportional to Q. Therefore,

$$\varepsilon'' = A\, \Delta T / f V_0^2 \qquad (3.14)$$

where A is a constant dependent on density and the specific heat of the sample and V_0 is the applied voltage. To simplify, a quantity ε''^* is defined to relate ΔT to Δh (height from baseline in the DTA data), yielding

$$\varepsilon''^* = B\, \Delta h / f V_0^2 \qquad (3.15)$$

In Fig. 3.13, ε''^* versus temperature is shown for the miscible blend of poly(vinyl nitrate) and an ethylene/vinyl acetate copolymer [30]. From the same data, contour surfaces of the dielectric loss can be obtained. In the contour surfaces, the dielectric loss, ε''^*, is taken perpendicular to the surface and $\log f$ is plotted against $1/T$. For miscible mixtures, a series of peaks occurs through which a single line can be constructed. For immiscible blends for which the constituents have different T_g's, two series of peaks are ob-

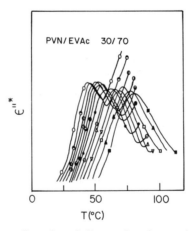

Fig. 3.13. Temperature dispersion of ε''^* at various frequencies for a miscible mixture of poly(vinyl nitrate with ethylene/vinyl acetate copolymer (86 wt% vinyl acetate) (30/70 wt ratio). [From S. Akiyama, Y. Komatsu, and R. Kaneko, *Polym. J.* **7**, 172 (1975).]

3.2. Glass Transition Temperature

served. This is illustrated in Figs. 3.14 and 3.15, respectively showing the miscible blend of poly(vinyl nitrate)–poly(vinyl acetate) and the immiscible blend of poly(vinyl acetate)–(ethylene/vinyl acetate) copolymer. This graphical technique yields a new method for determination of polymer miscibility by evaluation of the characteristic appearance of the contour surfaces.

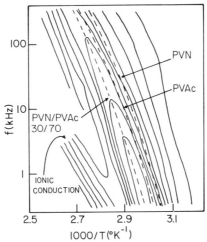

Fig. 3.14. Contour map of $\varepsilon''*$ for the miscible mixture of poly(vinyl nitrate) with poly(vinyl acetate) (30/70). The dashed lines represent crests corresponding to phase transitions. [From S. Akiyama, Y. Komatsu, and R. Kaneko, *Polym. J.* **7**, 172 (1975).]

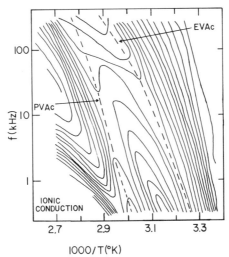

Fig. 3.15. Contour map of $\varepsilon''*$ for the immiscible mixture poly(vinyl acetate) and ethylene/vinyl acetate copolymer (86 wt% vinyl acetate) (40/60). [From S. Akiyama, Y. Komatsu, and R. Kaneko, *Polym. J.* **7**, 172 (1975).]

Reviews of dielectric characterization of polymers can be found in the cited references [31, 32].

3.2.3 Dilatometric Methods

Polymer glass transitions have many characteristics similar to a second-order thermodynamic transition. With respect to volume change, a discontinuity is observed in the rate of volume change with temperature in the region of the glass transition. Dilatometric methods to determine polymeric glass transitions were one of the most common techniques before mechanical methods became popular. Dilatometric techniques and experimental apparatus have been adequately discussed elsewhere [33, 34] and will not be reproduced here.

In a blend of two distinctly different polymers, two-phase behavior can be determined by two discontinuities in the derivative curve dV/dT corresponding to the T_g's of the respective phases. Experimental data exhibiting one-phase behavior for a polymer blend are shown in Fig. 3.16, illustrating the blend of syndiotactic poly(methyl methacrylate) ($T_g = 120°C$) and isotactic poly(methyl methacryate) ($T_g = 45°C$) [35]. Note that the major change in slope for the volume–temperature data occurs at the T_g (94°C) of the blend, well between the component T_g's. Dilatometric techniques are less sensitive than the dynamic mechanical methods previously discussed, and the presence of crystallinity hinders resolution.

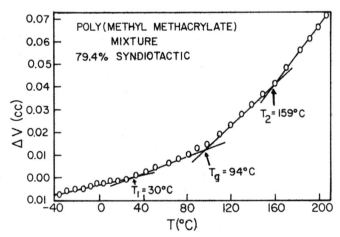

Fig. 3.16. Volume–temperature plot of a mixture of 79.4% (by wt) syndiotactic poly(methyl methacrylate) with 20.6% (by wt) isotactic poly(methyl methacrylate). [From S. Krause and N. Roman, *J. Polym. Sci.*, Part A **3**, 1631 (1965).]

3.2. Glass Transition Temperature

3.2.4 Calorimetric Methods

The utilization of calorimetric methods to determine the glass transition of polymers and their respective blends parallels that of dilatometric methods discussed in the previous section. The specific heat of polymers exhibits a change when passing through the glass transition, generating a maximum in the value of dC_p/dT as generalized in Fig. 3.17.

With the introduction of sensitive calorimeters within the last decade, the calorimetric technique has rapidly gained prominence. The most common instrument is the differential scanning calorimeter (DSC). The DSC measures the amount of heat required to increase the sample temperature by a value ΔT over that required to heat a reference material by the same ΔT. Through sophisticated instrumentation, controlled rates of heating or cooling are possible with high accuracy of heat input (or output) to small specimens (5–50 mg). More detailed description of the utility and design parameters of differential scanning calorimeters can be found elsewhere [36, 37].

This technique has successfully demonstrated polymer–polymer miscibility for the systems PPO–polystyrene [38], nitrile rubber–PVC [39], poly(vinyl methyl ether)–polystyrene [40], poly(vinylidene fluoride)–poly(methyl methacrylate) [41], and PVC–(ethylene/vinyl acetate/sulfur dioxide) terpolymer [42]. Differential scanning calorimetry has been particularly useful in studying the miscibility of the classic system: nitrile rubber–PVC. Using DSC, Zabrzewski [39] observed miscibility with PVC in all compositions at levels of 23 to 45% acrylonitrile in the nitrile rubber, in excellent agreement with dynamic mechanical data. Landi [43] investigated similar blends and observed single-phase behavior with a 34% acrylonitrile-content nitrile rubber blended with PVC. He noted that the DSC results could be more clearly illustrated by plotting the secant slope of the specific heat versus temperature,

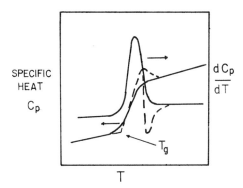

Fig. 3.17. Generalized behavior of specific heat versus temperature of polymers in the range of the glass transition temperature. Solid line = quenched; dashed line = annealed.

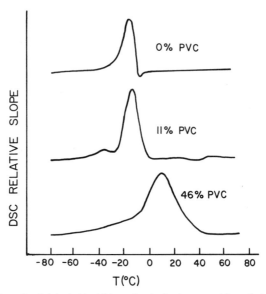

Fig. 3.18. Effect of poly(vinyl chloride) on the single glass transition of nitrile rubber (34% acrylonitrile). Data obtained on a differential scanning calorimeter (DSC). [From V. R. Landi, *Appl. Polym. Symp.* **25**, 223 (1974).]

as shown in Fig. 3.18. By using this data reduction technique, he clearly demonstrated that variations in acrylonitrile content as well as mixing techniques could be more clearly defined than by direct observation of the basic DSC thermogram.

3.2.5 Thermo-Optical Analysis

A technique termed thermo-optical analysis (TOA) has been employed by Shultz *et al.* [44–47] to investigate the miscibility of polymer blends. This technique involves scribing scratches onto a polymer or blend surface with a steel stylus. A polarizing microscope equipped with a hot stage capable of temperature programming is employed. Light transmitted through the film placed between crossed polarizer and analyzer is converted into voltage and plotted against temperature. The scratched surface is birefringent and thus light is only transmitted through the scratches. As the polymer (or constituents of the blend) pass through the glass transition temperature, the orientation produced by scratching the film disappears and the reduction in birefringence leads to a decrease in transmitted light.

Results for a miscible blend (styrene–*p*-chlorostyrene) copolymer–poly-(2,6-dimethyl-1,4-phenylene oxide) shown in Fig. 3.19 are compared with

3.2. Glass Transition Temperature

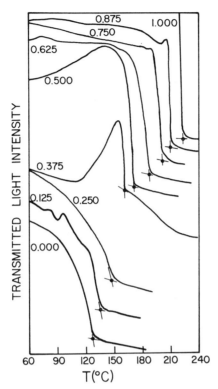

Fig. 3.19. Thermo-optical analysis curves for blends of poly(2,6-dimethyl-1,4-phenylene xoide) (PPO) and styrene–*p*-chlorostyrene copolymer (0.453 mole fraction styrene). Numbers on the plot represent the weight fraction of PPO in each blend. [Reprinted with permission from A. R. Shultz and B. M. Brach, *Macromolecules* **7**, 902 (1974). Copyright by the American Chemical Society.]

an immiscible blend of poly(*p*-chlorostyrene)–poly(2,6-dimethyl-1,4-phenylene oxide) shown in Fig. 3.20. Single transition temperatures monotonically increasing with the content of the higher T_g component are characteristic of the miscible blend (Fig. 3.19), whereas two transitions corresponding to the blend constituents are observed for the immiscible blend (Fig. 3.20). The conclusions [46] reached using thermo-optical analysis to characterize miscibility in polymer blends were in excellent agreement with more common techniques (e.g., dynamic mechanical and calorimetry).

3.2.6 Radioluminescence Spectroscopy

A unique method to measure the glass transition of polymer blends, termed radioluminescence spectroscopy, has been successfully utilized by Zlatkevich

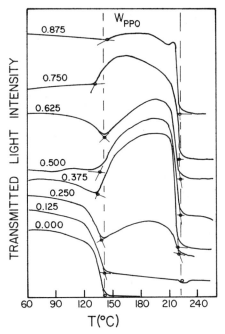

Fig. 3.20. Thermo-optical analysis curves for blends of poly(2,6-dimethyl-1,4-phenylene oxide) (PPO) and poly(p-chlorostyrene). Numbers on the plot represent the weight fraction of PPO in each blend. [Reprinted with permission from A. R. Shultz and B. M. Beach, *Macromolecules* **7**, 902 (1974). Copyright by the American Chemical Society.]

and Nikolskii [48] and Böhm et al. [49]. Irradiation (electron or γ ray) of the polymer or blend in the glassy state results in trapped secondary electrons which are rapidly released, yielding luminescence, once the sample temperature reaches the glass transition. Maximum luminescence is observed at a temperature quite close to T_g values reported by more common techniques. For two-phase blends, two distinct peaks can be observed in luminescence versus temperature, corresponding to the respective T_g's. Resolution of the T_g of a minor phase (as low as several volumes percent) is quite good, thus providing equal or superior sensitivity to mechanical or calorimetric methods. For a description of specific experimental procedures and equipment design, see Zlatkevich and Nikolskii [48] and Böhm [50].

3.3 MICROSCOPY

Direct visual confirmation of the presence of two phases has been used more often than any other method as a preliminary indication of the degree

3.3. Microscopy

of miscibility in a polymer–polymer system. Many have turned to microscopy to aid in determining not only the presence but the connectivities of the phases. Electron microscopy, with 50-Å resolution, has shown that heterogeneities exist even in miscible polymer systems. Such is the nature of solutions of 50-Å molecules.

3.3.1 Visible, Including Phase Contrast

Both transmitted-light and phase-contrast microscopy require as a minimum a difference in refractive index between the phases for contrast. Transmission contrast is best obtained with differences in opacity or color; however, with phase-contrast optics good contrast is obtained with transparent materials and is therefore the preferred method for polymer–polymer systems. Staining is another method of enhancing contrast and a limited staining technology has been developed for polymers; however, it pales in comparison with that known to biologists.

Typical phase-contrast techniques are summarized in the papers of Marsh et al. [51], Inoue et al. [52], Vasile and Schneider [53], and Walters and Keyte [54]. Generally, films of 5 μm or less are microtomed or cast for observations. Enhancement of contrast in mixtures of crystalline polymers can be obtained by use of polarized light [55]. Osmium tetroxide staining, well known in electron microscopy, has also been used to enhance contrast for optical work [52]. Stains for various polymers are listed by Brauer and Newman [56, 57]. A sample of the effects possible with stains is given in Table 3.1.

Optical microscopy on two-phase mixtures has revealed many types of structures, including interpenetrating phases. The fineness of the phases has been related to mixing intensity and viscosity ratio, but not often [58] to degree of solubility.

The use of scattering (dark-field) optics for the deduction of two phases is not common in polymer–polymer studies even though high intensity light at right angles to the optical axis (ultramicroscopy) can reveal the presence of scattering bodies far smaller than the resolving power of the microscope. Miyata and Hata [59] have described the use of the ultramicroscope on the system poly(methyl methacrylate)–poly(vinyl acetate).

3.3.2 Electron Microscope

Transmission electron microscopy (TEM) has been widely used in polymer–polymer studies. The necessary step of microtoming can be facilitated by cryogenic or chemical methods. Electron opacity differences are often achieved by selective chemical reaction [58, 60] or by annealing in the electron beam [58, 61]. Treatment of soluble polymer systems with any

TABLE 3.1

Colors Obtained with the Smith Stain, a Mixture of Methylene Blue and Sudan III Dyes, Applied to Various Polymers[a]

Material	Color
Colors of Hydrophilic Materials	
Cellophane	Bright blue
Cotton	Blue
Paper fibers	Very light to very deep blue depending on type of fiber and degree of hydration
Rayon (viscose)	Blue–green with lighter blue skin
Colors of Hydrophobic Materials	
Acrylate polymers	Orange
Butadiene/acrylonitrile copolymers	Orange to brownish red
Butadiene/styrene copolymers	Orange to brownish red
Ethyl cellulose	Dull red orange
Natural rubber, unfilled	Orange–yellow
Natural rubber, with hydrophilic filler	Greenish yellow
Polyamide resins	Orange–yellow
Polyethylene	Pale yellow
Polyisobutylene (Vistanex)	Pale pink
Vinyl chloride polymers	Pale pink
Vinyl pyridine copolymers (Gentac)	Orange
Colors of Mixed Hydrophilic–Hydrophobic Materials	
Cellulose acetate	Green
Cellulose nitrate, unplasticized	Colorless to light green
Cellulose nitrate, plasticized	Colorless to olive green
Any hydrophobic polymer with hydrophilic groups in the structure or with hydrophilic additives	Greenish orange
Materials Unaffected by Either Dye	
Dacron	—
Mylar	—
Nylon	—

[a] From S. B. Newman, in "Analytical Chemistry of Polymers" (O. M. Kline, ed.), Part III, p. 261, Wiley (Interscience), New York, 1962.

chemical agent should be regarded with caution as it may cause phase separation. Heating or cooling can have similar effects. The production of artifacts during microtoming, staining, replication, and exposure to the beam are well known but continue to cause difficulties.

Several polymer–polymer systems of reported miscibility have been shown

3.3. Microscopy

to contain domains by using the electron microscope. Smith and Andries [58] found that the system SBR–PB was immiscible even with as little as 3% styrene in the SBR, but that the phase size progressively decreased with styrene content. This is in conflict with the results of Marsh *et al.* [51], who found no evidence for multiple phases in SBR–PB or in SBR–(ethylene–butadiene) copolymer. Matsuo *et al.* [60] found some (400 Å) heterogeneity in the system PVC–NBR containing 40% acrylonitrile, although only one glass transition was observed.

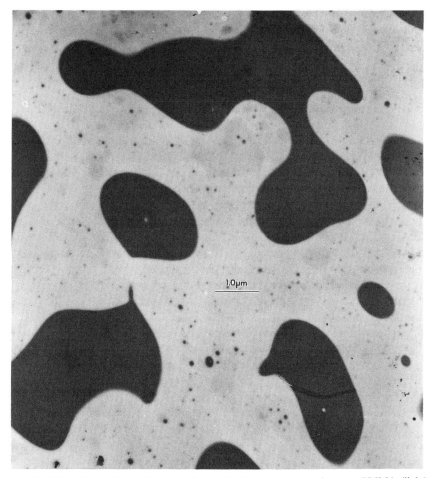

Fig. 3.21. Transmission electron micrograph showing contrast between PMMA (light) and SAN (dark) phases developed during exposure to the electron beam [Reproduced with permission from L. P. McMaster, *Adv. Chem. Ser.* **142**, 43 (1975). Copyright by the American Chemical Society.]

McMaster [61] found that TEM was useful for following the phase decomposition of the miscible system PMMA–SAN. The geometry of the phases correlated well with the expected occurrence of spinodal decomposition near the critical composition. Excellent contrast was achieved by prolonged exposure of the samples to the electron beam, as can be seen in Fig. 3.21. This technique has been studied in more detail by Thomas and Talmon [62], who have attributed the contrast development to differential thinning.

The technique of scanning electron microscopy (SEM) has found a niche in phase studies [54]. Contrast depends in this technique on differences in surface topography or texture and this can be emphasized by breaking the specimen in its glassy state. If vitrification requires cooling, there is again the danger of phase changes. Differential swelling [63] involves a similar hazard.

3.4 SCATTERING METHODS

3.4.1 Cloud-Point Method

By definition, a stable homogeneous mixture is transparent, whereas an unstable nonhomogeneous mixture is turbid unless the components of the mixture have identical refractive indexes [64]. Given a stable homogeneous mixture, the transition from the transparent to the turbid state can be brought about by variations of temperature, pressure, or composition of the mixture. The cloud point corresponds to this transition point—the point of incipient phase separation. It is not necessarily an equilibrium event, but the fact that the opalescence almost always disappears on reversal of the temperature–pressure–composition variation strongly indicates that the driving forces are thermodynamic in origin.

For polymer mixtures, the cloud-point curves are usually measured using a thin film made from a thoroughly mixed blend. The film is observed through a microscope illuminator for low-angle back or forward scattering relative to the incident light. The specimen is then heated at a very low rate such that the temperature increases at an infinitesimally slow rate. The first faint cloudiness appears, denoting the cloud point, and the temperature is recorded. A few degrees above this cloud point, the cycle is reversed; the sample is gradually cooled. The temperature at which the faintest opalescence just disappears is also recorded. This is repeated for a series of compositions and a temperature–composition plot is generated. The result is called the cloud-point curve (CPC). Generally, the CPCs found on heating and on cooling the sample do not agree. The reasons for this are many; they stem mostly from kinetic factors plus the fact that a phase transition point

3.4. Scattering Methods

can be observed only after big enough clusters have formed to create sufficient refractive index differences for scattering an observable quantity of light. This shortcoming in the CPC measurement is often corrected by presenting an average of the two CPCs.

A number of investigators [65–67] have made CPC measurements for several binary high-polymer mixtures. In each case, the CPCs are measured above the system's glass transition or melting point. Systems studied include polystyrene–poly(vinyl methyl ether) [65]; poly(ε-caprolactone)–poly(styrene-co-acrylonitrile) [66]; polycarbonate–poly(ε-caprolactone) [67]; and mixtures [67] of poly(vinylidene fluoride) with poly(methyl methacrylate), poly(ethyl methacrylate), poly(methyl acrylate), and poly(ethyl acrylate). All of these systems exhibit the lower critical solution temperature (lcst) behavior.

Other CPC measurements [68–72] on oligomeric and short chain length systems have exhibited upper critical solution temperature (ucst) behavior

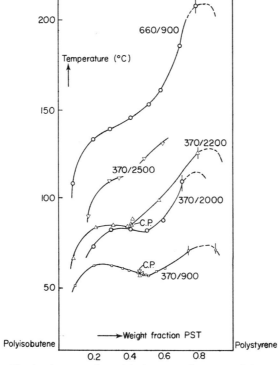

Fig. 3.22. Cloud-point curves for polyisobutene–polystyrene mixtures of various molecular weights. [From R. Koningsveld and L. A. Kleintjens, *J. Polym. Sci., Polym. Symp.* **61**, 221 (1977).]

and they have revealed unusual asymmetry and bimodality of the phase diagram. The asymmetry is found in the experimental CPCs of Allen, Gee, and Nicholson [68] for low molecular weight mixtures of polyisobutylene and poly(dimethyl siloxane). This asymmetry manifests itself as a shift of the maximum point from low concentrations of the high molecular weight component (with silicone) to high concentrations of that component. The bimodality has been demonstrated for a low molecular weight mixture of polystyrene with polyisoprene or polyisobutene [72]. This behavior has also been observed by Powers [70] for a low molecular weight α-methylstyrene–vinyl toluene copolymer mixed with a low molecular weight polybutene. Aharoni [71] observed a similar phenomenon with high molecular weight epoxy and copolyester comixed with 1,1′,2,2′-tetrachloroethane. The general observation for these systems was that the phenomenon was not related to the polydispersity of the polymers. In fact, the bimodality tended to increase as the polymer polydispersity was reduced.

Figure 3.22 illustrates the bimodal cloud-point curves for polyisobutene–polystyrene mixtures. For these measurements Koningsveld and Kleintjens [72] used a low-speed analytical centrifuge which allowed determination of the CPCs on the polymer melt within reasonable times. This also removed the necessity of making two CPC measurements, i.e., on heating and on cooling.

3.4.2 Conventional Light Scattering Method

Light scattering had its humble beginning with Lord Rayleigh's mathematical results [73] advanced in answering a seemingly innocent question regarding why the sky is blue. Later, Smoluchowski formulated [74] a fluctuations theory which extended Rayleigh's results to include liquid solutions; this was subsequently refined by Einstein [75]. According to these theories, if a light beam passes through a medium whose volume elements (containing the constituent particles) are small compared to the wavelength of the light, the light will be scattered. The scattered light intensity is proportional to the mean square of the concentration fluctuations in the small volume elements and, therefore, inversely related to the second derivative of the free energy with respect to concentration.

For multicomponent systems [76] and polydisperse polymers [77], the scattering is made up of two contributions: that due to density and that due to concentration fluctuations. At conditions far removed from the spinodal, the former can be eliminated merely by subtracting the scattering intensity of the pure solvent from that of the solution. In the region of the spinodal

3.4. Scattering Methods

this procedure breaks down. One now needs to allow for the coupling between the fluctuations as well as the additional energy required to establish the finite concentration gradient discussed earlier in Section 2.2.4. According to Debye [78], these further complications can be bypassed by making several measurements at a series of angles and extrapolating the scattering intensities to zero angle. Hence, so long as the measurements are made prior to actual phase separation, the scattered intensity depends on the mean square of the concentration fluctuations much the same way as in the original Rayleigh scattering [73].

For a multicomponent system, the scattered light extrapolated to zero scattering angle θ is given by Zernike [76] as

$$[R_\theta^c]_{\theta=0} = \frac{4\pi^2 n^2}{\lambda^4} kT \frac{\sum_i \sum_j \frac{dn}{dc_i} \frac{dn}{dc_j} B_{ij}}{|b|} \Delta V \qquad (3.16)$$

where R_θ^c, the Raleigh ratio, denotes the scattered intensity due to concentration fluctuation; λ, the wavelength of the light *in vacuo*; k, Boltzmann's constant; T, the absolute temperature; n, the refractive index; $|b|$, a determinant with elements $\partial^2(\Delta G)/\partial c_i \partial c_j$ in which G represents the Gibbs free energy of mixing for a volume element ΔV; c_i and c_j, the concentrations of the components i and j; and B_{ij}, the cofactor of the element i,j of the determinant.

In order to determine the elements of $|b|$, its cofactors, and hence an analytical expression for $[R_\theta^c]_{\theta=0}$, Scholte [79] made use of the Flory–Huggins [80, 81] free energy function which, for a binary mixture of polydisperse polymers, is expressed in weight fractions as [82]

$$\frac{\Delta G}{RT} = \left[\sum_i^m \frac{1}{M_{1,i}} W_{1,i} \ln W_{1,i} + \sum_j^m \frac{1}{W_{2,j}} \ln W_{1,j} + \Gamma \right] \rho \Delta V \qquad (3.17)$$

The Rayleigh ratio, after the necessary differentiations and substitution into Eq. (3.16), is given by [79]

$$[R_\theta^c]_{\theta=0} = \frac{4\pi^2 n^2}{N_A \lambda^4} \frac{1}{\rho} \left(\frac{dn}{dw_1}\right)^2 \left[\frac{1}{w_1 M w_1} + \frac{1}{(1-w_1) M w_2} + \frac{\partial^2 \Gamma}{\partial w_1^2} \right]^{-1} \qquad (3.18)$$

Hence, light scattering measurements at a number of concentrations for a known binary polymer mixture enable calculation of $\partial^2 \Gamma/\partial w_1^2$ and the determination, through double integration, of the polymer–polymer interaction function, Γ. The character of Γ yields information about the level of miscibility of the mixture.

Alternatively, one can generate the spinodal if it is recalled that Gibb's condition for the stability limit, expressed in Eq. (2.83) of Section 2.4.3 is equivalent to stating that

$$|b| = 0 \qquad (3.19)$$

This implies that, on approaching the spinodal from within the stable region, the reciprocal of $[R_\theta^c]_{\theta=0}$ should tend to zero. That is, if one performs light scattering measurements at constant concentration but various temperatures within the homogeneous stable region, a plot of $1/[R_\theta^c]_{\theta=0}$ versus temperature extrapolated to zero ordinate yields the spinodal temperature for the given concentration. Repeated for a series of concentrations, one is able to describe the spinodal locus. The procedure can be schematically represented as in Fig. 3.23 where the solid points are the scattering values obtained from measurements within the stable region. Using this technique for three polystyrene samples with M_w values of 51,000, 163,000, and 520,000, Scholte [79] successfully determined the spinodal envelope for polystyrene–cyclohexane.

Another light scattering method which permits determination of the spinodal curve is the van Aartsen quenching method [83]. The method relies on the fact that a highly concentrated solution can be thrust directly into the unstable region if a small, thin sample is used in a quenching medium of rather large heat capacity. Because of the resultant rapid exchange of heat, phase separation of the nucleation and growth type does not take place before the spinodal mechanism sets in. During the quenching period, the scattered intensity varies with time in a manner depicted in Fig. 3.24. In this

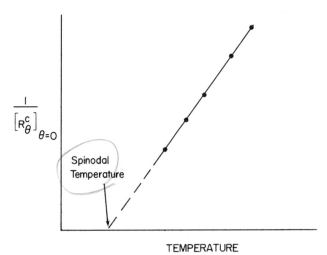

Fig. 3.23. Schematic of scattered light intensity at zero scattering angle as a function of temperature.

3.4. Scattering Methods

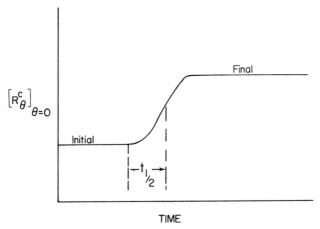

Fig. 3.24. Schematic of the time dependence of the scattered light intensity during the quenching period in the van Aartsen method.

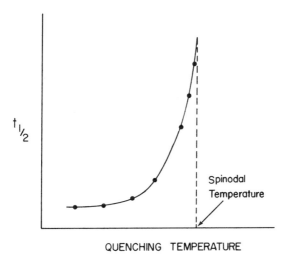

Fig. 3.25. Plot of $t_{1/2}$, as obtained from Fig. 3.24, against temperature, illustrating the determination of the spinodal temperature.

figure, $t_{1/2}$ is defined as the time necessary to reach half the maximum intensity. When measurements are made at a series of temperatures, $t_{1/2}$ plotted against the quenching temperature can be typically represented by Fig. 3.25.

The spinodal temperature is taken equal to that quenching temperature where $t_{1/2}$ increases to high values. For poly(2,6-dimethyl-1,4-phenylene oxide) in caprolactam, van Aartsen [84] determined the spinodal locus by taking $t_{1/2}$ to be 10 min. He also studied the solution of ethylene/vinyl acetate copolymer in caprolactam.

Kratochvil *et al.* [85, 86] have interpreted light scattering from *ternary* mixtures (polymer 1–polymer 2–mutual solvent) using Stockmayer's theory to obtain a parameter akin to the interaction parameter for the two polymers. Their results indicate with fair certainty that this parameter decreases as the calculated interaction parameter increases, an unexplained result. The values of the parameter did not depend on solvent—an important finding—but the number of solvents employed was limited.

3.4.3 Pulse-Induced Critical Scattering (PICS)

This elegant variation [87] of the conventional light scattering method was developed at the University of Essex by J. M. G. Cowie, M. Gordon, J. Goldsbrough, and B. W. Ready. The technique is essentially a hybrid of Scholte's method [79] of measuring scattered light in the stable region and van Aartsen's procedure [83, 84] of scattering measurements in the unstable region.

In the original design, the sample cell holds only a few microliters of the solution placed in a medium which is capable of delivering thermal pulses. The time scale of each thermal pulse is shorter than the time scale of the nucleation and growth mechanism; consequently, the scattering measurements can be made at temperatures within the metastable region and the

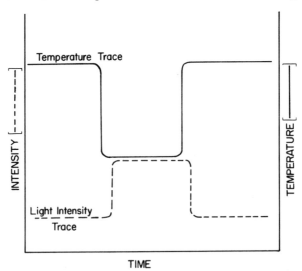

Fig. 3.26. Schematic representation of the principle behind the pulse-induced critical scattering (PICS) technique.

3.4. Scattering Methods

solution is stable and homogeneous during the period of the thermal pulse.

A diagrammatic representation of the principle behind the pulse-induced critical scattering method appears in Fig. 3.26. Light scattering measurements are performed at temperatures within the homogeneous stable phase as in Scholte's method [79]. However, by cooling and heating the sample very rapidly, the polymer mixture can be maintained within the metastable region and be brought back into the stable region before any phase separation occurs. Figure 3.27 illustrates the block diagram of one form of the apparatus used for pulse-induced critical scattering measurements. The light source is a low-power helium–neon laser and the system temperature is measured by a thermistor. The light is transmitted by means of a light guide to the phototransistor, which is the light-detecting system.

The sample-cell chamber contains a small heater which maintains the chamber at a slightly higher temperature than the surrounding, flowing water. The temperature pulse is produced by switching off the small heater; the sample cell and sample cool down rapidly (~ 3 sec) to the temperature of the flowing water. The scattering measurement is made and the heater is switched back on before phase separation ever begins. Considerable versatility is built into this system such that a stream of temperature pulses can be created for short periods of time at almost any base temperature.

Scholte's theoretical development [79] is, strictly speaking, no longer valid in this region. According to Debye's theory [78] of critical opalescence, there is a condition of mathematical singularity at the critical point. Beyond the

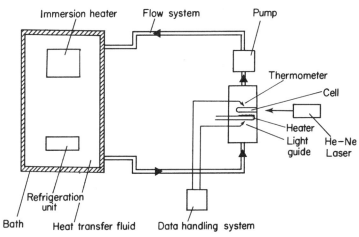

Fig. 3.27. Block diagram of one form of the apparatus for pulse-induced critical scattering measurements. [From K. E. Derham, J. Goldsbrough, and M. Gordon, *Pure Appl. Chem.* **38**, 97 (1974).]

immediate area of the singularity the theory predicts the scattered intensity, I, to obey the relation

$$I = TP(\theta)/[a(T - T_c) + b \sin^2 \theta] \qquad (3.20)$$

where $P(\theta)$ is the particle scattering factor; a and b are constants. When the temperature, T, equals the critical temperature, T_c, the scattered intensity at zero scattering angle, θ, diverges. Only at finite angle would the situation be saved. Equation (3.20) also indicates that a plot of $1/I$ would not be linear if $(T - T_c)$ is of the same order of magnitude as $b \sin^2 \theta$. This again limits how close a scattering measurement can be made near the critical point. Fortunately, however, theoretical estimation indicates that, for $\theta \sim 30°$, $b \sin^2 \theta$ is of the same order of magnitude as $T - T_c \leq 0.03°C$. That is, one could still go within $0.03°C$ of the critical point without approaching the nonlinear region. This has been essentially confirmed by experiments [87]. The implication of this is clear: Scholte's extrapolation technique [79] remains valid. Moreover, the spinodal temperature for each concentration is now located by means of a much shorter extrapolation, leading to a more accurately defined spinodal locus. A diagrammatic comparison of the conventional and the pulse-induced critical scattering is shown in Fig. 3.28.

Although the PICS method has been applied to a series of polymer–solvent systems, it has only recently been applied, by Koningsveld and Kleintjens [88], to measure the spinodal locus for polymer mixtures. Figures 3.29 and 3.30 represent the spinodal locus for polyisobutene–polystyrene. They constitute two of the systems in Fig. 3.22, namely, polyisobutene with M_w 370 and polystyrene of M_w 2200 and 2500. The curves are much better

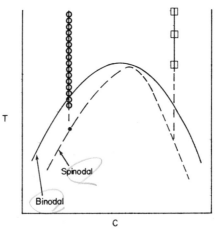

Fig. 3.28. Comparison of the conventional and PICS method, showing points where intensity measurements are made.

3.4. Scattering Methods

defined and they exhibit the now-familiar bimodality in agreement with the cloud-point curves of Fig. 3.22. Recently, Gordon et al. [89], at the University of Essex, developed a centrifugal homogenizer that allows small samples of high-polymer systems to be homogenized at elevated temperatures for pulse-induced critical scattering measurements. This new development promises to give a boost to the experimental studies in polymer–polymer miscibility.

3.4.4 Neutron Scattering Methods

While X-ray scattering is sensitive to density fluctuations, and light scattering to density and concentration fluctuations, neutron scattering

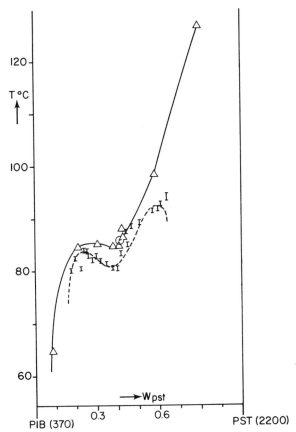

Fig. 3.29. Cloud-point curve ($\triangle$), spinodal by PICS (I), and critical point ($\bigcirc$) for a polyisobutene–polystyrene mixture with molecular weights of 370 and 2200, respectively. [From R. Koningsveld and L. A. Kleintjens, *Br. Polym. J.* **9**, 212 (1977).]

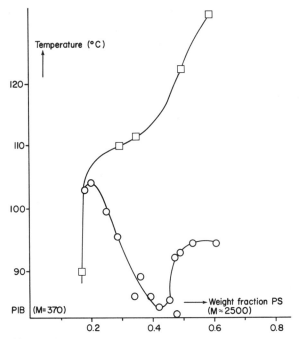

Fig. 3.30. Cloud-point curve (□) and spinodal (○) as determined by the PICS method for a polyisobutene–polystyrene mixture with molecular weights of 370 and 2500, respectively. [From R. Koningsveld and L. A. Kleintjens, *Br. Polym. J.* **9**, 212 (1977).]

measures the differential neutron scattering cross section of small concentrations of protonated polymer (tagged molecules) dispersed in a matrix of deuterated polymer. This allows a rather precise determination of the conformation of the tagged polymer, even in bulk. The three diffraction methods are analyzed in the same way.

If the Zernike relation [Eq. (3.16)] is written for a binary mixture, we have [76]:

$$[R_\theta^c]_{\theta=0} = \frac{4\pi^2 n^2 \, \Delta V \, (\partial n/\partial c_2)^2}{\lambda^4 (kT)^{-1}(\partial^2 G/\partial c_2^2)} \tag{3.21}$$

From standard thermodynamics [90],

$$\frac{1}{kT}\frac{\partial^2 G}{\partial c_2^2} = \frac{N_0 \, \Delta V}{c_2}\left[\frac{1}{M_w} + 2A_2 c_2 + 3A_3 c_2^2 + \cdots\right] \tag{3.22}$$

Substituting Eq. (3.22) into (3.21) gives

$$\frac{Kc_2}{[R_\theta^c]_{\theta=0}} = \frac{1}{M_w} + 2A_2 c_2 + 3A_3 c_2^2 + \cdots \tag{3.23}$$

3.4. Scattering Methods

where the constant K represents

$$K = \frac{4\pi^2 n \frac{\partial n}{\partial c}}{N_0 \lambda^4} \tag{3.24}$$

For a large particle such as a polymer chain, a dissymmetry correction $P(\theta)$ is introduced [90]:

$$P(\theta) = \frac{\text{scattered intensity for large particle}}{\text{scattered intensity without interference}} \tag{3.25}$$

The general expression for $P(\theta)$ in the limit of $\theta \to 0$ is

$$\lim_{\theta \to 0} \frac{1}{P(\theta)} = 1 + \frac{16\pi^2}{3\lambda^2} \langle R_G^2 \rangle_z \sin^2 \frac{\theta}{2} + \cdots + \tag{3.26}$$

and Eq. (3.23) can be written as

$$\frac{Kc_2}{[R_\theta^c]_{\theta=0}} = \frac{1}{M_w P(\theta)} + 2A_2 c_2 + 3A_3 c_2^2 + \cdots \tag{3.27}$$

It is to be noted that a plot of $Kc_2/[R_\theta^c]_{\theta=0}$ versus $\sin^2 \theta/2 + kc$, originally used by Zimm [90], gives two types of limiting results (k is an arbitrary constant chosen so as to provide a convenient spread of the data):

(i) When $\theta \to 0$

$$\frac{Kc_2}{[R_\theta^c]_{\theta=0}} = \frac{1}{M_w} + 2A_2 c_2 + 3A_3 c_2^2 + \cdots \tag{3.28}$$

one obtains a plot of $Kc/[R_\theta^c]_{\theta=0}$ versus kc, which gives $1/M_w$ as intercept and $2A_2/K$ as the limiting slope.

(ii) When $c \to 0$

$$\frac{Kc}{[R_\theta^c]_{\theta=0}} = \frac{1}{M_w} \left[1 + \frac{16\pi^2}{3\lambda^2} \langle R_G^2 \rangle_z \sin^2 \frac{\theta}{2} + \cdots \right] \tag{3.29}$$

one obtains a plot of $Kc/[R_\theta^c]_{\theta=0}$ versus $\sin^2 \theta/2$, which again gives $1/M_w$ as the intercept and $16\pi^2 \langle R_G^2 \rangle_z / 3\lambda^2$ as the limiting slope.

These are the sets of relations used by Kirste and co-workers [1] for analyzing the small-angle neutron scattering of a mixture of $\sim 1.5\%$ styrene/acrylonitrile copolymer in $\sim 98.5\%$ perdeuteropoly(methyl methacrylate) as well as the blend of 1.5% poly(α-methylstyrene) in 98.5% perdeuteropoly(methyl methacrylate). The major point of departure is in the calculation of K, which, for small-angle neutron scattering, is given by

$$K = (S_2 - v_2 S_1^*)/N_1 \tag{3.30}$$

where S_2 and S_1^* are the scattering length sum for polymer 2 and the deuterated polymer 1; v_2 is the partial specific volume of polymer 2.

Figure 3.31 illustrates the Zimm plot for the mixture of perdeuteropoly(methyl methacrylate) and styrene/acrylonitrile copolymer. Because of the linearity exhibited by the limiting curves and the correct value of M_w obtained from the diagram, it was concluded that the polymer mixture is

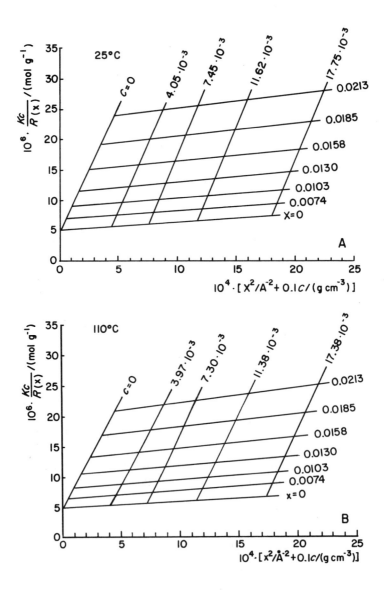

3.4. Scattering Methods

miscible on a molecular scale. Conversely, the skewed nature of the Zimm diagram for the second mixture, Fig. 3.32, indicates micelle formation and gross inhomogeneity.

Also calculated are the second virial coefficients for the homogeneous mixture of SAN–d-PMMA. These appear in Table 3.2. Two attempts were made at calculating A_2 directly from Flory–Huggins theory through the expression for the osmotic second virial coefficient

$$A_2 = v_2{}^2 V_1^{-1}(\tfrac{1}{2} - m_1\chi_{12}) \qquad (3.31)$$

The first assumes athermal mixing ($\chi_{12} = 0$); the second assumes a heat of mixing represented by the Hildebrand solubility parameter. Neither of the calculated results gave a satisfactory description of the experimental values, as can be verified from Table 3.2. An alternative method of testing the consequences of Eq. (3.31) is to calculate the Flory–Huggins interaction parameter from the experimental A_2 data. This was done by Koningsveld and Kleintjens [72], whose results appear in Table 3.2 (second column from the right).

Although the values are seemingly in the right direction, comparison with the χ_{12} values calculated directly from the Flory–Huggins theory (last column of Table 3.2) illustrates that the order of magnitude of the χ_{12} is not probable. The erroneous χ_{12} value obtained in the case of the SAN sample with 10% AN is a confirmation of the failure of the theory in correctly

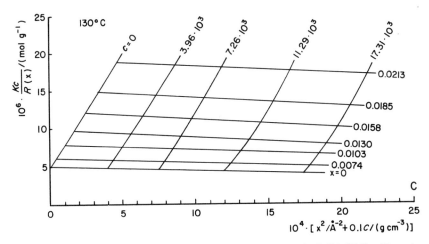

Fig. 3.31. Zimm diagram for the system poly(styrene–co-acrylonitrile) (28.7 wt% acrylonitrile) with deuterated poly(methyl methacrylate), both of approximately 200,000 weight average molecular weight (c = concentration of PSAN) (A = 25°C, B = 110°C, C = 130°C). [From W. A. Kruse, R. G. Kirste, J. Haas, B. J. Schmitt, and D. J. Stein, *Makromol. Chem.* **177**, 1145 (1976). Copyright by Hüthig and Wepf Verlag, Basel.]

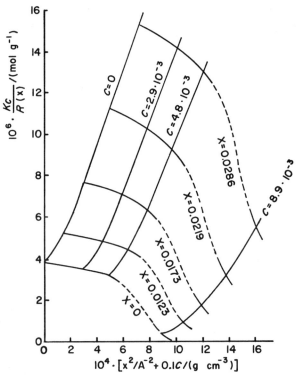

Fig. 3.32. Zimm diagram for poly(α-methylstyrene) mixed with deuterated poly(methyl methacrylate), both of approximately 250,000 weight average molecular weight [c = concentration of poly(α-methylstyrene)]. [From W. A. Kruse, R. G. Kirste, J. Haas, B. J. Schmitt, and D. J. Stein, *Makromol. Chem.* **177**, 1145 (1976). Copyright by Hüthig and Wepf Verlag, Basel.]

TABLE 3.2

Values for the Second Virial Coefficients from Neutron Scattering (Exp.) and Calculated Assuming Athermal Mixing (Athermal) and Solubility Parameters (δ) for Random SAN Copolymers Blended with d-PMMA[a]

% AN in PSAN	$M_w \times 10^{-5}$	A_2 (110°C) (cm^3 g^{-2} mole $\times 10^4$)			χ_{12} from $A_{2\,exp.}$ (FH)	$\chi_{12\,calc.}$(FH)
		Exp.	Athermal	δ		
19	2.7	1.15	0.02	−1.88	−0.0118	0.0007
19	4.4	1.15	0.01	−1.89	−0.0119	0.0005
10	0.7	−1.00	0.06	−0.08	+0.0110	0.0029
28.7	2.2	0.52	0.02	−5.98	−0.0052	0.0010

[a] The last two columns contain Flory–Huggins interaction parameters calculated from experimental neutron scattering results and directly from Flory–Huggins theory. Table from R. Koningsveld and L. A. Kleintjens, *J. Polym. Sci., Polym. Symp.* **61**, 221 (1977).

representing experimental evidence. A further attempt by Koningsveld and Kleintjens [72] in reconciling theory with experiment involved the use of the new theory of Huggins. With proper choice of the associated physical parameters, this theory did provide a correct interpretation of the experimental data.

Another investigation of low-angle neutron scattering from a miscible polymer mixture was carried out by Ballard, Rayner, and Schelten [91]. It involved poly(deutero-α-methylstyrene) mixed with polydeuterostyrene and/or polyprotostyrene. The weight percent compositions of the polymer mixtures investigated were 90/10/0, 90/5/5, 95/5/0, and 95/0/5, each with respect to the three polymers. Gunier plot procedure was used instead of the Zimm plot. It was found that the chain molecules are statistically distributed, indicating that the mixtures are miscible.

From the above, it is clear that the technique of low-angle neutron scattering will continue to find considerable application in the elucidation of the structure in miscible polymer mixtures. Particularly, it can provide answers to such questions as to whether the molecules in a mixture adopt their unperturbed configuration or a different dimension dictated by the neighboring, unlike interaction, and whether the molecules are randomly distributed or clustered. However, there is one point which conceivably may be a drawback of this technique. The question has been posed whether a deuterated matrix is thermodynamically different from a protonated one.

In order to answer this question, Koningsveld [72] used a slightly refined lattice expression in analyzing the chain length miscibility data of Kirste and Lehnen [92]. Kirste and Lehnen [92] performed low-angle neutron scattering experiments on mixtures of a normal, protonated poly(dimethyl siloxane) and a series of deuterated poly(dimethyl siloxanes) of varying chain lengths. The calculated second virial coefficient plotted as a function of the chain length of the deuterated PDMS did not agree with what is expected from a mixture of the same polymer. Indeed, the experimental data were satisfactorily represented by the expression derived for a binary mixture of two different polymers [72]. This finding supports the contention that a deuterated matrix is thermodynamically different from a protonated one. Further evidence in support of this conclusion has been found in the cloud-point and light scattering measurements on protonated and deuterated cyclohexane and polystyrene [93].

3.4.5 X-Ray Scattering and Other Methods

The physical structure of polymer mixtures can be characterized by the chain conformation, the local order, and the morphology. Small-angle neutron scattering elucidates the chain conformation [2]; the local order can be studied by means of electron and Rayleigh–Brillouin scattering [19], whereas the morphology can be studied by means of light scattering, small-angle X-ray scattering, magnetic birefringence, and visible, phase-contrast, and electron microscopy methods [2].

Small-angle neutron scattering has been used in the study of polymer miscibility (see Section 3.4.4). The use of electron scattering methods usually entails the derivation of pair distribution functions from the electron scattering curves [2]. No experiment of this nature has so far been done on polymer mixtures. The Rayleigh–Brillouin scattering technique has been applied successfully by Patterson and co-workers in the investigation of the miscibility of poly(vinylidene fluoride)–poly(methyl methacrylate) mixtures [94]. This involved measuring the polarized and depolarized light scattering spectra and analyzing them in terms of Rayleigh–Brillouin equations.

Probably the most extensive quantitative study of the morphology of polymer mixtures was the work of Stein and co-workers [95], who examined (in the solid state) the blends of poly(ε-caprolactone) (PCL) with poly(vinyl chloride) (PVC) by low-angle X-ray and small-angle light scattering. Because of the crystallinity of PCL, the authors not only had to describe the local order and the molecular distribution within the amorphous modular structures but also had to describe the spherulite size, the repeat period of the lamellar substructure, and the thickness of the crystalline in relation to the amorphous layers.

In the PVC concentration range from 0 to $\sim 60\%$, low-angle X-ray scattering measurements were interpreted using the Tsvankin–Buchanan technique [96, 97]. This provided a reasonable estimate of the repeat period of the PCL lamellar substructure, the thickness of the PCL crystalline layer, and the thickness of the amorphous layer containing both PCL and PVC segments. The spherulite sizes were measured by small-angle light scattering and, at $\sim 60\%$ PVC, no more crystallinity was observed. In this region, the small-angle X-ray scattering measurement was interpreted in terms of the Debye–Bueche theory [98] of scattering in heterogeneous media. The results were consistent with partial phase separation into statistical regions of two types: one made up of PCL domains containing dissolved PVC and the other made up of PVC domains containing dissolved PCL. The change in the intensity of scattering with concentration suggests a transition zone on the order of 30 Å between each of the two phases.

The significance of this sort of work cannot be overstated. However, it inherits a criticism usually leveled at methods used to identify miscible polymer blends in the solid state. Because changes of state took place in the preparation of the sample, one could not make a definite conclusion regarding the actual level of miscibility of the polymers involved. Observations made in the nonequilibrium glassy state are inadequate in rationalizing the thermodynamic aspect of polymer miscibility.

3.5 TERNARY-SOLUTION METHODS

3.5.1 Mutual-Solvent Method

Probably the oldest and most used method of determining polymer–polymer miscibility is the mutual-solvent approach. It consists of dissolving

TABLE 3.3

Polymers Used for the Ternary Solution Studies of Table 3.4[a]

Polymer No.	Material	Osmotic molecular weight	Intrinsic viscosity	Chemical characteristics
		Unfractionated		
1	Methyl cellulose	160,000	3.80^b	24% Methoxyl
2	Cellulose acetate	56,000	1.70^c	55.5% Acetic acid
3	Nitrocellulose	92,000	2.60^c	12.35% Nitrogen
4	Ethyl cellulose	35,000	1.10^c	47.6% Ethoxyl
5	Benzyl cellulose			46.5% Benzyl
6	Polystyrene	225,000	2.15^d	
7	Polyvinyl acetate			
7a	Rhodopas H	56,000	0.60^c	Totally acetylated
7b	Rhodopas HH	112,000	0.85^c	
8	Polyvinyl acetal	38,000	0.75^c	Copolymer of 10% polyvinyl alcohol, 2% polyvinyl acetate, 88% polyvinyl acetal
9	Methyl methacrylate	>2,000,000	3.65^c	
11	Rubber	Undetermined		
		Fractionated		
10	Polyvinyl alcohol			
10a	Fraction 1	60,000	1.10^b	
10b	Fraction 5		0.35^b	
12	Polystyrene			
12a	Fraction 1	700,000	4.2^d	
12b	Fraction 2	700,000	2.8^d	
13	Polyvinyl acetal			
13a	Fraction 1	97,000	1.10^c	Same product as for polyvinyl acetal above
13b	Fraction 3	39,000	0.58^c	
14	Cellulose acetate			
14a	Fraction 1	56,000	1.70^c	Commercial diacetate
14b	Fraction 3	16,000	0.39^c	

[a] From A. Dobry and F. Boyer-Kawenoki, *J. Polym. Sci.* **2**, 90 (1947).
[b] In water.
[c] In acetone.
[d] In chloroform.

TABLE 3.4

Results of Ternary Solution Studies Using Polymers Described in Table 3.3[a]

Mixture No.	Mixture of polymers (see Table 3.3)	Solvent	Limit of phase separation, dry content, %	Observations
1	1 + 10a	Water	3.2	
2	2 + 3	Acetone	5.5	Yield, 2:3 = 2:1[b]
3		Acetic acid	Miscible	
4	2 + 4	Acetone	2.8	
5		Acetic acid	5.5	
6	2 + 5	Ethyl acetate	2.0	
7	2 + 6	Methyl ethyl ketone	1.2	
8	2 + 7a	Acetone	5.5	7a:2 = 2:1
9	2 + 8	Acetone	2.1	
10	2 + 9	Acetone	1.5	2:9 = 3:1
11	2 + 10	No common solvent		
12	2 + 11	No common solvent		
13	3 + 4	Acetone	3.7	
14		Acetic acid	>20	Opaque film
15	3 + 5	Mesityl oxide	3.2	
16	3 + 6	Methyl ethyl ketone	0.85	
17	3 + 7b	Acetone	Miscible	
18		Methyl ethyl ketone	Miscible	
19		Acetic acid	Miscible	
20		Ethyl acetate	Miscible	
21		Amyl acetate	Miscible	
22	3 + 8	Acetone	1.8	
23		Methyl ethyl ketone	2.0	
24		Mesityl oxide	>5	Opaque film
25		Acetic acid	>20	Opaque film
26		Methyl acetate	2.6	
27		Ethyl acetate	1.8	
28		Propyl acetate	2.2	
29		Butyl acetate	1.8	
30		Amyl acetate + 10% absolute alcohol	2.2	
31	3 + 9	Acetone	Miscible	
32		Ethyl acetate	Miscible	
33	3 + 10	No common solvent		
34	3 + 11	No common solvent		
35	4 + 5	Chloroform	20	
36		Ethyl acetate	4.0	
37	4 + 6	Benzene	1.2	
38	4 + 7b	Chloroform	4.5	7b:4 = 3:1[b]
39	4 + 8	Chloroform	4.0	4:9 = 3:1[b]
40	4 + 9	Acetone	2.2	4:11 = 5:2[b]

3.5. Ternary-Solution Methods

TABLE 3.4 (*Continued*)

Mixture No.	Mixture of polymers (see Table 3.3)	Solvent	Limit of phase separation, dry content, %	Observations
41	4 + 10	No common solvent		
42	4 + 11	Benzene	1.3	
43	5 + 6	Chloroform	Miscible	
44	5 + 7b	Chloroform	2.5	$5:7b = 1:2^b$
45	5 + 8	Chloroform	10.5	
46	5 + 9	Dioxane	>10	Opaque film
47	5 + 10	No common solvent		
48	5 + 11	No common solvent		
49	6 + 7b	Chloroform	4.0	
50		Methyl ethyl ketone	1.5	
51	6 + 8	Chloroform	3.2	
52	6 + 9	Benzene	2.6	$6:9 = 5:2^b$
53	6 + 10	No common solvent		
54	6 + 11	Benzene	2.0	
55	7b + 8	Acetone	2.0	
56		Methyl ethyl ketone	3.5	
57		Chloroform	6.0	$7:8 = 3:1^b$
58	7a + 8	Acetic acid	7.2	
59		Mesityl oxide	12.0	$7:8 = 2:1^b$
60		Dioxane	7.0	
61		Mesityl acetate	3.8	
62		Ethyl acetate	3.2	
63		Propyl acetate	3.6	
64		Butyl acetate	3.9	
65		Amyl acetate	3.6	
66	7a + 9	Acetone	4.5	$4:9 = 12:1^b$
67		Ethyl acetate	8.5	$7:9 = 7:1^b$
68		Dioxane	>10	Opaque film
69		Acetic acid	>10	Opaque film
70	7 + 10	No common solvent		
71	7b + 11	Benzene	2.8	$7:11 = 5:1^b$
72	8 + 9	Acetone	2.2	$8:9 = 3:1^b$
73	8 + 10	No common solvent		
74	8 + 11	Benzene + 5% absolute alcohol	2.0	$8:9 = 5:2^b$
75	9 + 10	No common solvent		
76	9 + 11	Benzene	2.0	$9:11 = 2:3^b$
77	10 + 11	No common solvent		
78	7 + cellulose triacetate	Chloroform	7.5	$7:\text{triacetate} = 2:1^b$

[a] From A. Dobry and F. Boyer-Kawenoki, *J. Polym. Sci.* **2**, 90 (1947).
[b] Weight ratio of the amount of high polymers for which the limit of separation has been determined. For the other systems this yield is 1:1.

and thoroughly mixing a 50/50 mixture of two polymers at low to medium concentration in a mutual solvent. By allowing the mixture to stand, usually for a few days, miscibility is said to prevail if phase separation does not occur; if phase separation does occur the two polymers are said to be immiscible with each other. The method was first used in the field of paints, varnishes, and lacquers.

A varnish composition which on drying leaves a turbid, opaque, and usually brittle film is unacceptable; the occurrence was also known to be due to the immiscibility of the constituent polymers. It was not, however, until 1946 that a thorough and systematic study of polymer miscibility was undertaken. The study, reported in the classic paper by Dobry and Boyer-Kawenoki [99], involved 78 mixtures made up from 14 high polymers (cellulose, vinyl, and acrylic derivatives) dissolved in 13 solvents. The results of Dobry and Boyer-Kawenoki are represented in Tables 3.3 and 3.4. About a decade later, Kern and Slocombe [100] undertook a similar study on 27 other mixtures, the results of which appear in Table 3.5. The six general conclusions reached by Dobry and Boyer-Kawenoki [99], confirmed by the second study [100] stand to the present day almost without contradiction. They deserve direct quotation in part [99]:

(1) Of the 35 pairs of high polymers tested, only four do not show separation. Consequently, compatibility (miscibility) is the exception, *immiscibility is the rule.*

(2) When two high polymers are incompatible in one solvent, they are generally also incompatible in all other solvents. This rule represents the normal situation, but it is not always fulfilled.

(3) The limit of phase separation depends on the nature of the solvent.

(4) The molecular weight of the polymers is of great importance. The higher it is, the less compatible (miscible) are the samples and the more is the limit of phase separation shifted toward smaller (polymer) concentration.

(5) Theoretical considerations make it probable that not only the molecular weight but also the shape of the dissolved molecules influences their compatibility (miscibility).

(6) There is no obvious relationship between the compatibility (miscibility) of two polymers and the chemical nature of their monomers. The similarity of the principal chain is not sufficient to insure miscibility of two polymers.

Dobry and Boyer-Kawenoki's results [99] were also represented in triangular diagrams whose general nature is similar to Fig. 3.2.

Such phase diagrams support the spirit of the study, namely, that at rather high solvent concentration it is possible, by the mutual-solvent method, to tell that P_1 and P_2 are immiscible even though they are each completely miscible with the solvent.

3.5. Ternary-Solution Methods

TABLE 3.5

Results of Ternary Solution Studies[a]

System	Solvent and solute concentration	Volume ratio of upper/lower phase	Layer analysis	
			Upper	Lower
PMMA/PVAc[b]	Acetone, 20%	14/5	56% VAc	12.5% VAc
			60% VAc	10.6% VAc
PMMA/PVAc[c]	Acetone, 20%	14/5	71% VAC	6.3% VAc
PMMA/PMA	Acetone, 20%	16/9		
PMA/PVAc	Acetone, dioxane, benzene	—[d]		
PVAc/PMVK	Acetone, 20%	1/1	14.5% VAc	78% VAc
PVAc/PMVK	Ethyl acetate, 20%	7/5	68% VAc	4.5% VAc
PMMA/PS	Chloroform, 20%	2/3		
PMMA/PS	Methyl ethyl ketone, 15%	9/13		
PMMA/PMAN[e]	Acetone, 15%	9/13	92% MAN	13.7% MAN
PMMA/PMAN[f]	Acetone, 15%	9/10	88% MAN	1.0% MAN
PS/PVAc	Benzene, 20%	1/1		
PpClS/PVAc	Benzene, 20%	2/1	1.8% ClS	96% pClS
PpMeS/PVAc	Benzene, 20%	15/14		
PS/PpMeS	Benzene, 20%	14/15		
PS/PVC	Tetrahydrofuran, 15%	13/17	23% VC	86% VC
PS/PpClS	Benzene, 20%	2/1		
PpMeS/PpClS	Benzene, 20%	2/1	0.7% pClS	99% pClS
PS/PoMeS	Chloroform, 20%	—[d]		
PS/PmMeS	Chloroform, 20%	5/2		
PS/PpMeS	Chloroform, 20%	5/2		
PoMeS/PmMeS	Chloroform, 20%	5/3		
PoMeS/PpMeS	Chloroform, 20%	3/5		
PmMeS/PpMeS	Chloroform, 20%	—[d]		
PS/PE	Xylene, 10% (90°C)	25/14		
PS/PMA	Benzene, 20%	9/7		
PEA/PMA	Acetone, 20%	2/1		
PE/PBu	Xylene, 20% (90°C)	4/1		

[a] From R. J. Kern and R. J. Slocombe, *J. Polym. Sci.* **15**, 183 (1955).
[b] PVAc M_w ca. 50,000.
[c] PVAc M_w ca. 150,000.
[d] No phase separation.
[e] PMAN η_{sp} 0.034 0.1% in acetone.
[f] PMAN η_{sp} 0.35 0.1% in acetone.

All these experimental findings gained theoretical support from the Scott–Tompa ternary solution treatment of Flory–Huggins theory. The Scott–Tompa development [101, 102] was based on symmetric systems (systems where the Flory–Huggins polymer–solvent interaction parameters are equal), but it was surmised that the phase behavior for asymmetric systems would be similar. The analysis blames immiscibility on the unfavorable polymer–polymer interaction parameter. The solvent effect was considered rather positive; it merely dilutes the polymers so as to reduce the number of unfavorable contacts between the different polymers. Thus, it does not really matter what solvent is used; if two high polymers are immiscible in one solvent, they will be immiscible in all solvents.

This belief was held for a long time until contradictory experimental evidence began to appear. Hugelin and Dondos [103] were able to achieve maximum miscibility between polystyrene and poly(methyl methacrylate) only with solvents having comparable affinities for the polymers; solvents of disparate affinities fail, contrary to Scott–Tompa prediction. Bank et al. [40] observed that polystyrene is miscible with poly(vinyl methyl ether) in toluene, benzene, or perchloroethylene, but not in chloroform, methylene chloride, and trichloroethylene, while Kern [104] found that polystyrene is miscible with poly(methyl methacrylate) if the mixture is prepared in benzene or chlorobenzene, but immiscible if the solvent medium is ethyl acetate. The overriding conclusion of these experimental findings is that solvent effect is indeed significant and that Scott–Tompa results must be revised.

The first attempt at reexamining the Scott–Tompa treatment was undertaken by Zeman and Patterson [5], who calculated the spinodals for a number of ternary, polymer–polymer–solvent systems. Later, Hsu and Prausnitz [105] were able to make the notoriously difficult calculation of the binodals via numerical methods. Both studies arrived at identical conclusions:

(1) At low polymer concentration the difference between the two polymer–solvent interaction parameters is directly responsible for the polymer–polymer immiscibility.
(2) At high polymer concentration, the state of miscibility is controlled by the magnitude and sign of the polymer–polymer interaction parameter.
(3) When the interaction between the polymers is low or even negative a closed miscibility gap would result if the two polymer–solvent interaction parameters are different.

Even though these conclusions agree with experimental observations, they also cast serious doubt on the validity of the mutual-solvent method in identifying miscible polymer pairs. For instance, aside from the "normal" diagram of Fig. 3.2, three other types of ternary phase diagrams are possible, as shown in Fig. 3.33.

Based on the Patterson–Prausnitz [5, 105] ternary-solution treatment of Flory–Huggins theory, the qualitative deduction possible from mutual-solvent experiments falls far short. For a system whose phase behavior is similar to the "normal" one in Fig. 3.2, the common solvent method says nothing about the miscibility of polymers P_1 and P_2. As for a system with

3.5. Ternary-Solution Methods

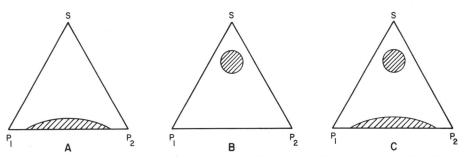

Fig. 3.33. Schematics of possible ternary phase diagrams involving two polymers (P_1, P_2) and a solvent (S).

a Fig. 3.33A phase diagram, a common solvent experiment performed at too high a solvent concentration would conclude erroneously that P_1 is completely miscible with P_2. Furthermore, the method might indicate complete immiscibility if Fig. 3.33B is representative of the ternary system, whereas the two polymers are miscible at all proportions. Examples of closed miscibility envelopes are found in the phase diagrams of benzene–butyl rubber–EPDM rubber and of diphenyl ether–atactic polypropylene–linear

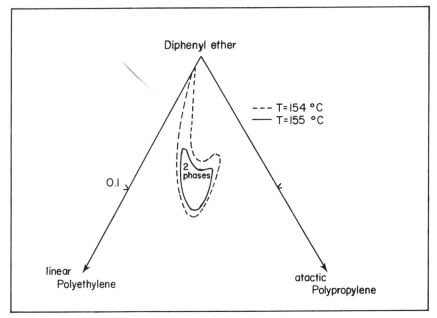

Fig. 3.34. Ternary phase diagram for the system polyethylene–polypropylene–diphenyl ether showing closed two-phase region. [From R. Koningsveld, L. A. Kleintjens, and H. M. Schoffeleers, *Pure Appl. Chem.* **39**, 1 (1974).]

polyethylene [82]. Figure 3.34 illustrates the phase diagram for the latter. The more complicated miscibility gap illustrated schematically in Fig. 3.33C is also found in nature; Fig. 3.35 illustrates the solvent-rich section of the phase diagram for diphenyl ether–isotactic polypropylene–linear polyethylene [82].

The more fruitful use of the Scott–Tompa ternary solution treatment is in calculating the polymer–polymer interaction parameter. The basic interpretative concept here is that a large positive value indicates unfavorable interaction, a low value indicates little interaction, and a negative value indicates a rather strong specific interaction.

The activity coefficient of the solvent in a ternary solution is

$$\ln a_1 = \ln \phi_1 + (1 - \phi_1) + (\chi_{12}\phi_2 + \chi_{13}\phi_3)(1 - \phi_1) - \chi'_{23}\phi_2\phi_3 \quad (3.32)$$

where the subscript 1 refers to the solvent and subscripts 2 and 3 refer to the two polymers; a_1, the solvent activity; ϕ, the volume fraction; χ, the binary interaction parameter; χ'_{23}, the polymer–polymer interaction parameter

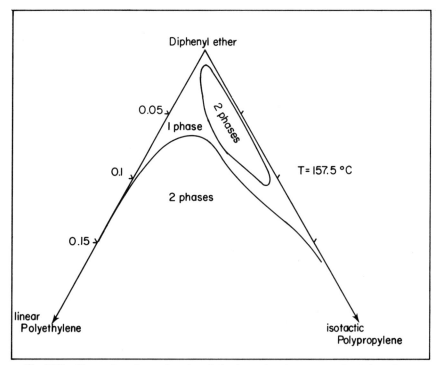

Fig. 3.35. Illustration of complex phase behavior, using the system polyethylene–isotactic polypropylene–diphenyl ether. [From R. Koningsveld, L. A. Kleintjens, and H. M. Schoffeleers, *Pure. Appl. Chem.* **39**, 1 (1974).]

3.5. Ternary-Solution Methods

per segment of polymer 2 [$\chi'_{23} = \chi'_{23}$ (Flory) x_1/x_2, where x is the number of segments in the molecule].

Kwei, Nishi, and Roberts [106] used vapor sorption measurements of polystyrene, poly(vinyl methyl ether), and their blends to successively calculate the χ_{12}, χ_{13}, and corresponding χ'_{23} as functions of temperature and composition. The χ'_{23} values appear in Table 3.6. Based on the sign of χ'_{23} and its temperature dependence, the authors concluded that the mixture is miscible (stable) and that it exhibits both the lower and upper cloud-point curves. This latter conclusion is based on Patterson's original analysis [107], which associates the occurrence of positive temperature coefficient of χ'_{23} with the presence of an upper critical solution temperature behavior, and a

TABLE 3.6

Interaction Parameters from Vapor Sorption Studies for Blends of Polystyrene and Poly(vinyl methyl ether)[a]

T (°C)	Wt% PVME in the film	ϕ_1	χ'_{23}
30	35.06	0.0967	−0.78
		0.1388	−0.73
		0.2006	−0.73
			av −0.75
30	45.30	0.0478	−0.67
		0.1104	−0.69
		0.1693	−0.72
			av −0.69
30	55.00	0.0389	−0.59
30	65.00	0.0698	−0.17
50	45.30	0.0367	−0.59
		0.0438	−0.60
		0.0516	−0.61
			av −0.60
50	65.00	0.0405	−0.47
		0.0498	−0.56
		0.0918	−0.36
			av −0.46

[a] Reprinted with permission from T. K. Kwei, T. Nishi, and R. F. Roberts, *Macromolecules* **7**, 667 (1974). Copyright by the American Chemical Society.

negative temperature coefficient of χ'_{23} with the presence of a lower critical solution temperature. The analysis may well be correct, but there is still some reservation regarding the validity of using solution measurements as a basis for deriving information about the state of the solvent-free polymer mixture. Furthermore, it is now well established that the state of thermodynamic stability of a mixture is not determined by the sign of the Gibbs free energy of mixing; rather, it is governed by the subtle details of the composition dependence of the free energy [72]. Consequently, the negative values calculated for the PS–PVME blends say nothing about the stability of the mixture.

3.5.2 Inverse Gas Chromatography Method

In the recent past, gas–liquid chromatography (GLC) has received general recognition as an effective, simple technique for rapid measurement of polymer interactions and solvent activity coefficients in molten homopolymers and their mixtures. It has been used in determining such properties as the glass transition temperature, crystallinity, adsorption isotherms, heats of adsorption, surface area, interfacial energy, diffusion coefficients, complex equilibria in solution, and curing processes in nonvolatile thermoset systems [108]. For these studies, its major advantages are (i) the simplicity, speed, and accuracy with which a large number of systems can be investigated, (ii) the wide range of easily controllable temperatures, and (iii) the ability to work at a single solution concentration.

In view of this unconventional usage of GLC, Guillet [109] has suggested the name "inverse gas chromatography" because traditional GLC determines the property of an unknown sample in the moving phase with a known stationary phase, while the inverse method determines the property of the stationary phase with the aid of a known vaporizable solute in the moving phase. He considers the latter as a molecular-probe experiment where the vaporizable molecules are designated probe molecules.

A schematic diagram of a gas chromatograph appears in Fig. 3.36. In operation, the polymer material, on a preferably inert support, is placed in the column maintained at a temperature which is at least 50°C above the system T_g for glassy material and T_m for a crystallizable system. A stream of an inert carrier gas continuously passes through the system at a known flow rate and under a predetermined pressure head, while the probe molecules are introduced in a pulse. The basic fundamental quantity of gas chromatography is the specific retention volume, defined as the volume of carrier gas per gram of stationary phase required to elute the probe molecule. Schematically, the interaction of the mobile probe solute with the stationary solvent phase (polymer) is illustrated in Fig. 3.37. The entering probe rup-

3.5. Ternary-Solution Methods

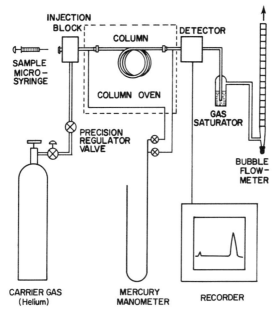

Fig. 3.36. Schematic of a typical gas chromatographic apparatus.

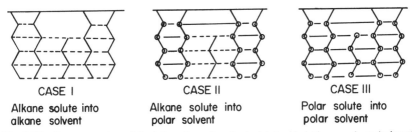

CASE I	CASE II	CASE III
Alkane solute into alkane solvent	Alkane solute into polar solvent	Polar solute into polar solvent

Fig. 3.37. Representation of the interaction of probe (solute) and stationary phase (solvent).

tures existing intermolecular forces and simultaneously forms new ones. Considering a probe of n segments and assuming that all the segments are randomly and completely absorbed by the stationary phase, n_{np} interactions would be gained by the system for case I, $2n_{np} - n_p$ for case II, and n_p for case III. Note that $n_{np} > 2n_{np} - n_p$; also, nonpolar probes in a polar stationary phase have a smaller V_g than if the stationary phase were nonpolar. It becomes apparent, therefore, that one could measure the polar, nonpolar, and specific interactions of a substrate by proper selection of the probe molecules. Furthermore, studies of two homopolymers and their blends could yield vital information about the polymer–polymer interaction.

The link between inverse chromatographic measurement and the interaction parameters of various solution theories is the infinite-dilution activity coefficient, which, from analysis of the dynamics of inverse chromatography, is obtainable directly from V_g data. For a probe in a polymeric stationary phase, the weight-fraction infinite-dilution activity coefficient is [108, 109]

$$\ln \Omega_1^\infty = \ln\left(\frac{a_1}{w_1}\right)^\infty = \ln\left[\frac{RT}{P_1^0 V_g M_1}\right] - \frac{P_1^0}{RT}(B_{11} - V_1) \quad (3.33)$$

This equation could alternatively be written in terms of V_g^0, the specific retention volume corrected to 0°C, where

$$V_g/T = V_g^0/273.2 \quad (3.34)$$

Now, in statistical thermodynamic theories, the probe activity, a_1, is generally written as the sum of two terms: a combinatorial entropy and a noncombinatorial free energy of mixing term. Written in the Flory–Huggins approximation and combined with the chromatographic expression, Eq. (3.33), the interaction parameter between the probe and the stationary phase is given by [108, 109]

$$\chi_{12} = \ln\left[\frac{RT v_2}{V_g V_1 P_1^0}\right] - \left(1 - \frac{V_1}{\overline{M}_2 v_2}\right) - \frac{P_1^0}{RT}(B_{11} - V_1) \quad (3.35)$$

In the newer equation-of-state solution theory, the noncombinatorial term is further broken down into two terms: the equation-of-state contribution due to the free-volume dissimilarity of the probe and the polymer, and the exchange energy term which reflects the energy involved when i–i or j–j contacts are replaced by i–j contacts. In the Flory approximation [110] the newly defined parameter, χ_{12}^*, a counterpart of χ_{12} based on conditions of a hypothetical liquid at 0°K, can be written in terms of V_g [108, 109]:

$$\chi_{12}^* = \ln\left[\frac{RT v_2^*}{V_g V_1^* P_1^0}\right] - \left(1 - \frac{V_1^*}{\overline{M}_2 v_2^*}\right) - \frac{P_1^0}{RT}(B_{11} - V_1) \quad (3.36)$$

The exchange energy parameter X_{12} can be calculated from χ_{12}^* by using Flory's expression [109]:

$$\chi_{12}^* = \frac{P_1^* V_1^*}{RT}\left\{\left[3\tilde{T}_1 \ln\left(\frac{\tilde{v}_1^{1/3} - 1}{\tilde{v}_2^{1/3} - 1}\right) + \tilde{v}_1^{-1} - \tilde{v}_2^{-1}\right] + \frac{X_{12}}{P_1^* \tilde{v}_2}\right\} \quad (3.37)$$

The development so far concerns the interaction of the probe with a homopolymer stationary phase. The extension to the case of the mixed stationary phase consisting of two high polymers has taken two forms,

3.5. Ternary-Solution Methods

both arising from the Scott–Tompa ternary-solution treatment. The activity of a solvent (as $\phi_1 \to 0$) in two polymers is given by

$$\ln a_1 = \ln \phi_1 + \left(1 - \frac{r_1}{r_2}\right)\phi_2 + \left(1 - \frac{r_1}{r_3}\right)\phi_3$$

$$+ \left(\chi_{12}\phi_2 + \chi_{13}\phi_3 - \frac{r_1}{r_3}\chi_{23}\phi_2\phi_3\right) \quad (3.38)$$

In applying this for a mixed stationary phase consisting of two high polymers and a probe used essentially at zero concentration, Patterson et al. [111] chose to use the infinite-dilution volume-fraction activity coefficient. Assuming that $r_i/r_j = v_i/v_j$, the resulting expression is

$$\ln\left(\frac{a_1}{\phi_1}\right)^\infty = \left(1 - \frac{V_1}{V_2}\right)\phi_2 + \left(1 - \frac{V_1}{V_3}\right)\phi_3$$

$$+ \left[\left(\frac{\chi_{12}}{V_1}\right)\phi_1 + \left(\frac{\chi_{13}}{V_1}\right)\phi_3 - \left(\frac{\chi_{23}}{V_2}\right)\phi_2\phi_3\right]V_1 \quad (3.39)$$

When equated to the corresponding chromatographic relation, we have

$$\left[\left(\frac{\chi_{12}}{V_1}\right)\phi_2 + \left(\frac{\chi_{13}}{V_1}\right)\phi_3 - \left(\frac{\chi_{23}}{V_2}\right)\phi_2\phi_3\right]V_1 = \ln\left(\frac{RT(w_2v_2 + w_3v_3)}{V_g V_1 P_1^0}\right)$$

$$- \left(1 - \frac{V_1}{V_2}\right)\phi_2 - \left(1 - \frac{V_1}{V_2}\right)\phi_3 - \frac{P_1^0}{RT}(B_{11} - V_1) \quad (3.40)$$

where χ_{ij}/V_i is symmetrical and dependent only on the nature of i,j regardless of the chain length. The new interaction parameters χ^*_{ij} are similarly described if the volume fractions are replaced by the segment fractions and the volumes by the hard-core values. These are related to the exchange energy parameters via

$$\left[\left(\frac{\chi^*_{12}}{V_1^*}\right)\phi_2^* + \left(\frac{\chi^*_{13}}{V_1^*}\right)\phi_3^* - \left(\frac{\chi^*_{23}}{V_2^*}\right)\phi_2^*\phi_3^*\right]V_1^*$$

$$= \left[\left(\frac{X_{12}}{S_1}\right)\theta_2 + \left(\frac{X_{13}}{S_1}\right)\theta_3 - \left(\frac{X_{23}}{S_2}\right)\theta_2\theta_3\right]\frac{s_1 M_1 v_1^*}{RT\tilde{v}}$$

$$+ \left[3\tilde{T}_1 \ln\left(\frac{v_1^{1/3} - 1}{v^{1/3} - 1}\right) + \frac{1}{\tilde{v}_1} - \frac{1}{\tilde{v}}\right]\frac{P_1^* M_1 v_1^*}{RT} \quad (3.41)$$

where X_{ij}/S_i is symmetrical and dependent only on the nature of i,j regardless of the chain length.

TABLE 3.7

Gas Chromatography Results for the Systems Tetracosane–Dioctyl phthalate and Tetracosane–Poly(dimethyl siloxane)[a]

	Interaction between solute (component 1) and pure stationary phase (component 3)								Interaction between two components (2 and 3) in the stationary phase			
	n-C$_{24}$			DOP		PDMS			n-C$_{24}$–DOP		n-C$_{24}$–PDMS	
	χ_{12}	$X_{12}/s_1 \times 10^8$ J cm^{-2}	χ_{12}	$X_{12}/s_1 \times 10^8$ J cm^{-2}	χ_{12}	$X_{12}/s_1 \times 10^8$ J cm^{-2}			$V_1\chi_{23}/V_2$ cm^{-3}	$X_{23}/s_2 \times 10^8$ J cm^{-2}	$V_1\chi_{23}/V_2$ cm^{-3}	$X_{23}/s_2 \times 10^8$ J cm^{-2}
Solute	60°C	60°C	75°C	75°C	75°C	75°C	60°C	60°C	75°C	75°C	60°C	60°C
n-Pentane	0.32	4.5	0.32	4.4	0.76	21.2	0.45	12.5	0.86	33.9	1.01	34.6
n-Hexane	0.24	4.6	0.24	5.0	0.67	19.2	0.43	13.4	0.72	25.5	0.48	11.7
n-Heptane	0.20	4.2	0.20	4.6	0.67	18.8	0.45	13.8	0.77	25.5	0.55	12.1
n-Octane	0.17	3.4	0.17	3.8	0.68	17.6	0.49	13.4	0.87	26.4	0.64	12.6
2-Methylpentane	0.26	5.0	0.27	5.9	0.69	20.1	0.42	13.0	0.72	26.8	0.57	15.9
3-Methylpentane	0.23	4.6	0.24	5.4	0.66	19.7	0.41	13.4	0.74	27.6	0.49	13.0
2,4-Dimethylpentane	0.26	5.0	0.26	5.4	0.70	21.3	0.42	13.4	0.77	28.5	0.49	13.0
Cyclohexane	0.17	5.4	0.17	5.9	0.48	19.2	0.44	19.2	0.62	28.5	0.42	10.9
Carbon tetrachloride	0.26	10.5	0.26	10.9	0.19	6.3	0.42	20.1	0.48	22.2		
Benzene	0.51	23.4	0.48	23.0	0.16	5.4	0.62	31.4	0.43	19.2	0.37	14.6
Toluene	0.35	15.5	0.36	16.3								

[a] Interaction parameters are listed. Reprinted with permission from D. D. Deshpande, D. Patterson, H. P. Schreiber, and C. S. Su, *Macromolecules* **7**, 530 (1974). Copyright by the American Chemical Society.

3.5. Ternary-Solution Methods

Patterson et al. [111] applied this formalism to the treatment of two mixed stationary phases, namely, tetracosane–dioctyl phthalate and tetracosane–poly(dimethyl siloxane). Table 3.7 summarizes some of the results. Within the alkane series, very little variation is observed for the X_{23}/S_2 and $V_1\chi_{23}/V_2$ except for pentane. This is in line with theory. However, the quantities are significantly different when polar probes are used, indicating that the interactions involved might not be correctly describable by the present theory. Such variation was not ascribed to experimental error because the check on data consistency was reasonably successful. For instance, the magnitude of X_{23}/S_2 for tetracosane–dioctyl phthalate is similar to that for the pure dioctyl phthalate with alkane probes—so also for tetracosane–poly(dimethyl siloxane) as compared to that of pure poly(dimethyl siloxane).

Patterson [112] also applied inverse chromatography to the study of thermodynamic interactions in poly(vinyl chloride) (PVC) plasticized by di-n-octyl phthalate (DnOP). Table 3.8 contains $V_1\chi_{23}/V_2$ values for PVC–DnOP (82:18) as affected by temperature. The concentration dependence of this quantity is represented in Fig. 3.38; each data point is an average of the results from four n-alkane probes. Variation of the interaction parameter with probe was much more evident here than in the previous study, and the authors ascribed this either to nonrandom mixing or to preferential solution of the probe in one of the components of the mixed stationary phase. From the composition dependence of the interaction parameter, the authors concluded that DnOP at low concentration (<0.25) is miscible with PVC, but immiscible when the volume fraction is higher than 0.55.

Olabisi's development [113] differs from Patterson's primarily in the

TABLE 3.8

Interaction Parameters for the System PVC–Dioctyl phthalate by Gas Chromatography[a]

	χ'_{23}[b]		
Probe	110°C	120°C	130°C
n-Heptane	−1.20	−1.04	−0.81
n-Octane	−1.17	−1.07	−0.94
n-Nonane	−1.24	−0.89	−0.43
n-Decane	−1.63	−0.66	−0.14
Toluene	−0.72	−0.66	−0.60
Chlorobenzene	−0.78	−0.68	−0.56

[a] From C. S. Su, D. Patterson, and H. P. Schreiber, J. Appl. Polym. Sci. **20**, 1025 (1976).
[b] Assumes no crystallinity in PVC.

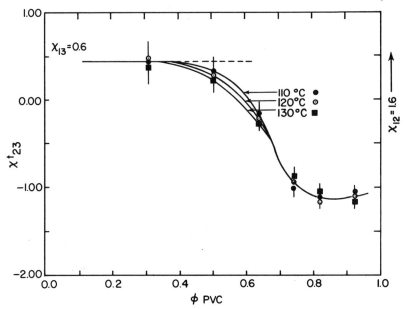

Fig. 3.38. The concentration dependence of the interaction parameter for the system PVC–dioctyl phthalate as measured by inverse gas chromatography. [From C. S. Su, D. Patterson, and H. P. Schreiber, *J. Appl. Polym. Sci.* **20**, 1025 (1976).]

level of assumptions made. For polymers with rather high degrees of polymerization and for a small enough probe such that $r_1 \sim 1$ [114], $r_1/r_2 \sim 0$, and $r_1/r_3 \sim 0$

$$\chi'_{23} = (\chi_{23})_{\text{Tompa}} = \frac{(\chi_{23})_{\text{Flory}}}{r_2} \sim \frac{r_1}{r_2}(\chi_{23})_{\text{Flory}} \quad (3.42)$$

When these are substituted into Eq. (3.36), the infinite-dilution weight-fraction activity coefficient is [113]

$$\ln \Omega^\infty_{1(23)} = \ln[v_1/(w_2 v_2 + w_3 v_2)]$$
$$+ (1.0 + \chi_{12}\phi_2 + \chi_{13}\phi_3) - \chi'_{23}\phi_2\phi_3 \quad (3.43)$$

When equated to the corresponding chromatographic expression, we have, after rearrangement [114],

$$\chi_{12}\phi_2 + \chi_{13}\phi_3 - \chi'_{23}\phi_2\phi_3 = \ln\left[\frac{RT(w_2 v_2 + w_3 v_3)}{P_1^0 V_g V_1}\right] - 1$$
$$- \frac{P_1^0}{RT}(B_{11} - V_1) \quad (3.44)$$

3.5. Ternary-Solution Methods

χ_{12} and χ_{13} are determined separately from the specific retention volume of the probe with the homopolymers as prescribed by Eq. (3.35) and χ'_{23} is unambiguously determined from the specific retention volume of the probe in the mixed stationary phase using Eq. (3.44). If the hard-core quantities are used, χ^*_{12}, χ^*_{13}, and χ^*_{23} are obtained. The exchange energy parameter χ_{ij}'s are calculated from an expression similar to Eq. (3.41) and estimated values of the segmental surface area ratio s_i/s_j.

By selecting probe molecules based on the relative magnitude of their dipole moments, polarizabilities, and hydrogen-bonding capabilities. Olabisi [113] investigated four types of polymer interactions:

(i) proton-acceptor strength with chloroform and ethanol as probes
(ii) proton-donor strength with methyl ethyl ketone and pyridine as probes
(iii) polar strength with acetonitrile and fluorobenzene as probes
(iv) nonpolar strength with hexane and carbon tetrachloride as probes

Recognizing that no such clear-cut division exists and that association complexes stabilized by electronic and electrostatic forces do exist even for nonpolar molecules, χ_{ij}, χ^*_{ij}, and X_{ij} were proposed merely as relative scales of interaction strengths between polymers. The data obtained for poly(ε-caprolactone), poly(vinyl chloride), and the mixture appear in Table 3.9. In calculating the exchange energy parameter, the segmental surface area ratio s_i/s_j was computed from the group contribution format of Bondi [16].

Based on the various interaction quantities obtained with chloroform as a probe and the fact that PCL–PVC mixtures are known to be stable over the

TABLE 3.9

Interaction Parameters for the System PVC–PCL at 120°C Using Gas Chromatography[a]

Solute	PCL		PVC		PCL–PVC (50:50)		
	χ^*_{12}	X_{12} (cal/cm³)	χ^*_{12}	X_{12} (cal/cm³)	χ_{23}	χ^*_{23}	X_{23} (cal/cm³)
Ethanol	1.15	21.4	2.35	40.6	0.21	−0.13	−2.8
Chloroform	−0.20	−4.20	1.38	16.6	0.33	−0.09	−2.4
Methyl ethyl ketone	0.533	2.75	1.00	4.41	−0.10	−0.61	−6.4
Pyridine	0.175	0.239	0.939	7.34	−0.17	−0.47	−5.4
Acetonitrile	1.11	19.3	1.85	29.3	−0.40	−0.98	−9.3
Fluorobenzene	0.127	−1.61	1.25	8.82	0.24	−0.15	−2.9
Carbon tetrachloride	0.391	1.06	1.49	10.2	1.07	0.63	3.0
Hexane	1.24	7.89	1.76	8.6	1.16	0.60	2.8

[a] Reprinted with permission from Olabisi, *Macromolecules* **8**, 316 (1975). Copyright by the American Chemical Society.

whole concentration range, it was noted that χ^*_{23} and χ_{23} present a clear picture of PCL–PVC miscibility and that complementary dissimilarity is responsible for the observed miscibility. Also observed is the fact that polar probes yield positive interaction indices representative of noncomplexing contributions, whereas specifically interacting probes sometimes give low or negative values. The variation of the polymer–polymer interaction with the probe molecules was ascribed to nonrandom absorption of the probe in the mixed stationary phase as well as to preferential solution of the probes in one of the constituents of the blend [113].

The foregoing chromatographic method has succeeded, by and large, in describing the miscibility state of solvent-free polymer mixtures by using the solvent at essentially zero concentration. It is capable of providing the interaction parameter at any given condition and may be able to provide some subtle details of the composition dependence of the free energy so paramount in defining the complete state of thermodynamic stability of polymer mixtures. The major uncertainty in the accuracy of this method can be found in the way chromatographic columns are prepared. Because the polymer mixture must first be dissolved in a mutual solvent prior to deposition on the inert support, the method would fail where the mutual-solvent method fails. The preparation of columns for systems with the sort of closed miscibility loop discussed earlier would introduce several uncertainties in the result. Furthermore, recent developments [88] in the field of polymer–polymer miscibility show that rather high accuracy is needed for proper definition of the state of thermodynamic stability, and it is doubtful that the inverse GLC can provide that level of significance. And, from a practical standpoint, the columns are extremely time consuming to prepare [114].

Nonetheless, the inverse GLC method has been applied successfully to the description of polymer miscibility in the liquid state. It has also been used in studying the concentration dependence of the glass transition temperature of polymer blends [108]. Because changes of state do occur during sample preparation, the applicability of this method may be limited to systems whose miscibility gap is known beforehand.

3.6 MISCELLANEOUS

3.6.1 Rheological Properties

a. Binary Studies. The determination of polymer–polymer miscibility by rheological measurements on binary systems is rare and indeed may be difficult to justify. But because the morphology of a two-phase system can

3.6. Miscellaneous

change with shearing rate, whereas the structure of a soluble system cannot, it is expected that the shear viscosity function of soluble systems will change monotonically with composition. Deviation from monotony can be taken as positive evidence of two phases.

Some examples of the use of melt viscosity as an indicator of miscibility are available. Kongarov and Bartenev [115] found a monotonic change of the viscosity function with composition for the system *cis*-1,4-polyisoprene–natural rubber, but completely unpredictable behavior for natural rubber–nitrile rubber. The system natural rubber–polybutadiene responded in an intermediate fashion.

Similar, but incomplete, results were obtained by Giniyatullin *et al.* [116] for poly(tetrahydrofuran)-based urethanes mixed with PVC. Linear polyurethane with PVC yielded melt viscosities which varied in a more regular fashion with concentration than those of branched polyurethanes with PVC. In no case was the variation completely monotonic, indicating limited solubility [117].

The viscoelastic properties of the PPO–PS system have been used to demonstrate its miscibility at a level corresponding to entanglement spacing [118]. Dynamic measurements at temperatures above the glass transition and extending into the flow region were found to change more smoothly with composition than the same measurements using a mixture of two PS samples of different molecular weights [119]. Viscosities of the former mixtures, corrected to the same free volume to account for the steady increase of T_g for the mixtures, were found to change smoothly with composition. This behavior was in accord with that predicted, based on an averaging of both the weight average molecular weights and entanglement molecular weights.

Similar conclusions concerning smooth changes with composition were reached by Schmidt [120] using blends of PPO with high-impact PS (HIPS). By measuring dynamic viscoelastic properties and steady-shear properties over a wide range of temperatures and frequencies for HIPS, PPO, and a 65/35 HIPS–PPO blend, it was concluded that the blend was intimately mixed on a segmental level.

A novel rheological technique for the detection of two phases in polymer mixtures has recently been suggested by Hubbel and Cooper [121]. This method presumes that the segmental orientation of the components in a miscible system will be the same, whereas the segmental orientation of the components in two-phase mixtures will differ significantly. While applied only to solids by Hubbel and Cooper, the technique should be equally applicable in any viscoelastic region. Results with the miscible systems nitrocellulose–PCL and PVC–PCL confirmed the presumptions of the method in that the orientation functions for both components were similarly related to the strain applied to the sample.

b. Ternary Studies (Polymer–Polymer–Solvent). The dependence of intrinsic viscosity $[\eta]$ on molecular weight can be used to estimate the interaction parameter via the Stockmayer–Fixman relationship [122]

$$[\eta] = K_\Theta M_2^{1/2} + 0.036\Phi[(1 - 2\chi_{12})/NV_1]M_2/\rho_2^2 \quad (3.45)$$

where K_Θ is the Mark–Houwink constant for a Θ-solvent, Φ is a universal constant of about 3.1×10^{24}, M_2 is the molecular weight of the polymer, χ_{12} is the interaction parameter for solvent 1 with polymer 2, V_1 is the molar volume of solvent, ρ_2 is the density of the polymer, and N is Avagadro's number. Thus, it is expected that viscosity, reflecting the size of polymer coil, will be influenced by the thermodynamics of these systems. Favorable interactions lead to higher intrinsic viscosities due to expansion of the polymer coil with solvent.

For ternary systems, it is not expected that $[\eta]$ will change much with changes in interactions between the two polymers because of the high dilution, but this does not always appear to be the case. Williamson and Wright [123] found large positive deviations from any average of the components' intrinsic viscosities for systems of highly interacting polymers such as PEO–poly(vinyl alcohol) and poly(acrylic acid)–poly(vinyl alcohol). The behavior of the latter is shown in Fig. 3.39. Apparently these systems consist of aggregates of several molecules. The discrepancies between average and observed intrinsic viscosity for most systems are very small [124, 125], and the method is not recommended for determining polymer–polymer miscibility.

A second type of experiment, lending itself to analysis by Eq. (3.45), has been performed [103]. In this the intrinsic viscosity of polymer 2 is determined

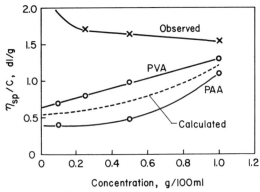

Fig. 3.39. Correlations of dilute solution viscosities of poly(vinyl alcohol) (PVA), poly(acrylic acid) (PAA), and their mixture ($\times$——$\times$), showing large deviation of the latter result from an average behavior (---). [From G. R. Williamson and B. Wright, *J. Polym. Sci., Part A* **3**, 3885 (1965).]

3.6. Miscellaneous

in a solvent containing a *constant* concentration of polymer 3. The value thus obtained is related to 2–3 miscibility. To obtain χ_{23}, the intrinsic viscosity of polymer 2 could be determined in this manner using several polymer 3 solutions of different concentrations. Extrapolation to a polymer 3 concentration of 1, followed by a Stockmayer–Fixman plot, would yield χ_{23}. Unfortunately, this would not be an accurate method because the second term of the Stockmayer–Fixman relationship would be small, due to the large value of V_1 as modified by the presence of polymer 3. All of these techniques depend heavily on the validity of the Stockmayer–Fixman relationship, which is in conflict with other theories [122].

A third type of solution viscosity experiment has been more widely used. In this experiment the Huggins constant K as defined by Eq. (3.46) below

$$\eta_{sp}/c = [\eta] + Kc[\eta]^2 \tag{3.46}$$

is examined [125]. Alternatively, the group $b = K[\eta]^2$ can be used [124]. Morawetz [122] has shown that interacting polymer systems in this experiment may show very high values of b compared with the average for each polymer. Böhmer et al. [124] have correlated directly the deviation of b from the average value with the polymer–polymer interaction parameter. They consider this more promising than deviations in K, which varies little, for determining polymer–polymer interactions. This method, and other adaptations [63, 103, 123, 125–129] of the Krigbaum and Wall treatment [130] are empirical in nature and should be used with caution.

As in any ternary experiment, it should be borne in mind that a dependence on *solvent* is likely and there is no certain method of eliminating this dependence, even at high solute concentration. As a minimum precaution, the experiment should be repeated in a wide variety of solvents.

3.6.2 Volume of Mixing

Blends of immiscible polymers and phase-separated block copolymers are generally expected to exhibit no volumetric deviation over that calculated utilizing an additivity relationship [131–133]. With miscible polymer blends, many experimental cases exist showing that the specific volumes are not additive [26, 106, 134, 135]. Generally, miscible-blend densities are higher than those calculated from volumetric additivity relationships, especially where specific interactions exist.

To make valid comparisons for the actual state in which the blend exists (i.e., glass or rubber), the calculated specific volume for the blend must employ pure-component volumes for the same state. In many experimental cases, such as a rubbery polymer blended with a glassy polymer, this will require extrapolated values, as shown in Fig. 3.40. Thus, the specific volume

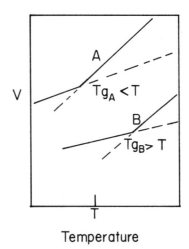

Fig. 3.40. Generalized volume–temperature response for polymers with different glass transition temperatures, showing extrapolation procedure for blends.

for the rubbery polymer A should be the extrapolated value of the volume–temperature data from the glassy state if the blend is in the glassy state. Conversely, if the blend T_g is below the testing temperature, extrapolation of the specific volume of the glassy polymer B from the rubbery state to the test temperature is required for determination of the specific volume to be employed in the calculations. Miscible blends of a rubbery and a glassy polymer would be expected to exhibit nonadditive specific volumes in the region between the T_g values of the respective components. Obviously, volume change (i.e., densification) should not be presented as evidence of polymer–polymer miscibility or interaction between the components unless this extrapolation is performed.

Studies by Kwei et al. [106] with polystyrene–poly(vinyl methyl ether) blends revealed significant densification, as shown in Table 3.10. These results were interpreted as being additional evidence for extensive mixing of the components.

Poly(vinylidene fluoride)–poly(methyl methacrylate) blends exhibited negative volume changes at high PVF_2 contents, but positive values for PMMA-rich blends [26]. However, extrapolation of the pure-component densities to the actual state of the blend was not attempted, which may explain these results.

Some work has been directed toward application of the free-volume concept to immiscible polymer mixtures. While not directly related to miscible polymer systems, these studies do indeed relate to volume change in polymer

3.6. Miscellaneous

TABLE 3.10

Densities of Mixtures of Polystyrene and Poly(vinyl methyl ether)[a]

			ρ (calcd)		
Wt% PVME	$\rho_j 23$	T_g (°C)	PS and PVME densities at 23°	Extrapolated PVME density	Extrapolated PS density
0	1.0505	102			
10.00	1.0562	80	1.0495	1.0508	
35.06	1.0661	18	1.0470		1.0654
45.30	1.0615	−18	1.0459		1.0614
70.00	1.0525	−23	1.0434		1.0520
100	1.0404	−29			

[a] Reprinted with permission from T. K. Kwei, T. Nishi, and R. F. Roberts, *Macromolecules* **7**, 667 (1974). Copyright by the American Chemical Society.

mixtures and may be of importance in understanding systems of partial miscibility. Lipatov and co-workers [136–139] have been primary contributors in this area. Starting with the Simha–Boyer relationship ($\Delta \alpha T_g = k$) [140], a verification of the following modification

$$\Delta \alpha_1 T_{g1} = k_1 \phi_1, \qquad \Delta \alpha_2 T_{g2} = k_2 \phi_2 \qquad (3.47)$$

was experimentally attempted. Using blends which exhibited T_g's of the respective components (thus indicating negligible phase mixing), dynamic dilatometry and isothermal compression data were obtained. While the pure-component $\Delta \alpha T_g$ data agreed reasonably well with the universal constant of the Simha–Boyer relationship ($k = 0.113$), values of the expression $\Delta \alpha_i T_{gi}/\phi_i$ were consistently higher than the predicted values. Lipatov and Vilenskii [138] concluded that this positive deviation implied that the density of molecular packing in the phase-separated mixture was lower than in the pure state, indicating higher molecular mobility or free volume. This excess free volume was proposed to be associated with the interphase region. Similar investigations may be of interest for systems exhibiting partial miscibility.

3.6.3 Heat of Mixing by Calorimetry

Calorimetry is one of the most direct methods of determining thermodynamic parameters. At constant pressure, the heat released by a mixing process is proportional to the enthalpy of mixing (ΔH) while the variation of enthalpy with temperature yields the free energy (ΔG) of mixing. The relevant relationship is

$$\left.\frac{\partial(\Delta H/T)}{\partial(1/T)}\right|_{P,N} = \Delta G \qquad (3.48)$$

For rather obvious reasons, no one has reported a calorimetric experiment involving the direct mixing of two polymeric components, although liquid oligomers have been successfully employed [68]. All determinations have used a solvent to aid the mixing process. The thermodynamic cycle used to calculate ΔH is

$$\begin{array}{c} \text{polymer 1 + polymer 2} \xrightarrow{\Delta H} \text{1-2 mixture} \\ \Delta H_1 \downarrow \quad \Delta H_2 \downarrow \quad \Delta H_4 \downarrow \\ \text{solution 1 + solution 2} \xrightarrow{\Delta H_3} \text{solution 1-2} \end{array} \qquad (3.49)$$

$$\Delta H = \Delta H_1 + \Delta H_2 + \Delta H_3 - \Delta H_4$$

An equivalent, simplified cycle is

$$\begin{array}{c} \overline{\text{polymer 1 + polymer 2}} \xrightarrow{\Delta H} \text{1-2 mixture} \\ \Delta H_{12} \downarrow \qquad \qquad \nearrow \Delta H_4 \\ \text{solution 1-2} \end{array} \qquad (3.50)$$

$$\Delta H = \Delta H_{12} - \Delta H_4$$

where ΔH_{12} is determined using a dry blend of the appropriate ratio of polymer 1 to polymer 2.

This cycle was used as early as 1958 by Struminskii and Slonimskii [141], who found a general agreement between ternary-phase behavior and heat of mixing. They were also the first to recognize the difficulties involved with calorimetric measurements on glassy polymers. Because the glass is not a thermodynamic state, the measured heats depend on how the glass is prepared.

Ichiara and co-workers [142, 143] used similar calorimetric techniques and reported similar problems with variation due to preparation of glassy samples. Their recommendation was that calorimetric techniques be confined to rubbery samples. Zverev et al. [144] experienced difficulties with crystallinity differences between components and mixtures. It is evident that crystallinity changes could lead to severe errors in the calculated heat of mixing.

Heat of mixing of selected polar polymers in the region of phase separation

3.6. Miscellaneous

can show unusual behavior with concentration, according to the results of Tager and co-workers [145, 146]. In the composition region corresponding to two phases, the heat of mixing was found to change signs, a behavior also cited by Patterson [147] for hydrocarbons.

Novakov *et al.* [148] have attempted to correlate "compatibility" with the heat effect of step 4 in Eq. (3.49). This might be possible if extrapolation to 100% solids were performed, but otherwise it appears to be an inadequate assumption because of the influence of solvent on the process.

No record has been found of an attempt to measure the temperature dependence of the heat of mixing to derive the free energy of mixing. This is not too surprising in view of the difficulty of the measurements and the compounding of error on taking derivatives. However, the free energy of mixing and, to some extent, its components have been determined with the aid of absorption studies. Tager and co-workers [145, 146] and Kwei *et al.* [106, 149] have used this technique.

Kwei *et al.* did not attempt to solve for the components of the free energy, they simply solved for the interaction parameter for the two polymers (components 2 and 3) using the traditional equations of regular solution theory. The sequence of relationships is

$$a_1 = P_1/P_1^0 \tag{3.51}$$

$$\Delta \mu_1 = RT \ln a_1 \tag{3.52}$$

$$= \ln \phi_1 + \phi_2 + \chi_{12}\phi_2^2 \quad \text{polymer 2-solvent 1}$$

$$= \ln \phi_1 + \phi_3 + \chi_{13}\phi_3^2 \quad \text{polymer 3-solvent 1}$$

$$= \ln \phi_1 + (1 - \phi_1) + (\chi_{12}\phi_2 + \chi_{13}\phi_3)(1 - \phi_1)$$

$$- \chi'_{23}\phi_2\phi_3 \quad \text{polymer 2-polymer 3-solvent 1}$$

where a_1 is the activity, P_1 is the partial pressure of the solvent, P_1^0 is the full vapor pressure of the solvent, $\Delta \mu_1$ is the partial molar free energy (chemical potential) of the solvent, χ_{12} and χ_{13} are the interaction parameters for polymer with solvent (based on solvent volume), and χ_{23} is the interaction parameter for the two polymers (also based on solvent volume). The result might be used to approximate ΔH^M through the relationship

$$\Delta M^M/V = RT\chi_{23}\phi_2\phi_3/v_1 \quad (\text{cal/cm}^3) \tag{3.53}$$

where v_1 is the molar volume of the solvent.

The approach of Tager and co-workers [145] to the analysis of absorption data for mixed polymer systems follows the more general thermodynamic route. The thermodynamic quantities are given on a convenient weight basis, the subscript 2 referring to either a polymeric component or the polymer mixture. The series of equations used is

$$\Delta\mu_1 = (1/M_1)RT\ln(P_1/P_1^0) \quad \text{(cal/g of solvent)} \quad (3.54)$$

$$\Delta\mu_2 = -\int_{-\infty}^{\Delta\mu}(w_1/w_2)\,d(\Delta\mu_1) \quad \text{(cal/g of polymer)} \quad (3.55)$$

$$\Delta g^M = w_1\Delta\mu_1 + w_2\Delta\mu_2 \quad \text{(cal/g of solution)} \quad (3.56)$$

$$\Delta G^M = \Delta g^M/w_2 \quad \text{(cal/g of polymer)} \quad (3.57)$$

where M_1 is the molecular weight of the solvent, w_1 and w_2 are weight fractions of solvent and polymer, and Δg^M is the average free energy of mixing. The limit of $\Delta g^M/w_2$ as w_2 goes to zero is the free energy of mixing the polymer (or polymer mixture) with an infinite amount of solvent. Because the absorption experiment cannot easily approach this limit, the result is very dependent on the assumption concerning the shape of the $\Delta g^M(w_2)$ relationship at high solvent content. The limiting values for ΔG^M, equal to ΔG, for both polymeric components and a $W_1:W_2$ mixture of the components may be combined using the relationship

$$\Delta G = W_1\Delta G_1 + W_2\Delta G_2 - \Delta G_4 \quad (3.58)$$

where the subscripts refer to the steps of the thermodynamic cycle in Eq. (3.49). The value of ΔG_3 is zero because of the infinite dilution of the solutes. The enthalpy of mixing may be calculated from the temperature variation of ΔG, using the analog of Eq. (3.48).

Qualitatively, the absorption isotherms for mixtures of low miscibility fall in a regular fashion between the isotherms for the pure components, while the isotherms for highly miscible systems fall below those of the components. Thus, an examination of the isotherms can provide a qualitative evaluation of the free energy function for the two components.

3.6.4 Melting Point Depression

The addition of low molecular weight soluble compounds to crystalline polymers results in a melting point depression. The melting point depression in this case can be determined by the expression

$$\frac{1}{T_m} - \frac{1}{T_m^\circ} = \frac{RV_2}{\Delta H_2 V_1}\{(1-\phi_2) - \chi_{12}(1-\phi_2)^2\} \quad (3.59)$$

where χ_{12} is the interaction parameter, T_m the experimental melting point, T_m° the equilibrium melting point, ΔH_2 the heat of fusion of 100% crystalline polymer per mole of repeat unit, V_1 the molar volume of diluent, V_2 the

3.6. Miscellaneous

molar volume of polymer repeat unit, and ϕ_2 the volume fraction of crystalline polymer. Melting point depression data for solute–polymer blends is an accepted method for the determination of the heat of fusion for the crystalline portion of semicrystalline polymers. Calorimetric data of the polymer yield ΔH_f; thus, with the ΔH_2 data of the solute–polymer blend from Eq. (3.59), the degree of crystallinity can be determined.

In polymer–polymer blends in which one component is crystalline, melting point depressions are also observed. Examples include poly(ε-caprolactone)–poly(vinyl chloride) [150], isotactic polystyrene–PPO [151], and poly(vinylidene fluoride)–poly(methyl methacrylate) [152, 153].

The utility of the melting point depression to calculate the interaction parameter was demonstrated by Nishi et al. [152, 153]. This method, which provides for calculation of χ_{12}, has definite importance and will be summarized here.

The general equation for melting point depression is

$$\frac{1}{T_m} - \frac{1}{T_m^\circ} = -\frac{RV_2}{\Delta H_2 V_1}\left[\frac{\ln \phi_2}{m_2} + \left(\frac{1}{m_2} - \frac{1}{m_1}\right)(1 - \phi_2) + \chi_{12}(1 - \phi_2)^2\right] \quad (3.60)$$

For polymer mixtures, m_1 and m_2 (the degree of polymerization for constituents 1 and 2) are very large, thus

$$\frac{1}{T_m} - \frac{1}{T_m^\circ} = -\frac{RV_2}{\Delta H_2 V_1}\chi_{12}(1 - \phi_2)^2 \quad (3.61)$$

Equation (3.61) indicates that a negative χ_{12} will yield a melting point depression as observed for experimental systems previously cited. With a positive interaction parameter, the theory predicts that a melting point elevation would result, as pointed out by Nishi and Wang [152]. Note that a positive χ_{12} will most probably result in phase separation due to the unfavorable thermodynamic situation for high molecular weight polymer mixtures.

In Eq. (3.61), χ_{12} and ΔH_2, present as a ratio, cannot be determined simultaneously from calorimetric measurements. In order to alleviate this experimental problem, Nishi and Wang suggested the following approach: The interaction parameter χ_{12} was assumed to be of the form

$$\chi_{12} = BV_1/RT \quad (3.62)$$

where B is the polymer–polymer interaction energy density. Equation (3.61) then reduces to

$$\frac{1}{\phi_1}\left[\frac{1}{T_m} - \frac{1}{T_m^\circ}\right] = -\frac{BV_2\phi_1}{\Delta H_2 T_m} \quad (3.63)$$

Recasting the data in the form of variables equal to $(1/T_m - 1/T_m^\circ)/\phi_1$ and ϕ_1/T_m allows B to be calculated from the slope of a plot of these variables; then χ_{12} can be determined. This procedure allows one to average experimental data graphically. Note that calculation of χ_{12} from data on a single blend is possible (with T_m° and ΔH_2 predetermined), but not as accurate.

This analysis indicates that a melting point depression of a crystalline polymer in a polymer blend implies miscibility and allows for the calculation of the interaction parameter. Some values cited are $\chi_{12} = -0.295$ (160°C) for poly(vinylidene fluoride)–poly(methyl methacrylate) mixtures [152] and $\chi_{12} = -0.34$ (160°C) for poly(vinylidene fluoride)–poly(ethyl methacrylate) mixtures [153]. In using the analysis for melting point depression to predict χ_{12}, it must be recognized that a miscible polymeric diluent in a crystalline polymer can alter the spherulite dimensions. As the melting point is influenced by the spherulite size, corrections for this variable will be necessary to obtain more accurate χ_{12} values.

3.6.5 Nuclear Magnetic Resonance

Proton nmr experiments on polymers are generally confined to studying the spin–spin and spin–lattice relaxation processes as a function of temperature and composition [154]. By convention, the latter is characterized by a relaxation time T_1 while the spin–spin relaxation time is called T_2. As with mechanical measurements, simpler results are expected with one-phase than with two-phase mixtures. But nmr has an advantage over mechanical measurements in that the signal should be independent of the shape and interconnectivity (but not the size) of the phases in a two-phase mixture. This allows one to decompose a multi-time relaxation process and analyze the phases thereby.

The magnitudes of T_1 and T_2 are influenced by molecular motions, and the changes with temperature can be analyzed in terms of the onset of such motions. The resulting T_1 versus temperature curve looks much like inverted mechanical loss response, while the T_2 versus temperature curve is quite reminiscent of an inverted modulus response. Although the origins of the *changes* with temperature of both mechanical and nmr responses are the same, the strengths or intensity of the changes may be quite different with the two methods. Also, the equivalent frequency of the nmr experiment is quite high: $\sim 10^{4.5}$ Hz for T_2 and $\sim 10^8$ Hz for T_1. When both mechanical and nmr methods do give information on a molecular motion, the agreement is quite good.

Nuclear magnetic resonance, as has been mentioned, has a particular advantage in two-phase systems. Here two times are often resolvable, one for each phase. This technique has been applied successfully to crystalline

3.6. Miscellaneous

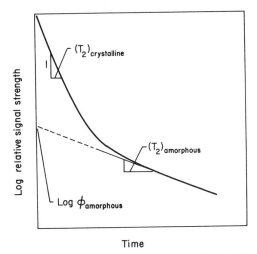

Fig. 3.41. Schematic of the decay of signal strength after a 90° pulse, showing the extraction of spin–spin relaxation times (T_2) for the rigid, crystalline phase and the flexible, amorphous phase.

systems where T_2 for the protons in the amorphous phase is much greater than that for the protons in the crystalline phase. A double exponential decay of induction signal, illustrated in Fig. 3.41, is ideally observed. The relative signal strength corresponding to these two processes reveals the crystallinity of the sample. The same effect occurs in glass–rubber phase mixtures if the T_g's are sufficiently far apart. The T_2's of a glass–crystalline mixture are too close to be resolved.

In quasi-binary polymer mixtures the T_2 relaxation has been analyzed in terms of the total proton content of each phase. One is able to determine the relative amounts of each phase if the components have about the same volume concentration of protons. Analyzing for the composition of the phases cannot be done without an assumption about the additivity of the pure components' relaxation times; however, this problem has been approached [106]. Figure 3.42 shows the decomposition process, using linear additivity of relaxation times for illustration.

Perhaps the most complete application of the nmr method to blends involves the PS–poly(vinyl methyl ether) (PVME) system. Using a 50:50 ratio, Kwei et al. [106] found multiple T_1's around 150°C, the temperature at which the system becomes opaque, but detected multiple T_2's at temperatures as low as 25°C. This is in accordance with the freedom of T_2 from spin diffusion, which tends to merge T_1's of closely associated regions. Thus, T_2 detects regions of relatively small size containing nearly pure material. The authors proposed that these regions are the natural result of geometrical

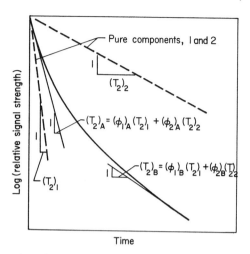

Fig. 3.42. Illustration of an analysis of the two phases in a binary blend using linear additivity of the T_2's for each component. The pure components must have different T_2's, as illustrated by the dashed lines. The relative amounts of each phase can be deduced from the y-intercept of straight-line fit to the long-time decay.

constraints in polymer systems. The reason for the disappearance of multiple T_2's below 25°C was not explained.

Using T_1 to follow phase separation upon annealing the same system, Nishi *et al.* [149] found that the volume portion of each phase remained nearly constant while the composition changed gradually with time, approaching that of the pure components. This behavior, expected for spinodal decomposition (see Section 2.2.3), is shown in Fig. 3.43.

Because of the sensitivity of T_2 to compositional variations, too much information may be obtained. Nishi *et al.* [155], using the system PVC–Hytrel, found single T_1's for each mixture but an unanalyzable T_2 signal. (Pure PVC itself has at least two T_2 components above the T_g, the shorter T_2 being associated with crystalline regions.) A single T_1 is an indication of the absence of aggregates greater than about 30 Å, according to these authors. Using the less miscible system PVC–PVAC, Elmqvist [156] detected a number of components of T_1, the values and intensities of which depended upon preparation and annealing procedures.

Nuclear magnetic resonance experiments other than relaxation are possible on solid polymers and polymer melts. Elmqvist and Svanson [157] showed that *broad line* nmr is a sensitive tool for the detection of small amounts of a soft phase imbedded in a hard matrix. The reason is that the resonance of protons in the soft phase is relatively sharp compared with the resonance band of the matrix protons. The intensity of the band due to the

3.6. Miscellaneous

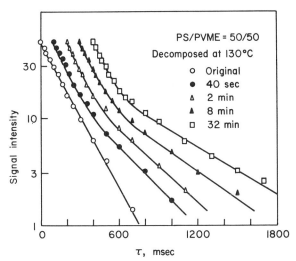

Fig. 3.43. Spin–lattice relaxation as measured from the signal decay following a 180°–τ–90° pulse sequence. The relative constancy of the intensity due to each phase demonstrates that the system PVME–PS is decomposing by the spinodal mechanism. [Reprinted with permission from T. Nishi, T. T. Wang, and T. K. Kwei, *Macromolecules* **8**, 227 (1975). Copyright by the American Chemical Society.]

soft segment is accordingly very high and plainly evident, even at concentrations as low as 1%.

High-resolution nmr is possible in polymer–polymer systems after slight swelling with a low viscosity solvent. Deuterated solvents avoid unnecessary complication of the spectrum. This technique proved useful in the study of alternating block copolymers of polycarbonate and poly(dimethyl siloxane) [158]. The introduction of a third component can influence the phase behavior, however, and caution must be exercised in the interpretation of such experiments.

3.6.6 Other Spectroscopic Techniques

In the previous section, nmr spectroscopy was discussed as a tool for analyzing the composition, amounts, and to some extent the sizes of the phases in a polymer mixture. But spectroscopy can also be used to investigate the solvation of molecules [159], i.e., the interaction of the molecule with its environment. Spectroscopy has proved particularly valuable in the interpretation of hydrogen bonding [160, 161], and it is natural to apply it to polymer–polymer systems.

The reasoning generally followed in the application of spectroscopy is that systems of high miscibility will produce spectra showing strong deviations from an average of the spectra of the two components. The degree of deviation as a function of miscibility cannot be satisfactorily predicted beforehand, however. This technique, therefore, can only substantiate the findings from other methods for demonstrating miscibility. It does provide valuable insight into the nature of the specific interactions between the macromolecules and can often provide clues for the improvement of miscibility (Section 4.5).

Infrared spectroscopy has most often been used in the analysis of polymer mixtures. Thus, the solubility of hard segments of aromatic polyurethane in soft segments of various polyesters and polyethers has been investigated by frequency shifts due to hydrogen bonding of the urethane NH group [162, 163]. Specific interactions in the systems poly(acrylic acid)–poly(ethylene imine) and poly(methacrylic acid)–poly(ethylene imine) were demonstrated via infrared spectroscopy by Zezin *et al.* [164]. The system PMMA–poly(vinylidene fluoride) exhibits specific interactions involving the carbonyl group, according to infrared spectroscopy performed by Coleman and co-workers [165].

Infrared and ultraviolet spectroscopy on the well-known blend PS–PPO by Wellinghoff and co-workers [166] provided evidence for the following conclusions: PPO is loosely packed in the glassy state and the addition of PS reduces the free volumes. The chains of the two components interpenetrate significantly. The reason for the high miscibility is a strong interaction between the phenyl group of the PS and the phenylene group of the PPO.

Ultraviolet emission intensity has been suggested as a tool for quantifying the degree of miscibility of two polymeric components [167, 168]. To employ this technique the components of the blend must contain chromophoric structures active in the ultraviolet, or they must be modified with appropriate groups (e.g., naphthyl, anthryl). This is a possible disadvantage because any modification of structure can change phase relationships in the region of the modification.

In the technique proposed by Morawetz [167], two *different* chromophores are incorporated, one on each polymeric component at a level of about 1%. These groups are selected so that a radiationless transfer can occur between the two. This transfer is assumed to be more efficient as the miscibility increases because close proximity (e.g., 4 Å) of the groups is critical to the transfer process. The measured emission reflects this efficiency; that is, less radiation is emitted by the transfer donor as the probability of a radiationless transfer increases. In one study using this method, naphthyl- and anthryl-tagged PMMA and poly(methyl methacrylate–co–butyl methacrylate) were found to show steadily decreasing miscibility as the butyl content of the copolymer was increased from 0 to 40%.

While the preceding method relies on different groups capable of radiationless transfer, the method proposed by Frank [168] employs the emission from an eximer complex formed by two identical groups. This eximer complex, with a lifetime of about 100 nsec, requires a specific geometry for formation; its concentration is therefore sensitive to conformational aspects of the macromolecular components as well as their phase relationships. Temperature, as well, can independently influence the concentration. In work conducted by Frank using a wide variety of polymeric components, the eximer emission intensity was found to correlate with the solubility parameter difference of the components.

REFERENCES

1. W. A. Kruse, R. G. Kirste, J. Haas, B. J. Schmitt, and D. J. Stein, *Makromol. Chem.* **177**, 1145 (1976).
2. E. W. Fischer, J. H. Wendorff, M. Dettenmaier, G. Lieser, and I. Voigt-Martin, *J. Macromol. Sci., Phys.* **12**, 41 (1976).
3. H. Yu, *Adv. Chem. Ser.* **99**, p. 1 (1971).
4. H. Berghmans and N. Overbergh, *J. Polym. Sci., Polym. Phys. Ed.* **15**, 1757 (1977).
5. L. Zeman and D. Patterson, *Macromolecules* **5**, 513 (1972).
6. A. Robard, D. Patterson, and G. Delmas, *Macromolecules* **10**, 706 (1977).
7. Y. Y. Tan and G. Challa, *Polymer* **17**, 739 (1976).
8. G. Rehage and W. Borchard, *in* "The Physics of Glassy Polymers" (R. N. Haward, ed.), Chapter 1. Appl. Sci. Publ., London, 1973.
9. A. K. Doolittle, *J. Appl. Phys.* **22**, 471 (1951).
10. H. H. Cohen and D. Turnbull, *J. Chem. Phys.* **31**, 1164 (1959).
11. A. A. Miller, *J. Phys. Chem.* **67**, 1031 (1963).
12. M. I. Williams, R. F. Landel, and J. D. Ferry, *J. Am. Chem. Soc.* **77**, 3701 (1955).
13. J. H. Gibbs and E. A. DiMarzio, *J. Chem. Phys.* **28**, 373 (1958).
14. T. Nose, *Polym. J.* **2**, 124, 428, 437, and 445 (1971).
15. R. N. Haward, ed., "The Physics of Glassy Polymers." Appl. Sci. Publ., London, 1973.
16. A. Bondi, "Physical Properties of Molecular Crystals, Liquids, and Glasses." Wiley, New York, 1968.
17. R. N. Haward, *in* "Molecular Behavior and the Development of Polymeric Materials" (A. Ledwith and A. M. North, eds.), Chapter 12. Wiley, New York, 1975.
18. G. D. Wignall, D. G. H. Ballard, and J. Schelter, *J. Macromol. Sci., Phys.* **12**, 75 (1976).
19. G. D. Patterson, *J. Macromol. Sci., Phys.* **12**, 61 (1976).
20. P. Geil, *Ind. Eng. Chem., Prod. Res. Dev.* **14**, 59 (1975).
21. D. S. Kaplan, *J. Appl. Polym. Sci.* **20**, 2615 (1976).
22. J. Gillham, *AIChE J.* **20**, 1066 (1974).
23. L. E. Nielsen, "Mechanical Properties of Polymers." Van Nostrand-Reinhold, Princeton, New Jersey, 1962.
24. I. M. Ward, "Mechanical Properties of Solid Polymers." Wiley, New York, 1971.
25. M. Takayanagi, H. Harima, and Y. Iwata, *Mem. Fac. Eng., Kyushu Univ.* **23**, 1 (1963).
26. D. R. Paul and J. O. Altamirano, *Adv. Chem. Ser.* **142**, 371 (1975).
27. M. Matzner, D. L. Schober, and J. E. McGrath, *Polym. Prepr., Am. Chem. Soc., Div. Polym. Chem.* **13**, 754 (1972).

28. M. T. Shaw, *J. Appl. Polym. Sci.* **18**, 449 (1974).
29. W. J. MacKnight, J. Stoelting, and F. E. Karasz, *Adv. Chem. Ser.* **99**, 26 (1971).
30. S. Akiyama, Y. Komatsu, and R. Kaneko, *Polym. J.* **7**, 172 (1975).
31. N. G. McCrum, B. E. Read, and G. Williams, "Anelastic and Dielectric Effects in Polymer Solids." Wiley, New York, 1967.
32. T. G. Parker, *in* "Polymer Science" (A. D. Jenkins, ed.), Vol. 2, Chap. 19. Elsevier, New York, 1972.
33. N. Bekkedahl, *J. Res. Natl. Bur. Stand.* **43**, 145 (1949).
34. R. E. Gibson and O. H. Loeffler, *J. Am. Chem. Soc.* **61**, 2515 (1939).
35. S. Krause, *J. Polym. Sci., Part A* **3**, 1631 (1965).
36. E. P. Manche and B. Carroll, *in* "Physical Methods in Macromolecular Chemistry" (B. Carroll, ed.), Vol. 2, Chap. 4. Dekker, New York, 1972.
37. B. Ke, "Newer Methods of Polymer Characterization." Wiley, New York, 1964.
38. J. Stoelting, F. E. Karasz, and W. J. MacKnight, *Polym. Eng. Sci.* **10**, 133 (1970).
39. G. A. Zabrzewski, *Polymer* **14**, 347 (1973).
40. M. Bank, J. Leffingwell, and C. Thies, *Macromolecules* **4**, 32 (1971).
41. J. S. Noland, N. N.-C. Hsu, R. Saxon, and J. M. Schmitt, *Adv. Chem. Ser.* **99**, 15 (1971).
42. J. J. Hickman and R. M. Ikeda, *J. Polym. Sci., Polym. Phys. Ed.* **11**, 1713 (1973).
43. V. R. Landi, *Appl. Polym. Symp.* **25**, 223 (1974).
44. A. R. Shultz and B. M. Gendron, *J. Appl. Polym. Sci.* **16**, 461 (1972).
45. A. R. Shultz and B. M. Beach, *J. Appl. Polym. Sci.* **21**, 2305 (1977).
46. A. R. Shultz and B. M. Beach, *Macromolecules* **7**, 902 (1974).
47. A. R. Shultz and B. M. Gendron, *J. Polym. Sci., Polym. Symp.* **43**, 89 (1973).
48. L. Y. Zlatkevich and V. G. Nikolskii, *Rubber Chem. Technol.* **46**, 1210 (1973).
49. G. G. A. Böhm, K. R. Lucas, and W. G. Mayes, *Rubber Chem. Technol.* **50**, 714 (1977).
50. G. G. A. Böhm, *J. Polym. Sci., Polym. Phys. Ed.* **14**, 437 (1976).
51. P. A. Marsh, A. Voet, L. D. Price, and T. J. Mullens, *Rubber Chem. Technol.* **41**, 344 (1968).
52. T. Inoue, T. Soen, T. Hashimoto, and H. Kawai, *Polym. Prepr., Am. Chem. Soc., Div. Polym. Chem.* **10**, 538 (1969).
53. C. Vasile and L. A. Schneider, *Eur. Polym. J.* **7**, 1205 (1965).
54. M. H. Walters and D. N. Keyte, *Rubber Chem. Technol.* **38**, 62 (1965).
55. M. Kryszewski, A. Galeski, T. Pakula, and J. Grebowicz, *J. Colloid Interface Sci.* **44**, 85 (1973).
56. G. M. Brauer and S. B. Newman, *in* "Analytical Chemistry of Polymers" (G. M. Kline, ed.), Vol. 3, p. 141. Wiley, New York, 1962.
57. S. B. Newman, *in* "Analytical Chemistry of Polymers" (G. M. Kline, ed.), Vol. 3, p. 261. Wiley, New York, 1962.
58. R. W. Smith and J. C. Andries, *Rubber Chem. Technol.* **47**, 64 (1974).
59. S. Miyata and T. Hata, *Proc. Int. Congr. Rheol., 5th, 1968* Vol. 3, p. 71 (1970).
60. M. Matsuo, C. Nozaki, and Y. Jyo, *Polym. Eng. Sci.* **9**, 197 (1969).
61. L. P. McMaster, *Adv. Chem. Ser.* **142**, 43 (1975).
62. E. L. Thomas and Y. Talmon, *Polymer* **19**, 225 (1978).
63. K. Dimov and Ye. Dimova, *Polym. Sci. USSR (Engl. Transl.)* **16**, 1476 (1974).
64. J. W. Gibbs, "Collected Works," Vol. 1. Longmans, Green, New York, 1928.
65. M. Bank, J. Leffingwell, and C. Thies, *J. Polym. Sci., Part A-2* **10**, 1097 (1972).
66. L. P. McMaster, *Macromolecules* **6**, 760 (1973).
67. R. E. Bernstein, C. A. Cruz, D. R. Paul, and J. W. Barlow, *Macromolecules* **10**, 681 (1977).
68. G. Allen, G. Gee, and J. P. Nicholson, *Polymer* **2**, 8 (1961).
69. D. McIntyre, N. Rounds, and E. Campos-Lopez, *Polym. Prepr., Am. Chem. Soc., Div. Polym. Chem.* **10**, 531 (1969).

70. P. O. Powers, *Polym. Prepr., Am. Chem. Soc., Div. Polym. Chem.* **15,** 528 (1974).
71. S. M. Aharoni, *Macromolecules* **11,** 277 (1978).
72. R. Roningsveld and L. A. Kleintjens, *J. Polym. Sci., Polym. Symp.* **61,** 221 (1977).
73. J. W. Strutt (Lord Rayleigh), *Philos. Mag.* [4], **41,** 107 (1871); [4], **41,** 447 (1871).
74. M. Smoluchowski, *Ann. Phys. (Leipzig)* [4] **25,** 205 (1908).
75. A. Einstein, *Ann. Phys. (Leipzig)* [4] **33,** 1275 (1910).
76. F. Zernike, *Arch. Neerl. Sci. Exactos Nat., Ser. 3A* **4,** 74 (1918).
77. H. C. Brinkman and J. J. Hermans, *J. Chem. Phys.* **17,** 574 (1949).
78. P. Debye, *J. Chem. Phys.* **31,** 680 (1959).
79. T. G. Scholte, *J. Polym. Sci., Part C* **39,** 281 (1972).
80. P. J. Flory, *J. Chem. Phys.* **10,** 51 (1942).
81. M. L. Huggins, *Ann. N. Y. Acad. Sci.* **43,** 1 (1942).
82. R. Koningsveld, L. A. Kleintjens, and H. M. Schoffeleers, *Pure Appl. Chem.* **39,** 1 (1974).
83. J. J. van Aartsen, *Eur. Polym. J.* **6,** 919 (1970).
84. C. A. Smolders, J. J. van Aartsen, and A. Steenbergen, *Kolloid Z. & Z. Polym.* **243,** 14 (1971).
85. P. Kratochvil, J. Vorlicek, D. Strakova, and Z. Tuzar, *J. Polym. Sci., Polym. Phys. Ed.* **13,** 2321 (1975).
86. P. Kratochvil, D. Strakova, and Z. Tuzar, *Br. Polym. J.* **9,** 217 (1977).
87. K. E. Derham, J. Goldsbrough, and M. Gordon, *Pure Appl. Chem.* **38,** 97 (1974).
88. R. Koningsveld and L. A. Kleintjens, *Br. Polym. J.* **9,** 212 (1977).
89. M. Gordon, L. A. Kleintjens, B. W. Ready, and J. A. Torkington, *Brit. Polym. J.* **10,** 170 (1978).
90. B. H. Zimm, *J. Chem. Phys.* **16,** 1099 (1948).
91. D. G. H. Ballard, M. G. Rayner, and J. Schelten, *Polymer* **17,** 640 (1976).
92. K. G. Kirste and B. R. Lehnen, *Makromol. Chem.* **177,** 1137 (1976).
93. C. Strazielle and H. Benoit, *Macromolecules* **8,** 203 (1975).
94. G. D. Patterson, T. Nishi, and T. T. Wang, *Macromolecules* **9,** 603 (1976).
95. F. B. Khambatta, F. Warner, T. Russell, and R. S. Stein, *J. Polym. Sci., Polym. Phys. Ed.* **14,** 1391 (1976).
96. D. Ya. Tsvankin, *Polym. Sci. USSR (Engl. Transl.)* **6,** 2304 and 2310 (1964).
97. D. R. Buchanan, *J. Polym. Sci., Part A-2* **9,** 645 (1971).
98. P. Debye and A. Bueche, *J. Appl. Phys.* **20,** 518 (1944).
99. A. Dobry and F. Boyer-Kawenoki, *J. Polym. Sci.* **2,** 90 (1947).
100. R. J. Kern and R. J. Slocombe, *J. Polym. Sci.* **15,** 183 (1955).
101. R. L. Scott, *J. Chem. Phys.* **17,** 279 (1949).
102. H. Tompa, *Trans. Faraday Soc.* **45,** 1142 (1949).
103. C. Hugelin and A. Dondos, *Makromol. Chem.* **126,** 206 (1969).
104. R. J. Kern, *J. Polym. Sci.* **21,** 19 (1956).
105. C. C. Hsu and J. M. Prausnitz, *Macromolecules* **7,** 320 (1974).
106. T. K. Kwei, T. Nishi, and R. F. Roberts, *Macromolecules* **7,** 667 (1974).
107. G. Delmas and D. Patterson, *J. Paint Technol.* **34,** 677 (1962).
108. J. M. Braun and J. E. Guillet, *in* "Progress in Gas Chromatography" (J. H. Purnell, ed.), p. 107. Wiley (Interscience), New York, 1976.
109. J. E. Guillet, *in* "Progress in Gas Chromatography" (J. H. Purnell, ed.), p. 187. Wiley (Interscience), New York, 1973.
110. P. J. Flory, *J. Am. Chem. Soc.* **87,** 1833 (1965).
111. D. D. Deshpande, D. Patterson, H. P. Schreiber, and C. S. Su, *Macromolecules* **7,** 530 (1974).
112. C. S. Su, D. Patterson, and H. P. Schreiber, *J. Appl. Polym. Sci.* **20,** 1025 (1976).

113. O. Olabisi, *Macromolecules* **8**, 316 (1975).
114. A. B. Littlewood, "Gas Chromatography." Academic Press, New York, 1970.
115. G. S. Kongarov and G. M. Bartenev, *Rubber Chem. Technol.* **47**, 1188 (1974).
116. M. Kh. Giniyatullin, R. G. Timergaleev, M. Kh. Khasanov, and V. A. Voskresenskii, *Sov. Plast.* (*Engl. Transl.*) No. 6, p. 8 (1973).
117. E. R. Galimov, G. G. Ushakova, R. G. Timergaleev, and V. A. Voskresenskii, *Int. Polym. Sci. Technol.* **2** (5), T.7 (1975).
118. W. M. Prest, Jr. and R. S. Porter, *J. Polym. Sci., Part A-2* **10**, 1639 (1972).
119. W. M. Prest, Jr. and R. S. Porter, *Polym. J.* **4**, 154 (1973).
120. L. R. Schmidt, *J. Appl. Polym. Sci.* **23**, 2463 (1979).
121. D. S. Hubbell and S. L. Cooper, *J. Polym. Sci., Polym. Phys. Ed.* **15**, 1143 (1977).
122. H. Morawetz, "Macromolecules in Solution." Wiley (Interscience), New York, 1965.
123. G. R. Williamson and B. Wright, *J. Polym. Sci., Part A* **3**, 3885 (1965).
124. B. Bohmer, D. Berek, and S. Florian, *Eur. Polym. J.* **6**, 471 (1970).
125. A. Rudin, H. L. W. Hoegy, and H. K. Johnston, *J. Appl. Polym. Sci.* **16**, 1281 (1972).
126. V. I. Aleksieyenko and I. U. Misrustin, *Polym. Sci. USSR* (*Engl. Transl.*) **2**, 63 (1961).
127. V. N. Kuleznev and L. S. Krokhina, *Polym. Sci. USSR* (*Engl. Transl.*) **15**, 1019 (1973).
128. C. Vasile and I. A. Schneider, *Makromol. Chem.* **141**, 127 (1971).
129. S. M. Liguori, M. De Santis Savino, and M. D'Alagni, *Polym. Lett.* **4**, 943 (1966).
130. W. R. Krigbaum and F. T. Wall, *J. Polym. Sci.* **5**, 505 (1950).
131. Y. J. Shur and B. Ranby, *J. Appl. Polym. Sci.* **20**, 3121 (1976).
132. Y. J. Shur and B. Ranby, *J. Appl. Polym. Sci.* **20**, 3105 (1976).
133. Y. J. Shur and B. Ranby, *J. Appl. Polym. Sci.* **19**, 1337 (1975).
134. B. G. Ranby, *J. Polym. Sci., Polym. Symp.* **51**, 89 (1975).
135. J. J. Hickman and R. M. Ikeda, *J. Polym. Sci., Polym. Phys. Ed.* **11**, 1713 (1973).
136. Y. S. Lipatov, *J. Polym. Sci., Polym. Symp.* **42**, 855 (1973).
137. Y. S. Lipatov, *Polym. Sci. USSR* (*Engl. Transl.*) **17**, 2717 (1975).
138. Y. S. Lipatov and V. A. Vilenskii, *Polym. Sci. USSR* (*Engl. Transl.*) **17**, 2389 (1975).
139. Y. S. Lipatov, V. F. Babich, and V. F. Rosovizky, *J. Appl. Polym. Sci.* **18**, 1213 (1974).
140. R. Simha and R. F. Boyer, *J. Chem. Phys.* **37**, 1003 (1962).
141. G. V. Struminskii and G. L. Slonimskii, *Rubber Chem. Technol.* **31**, 250 (1958).
142. S. Ichihara and T. Hata, *Kobunshi Kagaku* **26**, 249 (1969).
143. S. Ichihara, A. Komatsu, and T. Hata, *Polym. J.* **2**, 640 (1971).
144. M. P. Zverev, L. A. Polovikhina, A. N. Barash, L. P. Mal'kova, and G. D. Litovchenko, *Polym. Sci. USSR* (*Engl. Transl.*) **16**, 2098 (1974).
145. A. A. Tager, T. I. Scholokhovich, and Yu. S. Bessenov, *Eur. Polym. J.* **11**, 321 (1975).
146. A. A. Tager and Yu. S. Bessonov, *Polym. Sci. USSR* (*Engl. Transl.*) **17**, 2741 (1975).
147. D. Patterson, *J. Polym. Sci., Part C* **16**, 3379 (1968).
148. P. Novakov, Ch. Konstantinov, and P. Mitanov, *J. Appl. Polym. Sci.* **16**, 1827 (1972).
149. T. Nishi, T. T. Wang, and T. K. Kwei, *Macromolecules* **8**, 227 (1975).
150. C. J. Ong, Ph.D. Thesis, University of Massachusetts, Amherst (1974).
151. W. Wenig, F. E. Karasz, and W. J. MacKnight, *J. Appl. Phys.* **46**, 4194 (1975).
152. T. Nishi and T. T. Wang, *Macromolecules* **8**, 909 (1975).
153. T. K. Kwei, G. D. Patterson, and T. Wang, *Macromolecules* **9**, 780 (1976).
154. D. W. McCall, *Acc. Chem. Res.* **4**, 223 (1971).
155. T. Nishi, T. K. Kwei, and T. T. Wang, *J. Appl. Phys.* **46**, 4157 (1975).
156. C. Elmqvist, *Eur. Polym. J.* **13**, 95 (1977).
157. C. Elmqvist and S. E. Svanson, *Eur. Polym. J.* **11**, 789 (1975).
158. D. G. LeGrand, *Trans. Soc. Rheol.* **15**, 541 (1971).

References

159. J. R. Dyer, "Applications of Absorption Spectroscopy of Organic Compounds." Prentice-Hall, Englewood Cliffs, New Jersey, 1965.
160. G. C. Pimentel and A. L. McClellan, "The Hydrogen Bond." Freeman, San Francisco, California, 1960.
161. J. D. Crowley, G. S. Teague, Jr., and J. W. Lowe, Jr., *J. Paint Technol.* **38,** 269 (1966).
162. R. W. Seymour, G. M. Estes, and S. L. Cooper, *Macromolecules* **3,** 579 (1970).
163. C. S. P. Sung and N. S. Schneider, *Macromolecules* **8,** 68 (1975).
164. A. B. Zezin, V. B. Rogacheva, V. S. Komarov, and Ye. F. Razvodovskii, *Polym. Sci. USSR (Engl. Transl.)* **17,** 3032 (1975).
165. M. M. Coleman, J. Zarian, D. F. Varnell, and P. C. Painter, *J. Polym. Sci., Polym. Lett. Ed.* **15,** 745 (1977).
166. S. T. Wellinghoft, J. L. Koenig, and E. Baer, Jr, *J. Polym. Sci., Polym. Phys. Ed.* **15,** 1913 (1977).
167. H. Morawetz and F. Amrani, *Macromolecules* **11,** 281 (1978).
168. C. Frank, private communication.

Chapter 4

Methods of Enhancing Miscibility

In this chapter, various methods for enhancing the miscibility of polymer blends will be discussed. One situation which will be covered in depth concerns the change in structure of the polymer or polymers involved in the blend in question. In cases where the polymer constituents of a blend are close to miscibility, minor changes in structure may render the system miscible. The introduction of strong specific interactions (e.g., acid–base) is an extreme of such a change. Other variations of chemical structure can result in the permanent attachment of these polymers to each other via block and graft copolymers, formation of interpenetrating networks (IPN), and cross-linking. The thermodynamic theories for the phase separation of block copolymers have been previously discussed in Section 2.6 and have clearly pointed out the feasibility of covalent bonding of polymer components of a blend for enhancing the degree of miscibility. While these theories have primarily dealt with block copolymers, the general results translate to graft copolymers, cross-linked systems, and interpenetrating networks. With block copolymers, the application of the theory will be discussed and several experimental cases will be presented which substantiate the essence of the various theoretical approaches. As no complete theoretical treatment of the phase behavior of interpenetrating networks has been developed, other than a semiempirical approach by Donatelli *et al.* [1], documentation of the IPN approach to improving miscibility will be discussed in terms of experimentally reported results. The ramifications of structural modification of the components of the blend to improve miscibility are quite important in designing miscible polymer blends, and these will be discussed in more detail than the other techniques.

4.1 MINOR MODIFICATIONS OF STRUCTURE

If two polymers form an insoluble system, but it is suspected that the free energy of mixing is not greatly positive, it may be possible to modify the system slightly to increase miscibility significantly. One of the most direct methods of achieving enhanced miscibility is to modify the structure of one or both components. Modification of the monomer unit itself and copolymerization are the two most commonly followed routes. Postpolymerization modification is also possible; for example, the chlorination of polyethylene or poly(vinyl chloride). Modification to introduce strongly interacting groups will be discussed separately in Section 4.5.

The behavior of copolymer systems may not always be as expected because of the dependence of sequence distribution on monomer feed ratio or on extent of reaction. Thus, the preparation of copolymer systems must be carefully controlled to provide a pure, single polymeric component for study. The necessity of this care is best illustrated by the possibility of forming two-phase mixtures with a single copolymer, the components of the system being the copolymer with different monomer ratios. Scott [2] demonstrated with Flory–Huggins thermodynamics that the maximum allowable standard deviation of the solubility parameters of molecules making up a single-phase copolymer mixture is $(RT\rho/2M)^{1/2}$. [For two homopolymers with equal molecular weights the maximum difference in solubility parameter is $(2RT\rho/M)^{1/2}$.] Molau [3] confirmed this experimentally using styrene/acrylonitrile copolymers, the mixtures of which phase-separated at a 4% (average) compositional difference. A reasonable solubility parameter difference of 3 between polystyrene and polyacrylonitrile (see Chapter 2, Table 2.1) would be adequate for explaining Molau's findings using the above-mentioned relationship due to Scott.

Another, more recent, study was done by Kollinsky and Markert [4] with copolymers of methyl methacrylate and n-butyl acrylate. They found that compositional differences of up to 25 mole% still allowed miscibility in this copolymer system where the monomers are structurally more similar than the styrene and acrylonitrile used by Molau. The allowable compositional differences fell, however, as the molecular weight of the components increased or as the overall mole fraction of methyl methacrylate increased (Fig. 4.1). The latter effect may be explained by the size difference of the monomers [5]. Neither Molau nor Kollinsky and Markert noted any improvements in miscibility for mixtures of these copolymers after adding components of intermediate composition.

Kollinsky and Markert found that complete polymerization of their acrylate copolymer system could lead to phase separation because of compositional changes during reaction. Grafting was found to influence the

4.1. Minor Modifications of Structure

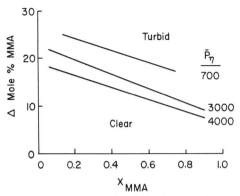

Fig. 4.1. Boundaries between clear and turbid mixtures of copolymers of methyl methacrylate and n-butyl acrylate. The boundaries for various molecular weights, denoted by the viscosity-average degree of polymerization, $\bar{P}_\eta$, are plotted in the plane defined by the overall mole fraction of methyl methacrylate in the system, X_{MMA}, and the difference in methyl methacrylate content of each copolymer, Δmole% MMA. [Reprinted by permission from Norbert A. S. Platzer, ed., "Multicomponent Polymer Systems," Adv. Chem. Ser. No. 99. Copyright 1971 by the American Chemical Society.]

observed miscibility under these conditions. Similar behavior has been noted by Ambler [6] for NBR (copolymer of acrylonitrile and butadiene), where compositional differences during normal polymerization led to phase separations. As an extreme, Landi [7] found that two clearly distinguishable glass transitions could be achieved with NBR containing 29% acrylonitrile. The material with the higher T_g (higher acrylonitrile content) was miscible with PVC, whereas the material with the higher butadiene content was not. If the polymerization was carried out with 34% acrylonitrile, then one glass transition for the PVC blends was found. The rather broad compositional range of NBR which will produce miscibility with PVC is undoubtedly due to weak hydrogen bonding between the two components. A similar result was reported by Robeson and McGrath [8], who found that terpolymers of ethylene, carbon monoxide, and either ethyl acetate or vinyl acetate were miscible if the carbon monoxide content was in the range of 8 to 18%. In still another study by Kruse and co-workers [9], the solubility of PMMA in SAN copolymers was shown to be possible over a broad compositional range of the SAN, due to an undefined interaction which gave an exothermic heat of mixing.

One of the more subtle of possible changes in polymer structure is tacticity. Not only do mixtures of stereoisomeric polymers show profound effects including complexation [10–13], but the tacticity of a homopolymer component may influence its phase behavior with other structures. Thus, Schurer et al. [14] found the behavior shown in Fig. 4.2 for the system

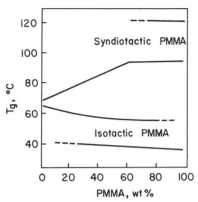

Fig. 4.2. Glass transition temperatures of mixtures of PVC and isotactic or syndiotactic PMMA. With the PVC–(i-PMMA) mixtures, two transitions are observed over the entire composition range, while with the PVC–(s-PMMA) system a single transition was observed with mixtures containing up to 60% PMMA. [Reprinted by permission from J. W. Schurer, A. de Boer, and G. Challa, *Polymer* **16**, 201–204 (1975). Copyright by IPC Business Press Ltd.]

PMMA–PVC. Because the breakpoint for the soluble s-PMMA–PVC system occurs at a 1:1 mole ratio, a rather specific interaction, incapable of occurring with i-PMMA, was suggested.

Configurational isomers containing the same functional groups can be considered to represent slightly differing structures, as most additive schemes

TABLE 4.1

Polymer Structural Isomers Arranged by Structural Classes

Structural family	Commercially available isomers
Aliphatic hydrocarbon	(1) PE, PP, PIB, Poly(butene-1)
Aliphatic ether	(2) Poly(vinyl methyl ether), poly(propylene oxide)
	(3) Poly(vinyl ethyl ether), poly(THF), poly(isobutylene oxide)
	(4) Poly(vinyl propyl ethers)
	(5) Poly(vinyl butyl ethers)
Aromatic hydrocarbon	(6) Poly(α-, o-, m-, p-, methylstyrenes)
	(7) Poly(p-xylylene), polystyrene
Polyester	(8) PVA, PMA
	(9) PEA, PMMA
	(10) Poly(propyl acrylates), polycaprolactone, PEMA
	(11) Poly(butyl acrylates)
Chlorinated hydrocarbon	(12) PVC, chlorinated polyethylene ($\sim 50\%$)

4.1. Minor Modifications of Structure

for solubility parameter will predict insignificant differences in cohesive energy for these materials (see Section 2.3). Several readily available series of configurational isomers exist, as shown in Table 4.1.

Some work has been done on these materials. Kern [15] found that members of series 6 had considerably different phase behavior (ternary) with polystyrene. The same author [16] found that the members of series 8 were not miscible with PVC, while in series 9, 10, and 11 only the methacrylates exhibited PVC miscibility. (Polycaprolactone was not included in the study, which was confined to acrylics.) This strongly suggests a specific interaction of some kind and demonstrates the importance of fine changes in structure.

In a study of the isomers of series 5, Hughes and Britt [17] discovered that poly(vinyl *tert*-butyl ether) and poly(vinyl isobutyl ether) were *not* miscible, nor were any combinations in series 8–11.

Small structural changes could also be extended to include the slight modification of a large repeat unit by substitution. Thus, α-methylstyrene is a slight modification of polystyrene and the following,

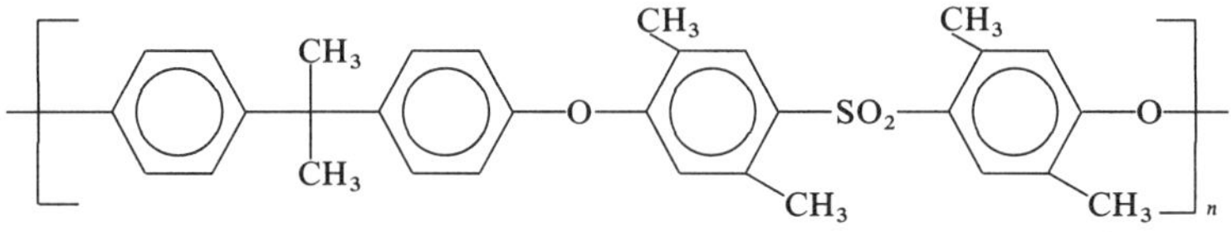

[chemical structure]

is a slight modification of

[chemical structure]

| 0 | 8 | 12 | 18 | 20 | 24 |

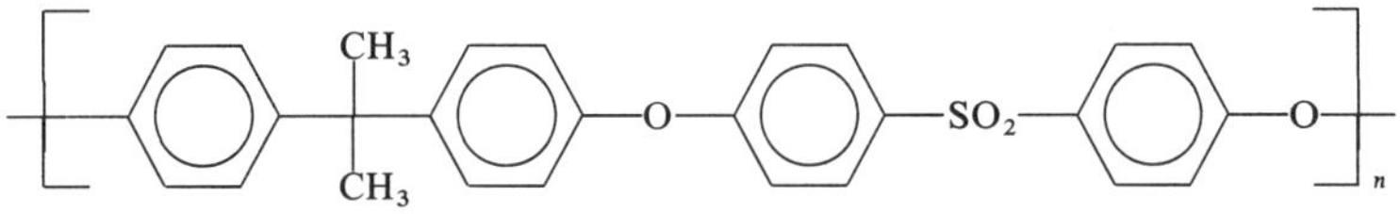

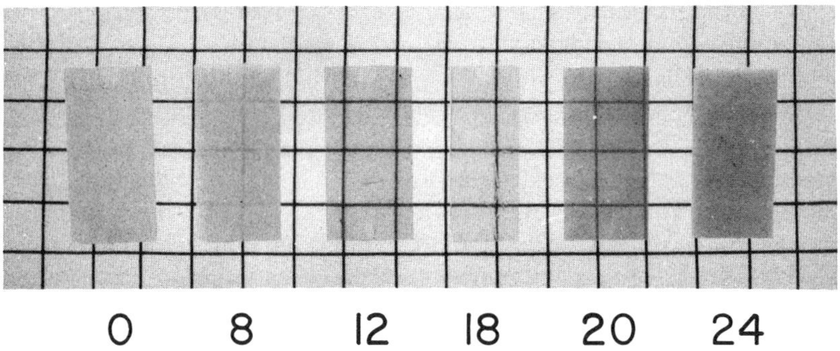

Fig. 4.3. Photograph of mixtures of a methyl-substituted polysulfone with poly(styrene–co-acrylonitrile) showing the minimum in opacity as the acrylonitrile content on the SAN copolymer is increased from 0 to 24%. [Reprinted by permission from M. T. Shaw, *J. Appl. Polym. Sci.* **18**, 449 (1974). Copyright by John Wiley and Sons, Inc.]

The most extensive study of this type was by Shaw [18], who reported on the simultaneous, incremental variation of the above structures and styrene/acrylonitrile copolymers to achieve miscibility. The appearance of the methyl-substituted polysulfone with various SAN copolymers is shown in Fig. 4.3, where miscibility was found over the range of 12 to 18% AN in the SAN copolymer. The unsubstituted polysulfone showed little miscibility with any of these copolymers, demonstrating the sensitivity of this system to slight modifications in structure.

4.2 BLOCK AND GRAFT COPOLYMER FORMATION

The phase separation of block copolymers has been discussed in Section 2.6, which considered the thermodynamics of the phase behavior of block copolymer systems. The various theories predict that covalent bonding (even with only one covalent bond per molecule, as in an AB block copolymer) can significantly increase the degree of miscibility of the polymeric components of the block copolymer over that which would be expected for simple blends of the polymers with molecular weights comparable to the component block molecular weights. Experimental verification of this has been shown in many examples, several of which will be described here.

The most rigorous experimental test of the block copolymer theories of phase separation has been with the AB block copolymer of poly(α-methylstyrene) and polystyrene. This system approximates the parameters desired in a model experimental system in that dispersive interactions are dominant due to the expected negligible polar or specific interactions. This translates into improved predictive capabilities for the interaction parameter χ_{12}. The polymers are also sufficiently similar that reasonably high molecular weights are required for phase separation, even in simple mixtures. Block copolymers of high purity with M_w/M_n values of less than 1.1 can be prepared using well-established techniques of anionic polymerization [19]. Poly(α-methylstyrene) and polystyrene blends exhibit two-phase behavior at molecular weights above $\sim$100,000. However, in block copolymers, one-phase behavior is observed at considerably higher molecular weights. An example illustrating this was given in Chapter 2 (Fig. 2.36) for the poly(α-methylstyrene)–polystyrene AB block copolymer with molecular weights of 200,000 for the individual blocks. By comparison, simple mixtures of the same components at a similar range in molecular weights exhibit two-phase behavior, as clearly illustrated in Fig. 4.4 [20]. Other investigators have observed similar results with this block copolymer [21–25].

The phase separation of the styrene–diene (diene = butadiene or isoprene) AB or ABA block copolymers has also been studied by many investi-

4.2. Block and Graft Copolymer Formation

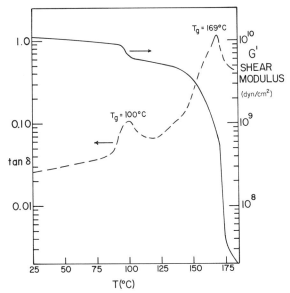

Fig. 4.4. Mechanical loss and shear modulus versus temperature data for blend of polystyrene–poly(α-methylstyrene), 25/75. Polystyrene $M_w = 270,000$; poly(αmethylstyrene) $M_w = 310,000$. [Reprinted from L. M. Robeson, M. Matzner, L. J. Fetters, and J. E. McGrath, in "Recent Advances in Polymer Blends, Grafts, and Blocks" (L. H. Sperling, ed.), p. 281. Copyright 1974 by Plenum Press.]

gators. The mechanical properties (e.g., tensile strength) require phase separation in order for the hard block (polystyrene) to have the rigidity required to achieve the "cross-link" character necessary for the thermoplastic elastomer behavior. The existence of an interfacial region between the phase-separated blocks is predicted by Leary and Williams [26, 27] and utilized in their theoretical treatment of block copolymer phase separation. This interfacial region has been experimentally verified by several investigators [28–30]. Miyamoto *et al.* [28] observed an intermediate transition for styrene–butadiene–styrene ABA block copolymers when cast from ethyl acetate and methyl ethyl ketone. However, when the same materials were cast from toluene and CCl_4, intermediate peaks were not observed, as illustrated in Fig. 4.5. They proposed that the aggregation of styrene blocks was not sufficient to yield distinct phase separation with specimens cast from poor solvents. Beecher *et al.* [30] observed an intermediate transition for the styrene–butadiene–styrene ABA block copolymer with films cast from CCl_4. This transition was considered to be a diffuse interfacial region in which the transition peak may not represent a composition identical to the overall block copolymer composition. Inoue *et al.* [31] studied the domain formation

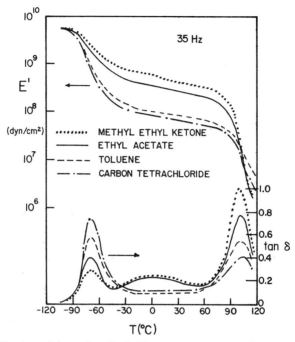

Fig. 4.5. Elastic modulus and mechanical loss versus temperature for a styrene–butadiene–styrene ABA block copolymer (polystyrene content = 30 wt%; polystyrene block molecular weight = 15,100). [Reprinted from T. Miyamoto, K. Kodama, and K. Shibayama, *J. Polym. Sci.*, Part A-2 **8**, 2095 (1970).]

of the styrene–isoprene AB block copolymer and the ternary blend of the block copolymer with the homopolymer components. The results indicated that solubilization of the homopolymer into respective block domains was limited to molecular weights of the homopolymers equal to or less than that for the specific blocks. Fedors [32] predicted minimum molecular weights for butadiene segments of 6000 and for polystyrene segments of 2500 to yield phase separation in simple mixtures. Meier [33] determined from his theoretical treatment that domain formation for the styrene–butadiene AB block copolymer would occur with a polystyrene block molecular weight in the range of 5000–10,000, if the polybutadiene block molecular weight were equal to 50,000. He used tensile strength data of Holden *et al.* [34], which showed a tensile strength increase of 20-fold with an increase in polystyrene molecular weight of 6000 to 10,000, as experimental evidence to support his theory. For simple mixtures, phase separation occurs at molecular weights 3 to 5 times less than the block copolymer values, according to this analysis.

4.3. Interpenetrating Network Formation

Phase separation in block copolymers composed of rigid blocks was experimentally investigated by McGrath et al. [35]. Polysulfone–bisphenol A polycarbonate $(AB)_n$ (5000/5000 M_n's) block copolymer exhibited single-phase behavior, whereas high molecular weight mixtures exhibited phase separation. Polysulfone–poly(butylene terephthalate) $(AB)_n$ block copolymers also exhibited single-phase behavior. Further investigations by McGrath et al. [36] showed single-phase behavior for the polysulfone–bisphenol A polycarbonate $(AB)_n$ (17,000/16,000 M_n's), whereas cast films of the oligomers comprising the block were phase-separated at much lower molecular weights.

Graft copolymers can be expected to show many of the same miscibility-enhancing features of block copolymers. In one study, nylon-6 grafted onto an ethylene–acrylic acid copolymer exhibited a single T_g with permeability as a function of composition similar to that expected of miscible systems [37]. The ungrafted blends of the components were stated to have single T_g's although broader in nature than the grafted counterparts.

Ethylene–propylene rubber grafted with polystyrene and polyisobutylene exhibited a single low-temperature relaxation, indicating intimate mixing of the ethylene–propylene rubber and polyisobutylene, in studies by Kennedy [38]. There are many other examples of block and graft copolymer miscibility which will be listed and discussed in Chapter 5.

4.3 INTERPENETRATING NETWORK FORMATION

The utility of *in situ* polymerization and cross-linking techniques to yield interpenetrating networks has been shown to lead to higher levels of molecular mixing than for simple polymer mixtures. Many polymeric systems have been combined using the numerous variations of this approach. Several combinations have exhibited either transitions sufficiently shifted to denote some level of molecular mixing or even a microheterogeneous level of miscibility as defined by a single broad glass transition. Theoretical treatments (other than a semiempirical approach by Donatelli et al. [1]) specifically related to interpenetrating networks have not been developed, and even experimental methods for defining the level of interpenetration are quite limited and merely qualitative. In fact, experimental determination of transitional behavior, relative to that for the simple polymer mixtures, is commonly used as a qualitative measure of the degree of interpenetration, and even this measure does not distinguish clearly between the desired formation of loops and the possible formation of grafts. For this reason, experimental examples of IPNs will be discussed only to demonstrate the potential in improving the miscibility of polymer blends by the techniques employed.

Frisch et al. [39] observed single transitions from calorimetric measurements for interpenetrating network combinations of (1) polyurethane–polyester, (2) polyurethane–epoxy, and (3) polyurethane–polyacrylate. Electron microscopy on the polyurethane–polyacrylate IPN did not reveal any evidense of phase separation. Intermediate IPN compositions exhibited tensile strength values higher than those of the unblended constituents. This improvement in tensile strength was attributed to network interpenetration resulting in catenane structures—thus, higher apparent "cross-link density."

IPNs of poly(ethyl acrylate) and poly(styrene–co-methyl methacrylate) were studied by Huelck et al, [40] as a function of methyl methacrylate content. Increasing the methyl methacrylate content of the copolymer led to improved molecular mixing, as determined by dynamic mechanical measurements. This trend is not surprising in view of the structural similarity of ethyl acrylate and methyl methacrylate. The interesting aspect is, however, that single broad glass transitions are observed with the PEA–PMMA IPN, whereas other literature references [41, 42] on the simple mixture of PEA and PMMA indicate phase separation.

Sperling et al. [43] prepared latex interpenetrating polymer networks by swelling cross-linked latex particles of polymer A with monomer B plus cross-linking agent. This was followed by the polymerization of monomer B. With polybutadiene–polystyrene IPNs, two-phase systems were inherent. Poly(ethyl acrylate)–poly(n-butyl acrylate) IPNs were found to exhibit a certain level of miscibility with single broad glass transitions for several combinations. Reversing the monomers gave results different enough to imply phase separation for the "semicompatible" IPNs.

As with the block and graft copolymers, many other examples of IPNs showing at least a small level of miscibility will be listed and discussed in Chapter 5.

4.4 CROSS-LINKING

It would appear that the "last resort" method for forcing miscibility would be to form covalent links between the components by cross-linking. Unfortunately, this is neither straightforward nor consistently desirable, but it has been tried in some instances. The principal opportunities for beneficially using this approach occur naturally in elastomer technology.

The preferred method for cross-linking depends heavily on the system. Many common polymers are difficult or impossible to cross-link by conventional means, such as peroxides or radiation. A promising method for some of these is the polymerization of monomer in the presence of the

4.4 Cross-Linking

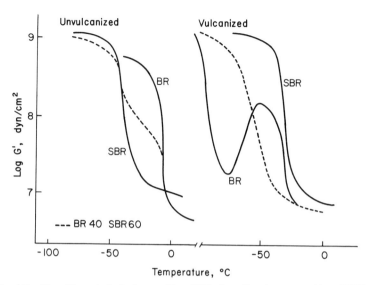

Fig. 4.6. Transitions in butadiene rubber (BR), butadiene/styrene rubber (SBR), and a 40/60 mixture by dynamical mechanical measurements. The influence of vulcanization of the transitions is shown. [Adapted from N. Yoshimura and K. Fujimoto, *Rubber Chem. Technol.* **42**, 1009 (1969).]

second polymer component, leading to adventitious cross-linking, or IPN formation. This subject is covered separately in Section 4.3.

Straight blending, followed by cross-linking, has apparently led to increased miscibility in the case of systems SBR–BR [44] and NR–BR [45]. The dynamical–mechanical data in the case of the former demonstrates clearly that cross-linking can force increased solubility in polymer–polymer systems, as can be seen in Fig. 4.6 [44].

It is evident from the figure that the unvulcanized 40/60 mixture contains mixed phases of nearly pure SBR and (crystalline) BR. Vulcanization, which removes much of the BR crystallinity in the case of the pure polymer, has removed all evidence of it in the 40/60 mixture and has yielded a material with a single T_g located between the T_g's for BR and SBR.

Understanding the influence of covalent bonds on the phase behavior of multicomponent polymer systems is probably best approached through the preparation of the appropriate block copolymers, followed by addition of the homopolymers. This technique, which attempts to introduce bonds in a controlled fashion, is discussed in Sections 2.6 and 4.2. Another possibility would be the random inter-cross-linking of polymer components with a characterizable reaction. The reaction below, with collection of ethanol, is an example.

$$\begin{array}{c}\text{CH}_3\\ |\\ -\text{CH}_2-\text{C}-\\ |\\ \text{C}=\text{O}\\ |\\ \text{O}-\text{CH}_2\text{CH}_2\text{OH}\end{array} + \begin{array}{c}-\text{CH}_2-\text{CH}-\\ |\\ \text{C}=\text{O}\\ |\\ \text{O}-\text{CH}_2\text{CH}_3\end{array} \longrightarrow \begin{array}{c}\text{CH}_3\\ |\\ -\text{CH}_2-\text{C}-\\ |\\ \text{C}=\text{O}\\ |\\ \text{O}\\ |\\ \text{CH}_2\\ |\\ \text{CH}_2\\ |\\ \text{O}\\ |\\ \text{C}=\text{O}\\ |\\ -\text{CH}_2-\text{CH}-\end{array} + \text{CH}_3\text{CH}_2\text{OH}$$

No study of this type is known, but the work of Robeson and Furtek [46], involving adventitious grafting of polyesters, does provide a start on the path toward an understanding of the effect of intercomponent cross-linking on phase behavior.

4.5 INTRODUCTION OF INTERACTING GROUPS

The most important method for enhancing the miscibility of polymers is via the introduction of specific interactions between constituents of the individual polymer chains. The importance of this is well illustrated by the necessary free energy of mixing criterion for miscibility: $\Delta G^M \leq \Delta H^M - T\Delta S^M$. With high molecular weight polymers, $\Delta S^M \rightarrow 0$ as the $M_w \rightarrow \infty$; thus, ΔH^M must be zero or negative for miscibility to occur. For ΔH^M to be negative, specific interactions must be present. The importance of specific interaction has been previously pointed out by many investigators [47–50]. Olabisi [51] proposed the concept of "complementary dissimilarity" in enhancing the miscibility of polymer blends wherein constituents of the individual polymer chains are quite different but the interactions (electronic, hydrogen bonding) between these groups provide the necessary driving force for miscibility. In his study, the sorption behavior (extrapolated to infinite dilution) of a series of penetrants in the miscible blend of poly(vinyl chloride)–poly(ε-caprolactone) revealed the interaction behavior for the blend by comparison of the blend behavior with the unblended constituent polymers. This gas chromatography method appears to be an excellent experimental tool for probing the molecular interactions of polymer blends.

The potentially useful specific interactions range from strong ionic inter-

4.5. Introduction of Interacting Groups

TABLE 4.2

Hydrogen Bonding Classification of Various Structural Groups[a]

H-bond strength[b]	Electron donors (proton acceptors)	Electron acceptors (proton donors)	Electron/proton donors–acceptors
Strong	Pyridine		Water Alcohols Carboxylic acids
Moderate	Ketones Ethers Esters		Aniline
Weak	Nitriles Nitros Olefins Aromatic hydrocarbons	Halogenated hydrocarbons	
Unclassified	Aldehydes Tertiary amines Sulfones Sulfoxides		Primary and secondary amines

[a] Adapted from H. Burrell, *Interchem. Rev.* **14**, 3 (1955), and E. P. Lieberman, *Off. Dig., Fed. Soc. Paint Technol.* **34**, 30 (1962).

[b] Mixtures of strong acceptors and donors may result in proton transfer, giving an acid–base (electrostatic) interaction, rather than a resonance interaction.

actions to weak dispersive interactions (see Section 2.1.6). Hydrogen bonding is an important specific interaction which has been proposed for many miscible polymer systems. Other types of specific interactions include charge (electron) transfer, acid–base (Lowry–Brönsted or Lewis), dipole–aromatic ring induction, ion pairing, and coordination with metallic ions.

Burrell [52] and Lieberman [53] have qualitatively classified the hydrogen bonding capability of polar units. This classification is illustrated in Table 4.2. Many miscible polymer blends can be hypothetically rationalized to exhibit miscibility based on the specific interactions chosen from this list. The experimental evidence to support the existence of these interactions in polymer blends, however, is often lacking.

Ionic interactions between polymers constitute the strongest potential for specific interaction short of direct covalent bonding. Polyelectrolyte complexes are classic examples of the utility of this specific interaction. Combinations of two oppositely charged polyelectrolytes lead to salt or complex formation with properties quite different than the individual components.

A typical example of a strong polyelectrolyte complex studied in detail [54] is illustrated below:

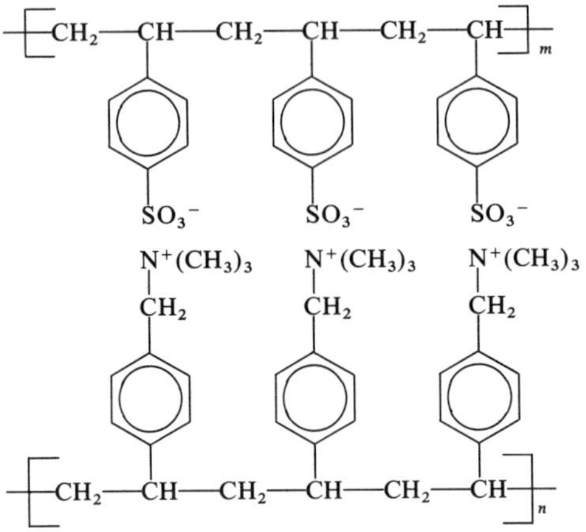

$$-[CH_2-CH-CH_2-CH-CH_2-CH]_m-$$

with pendant phenyl–SO_3^- groups (Sodium polystyrene sulfonate polyanion)

and paired with $N^+(CH_3)_3$–CH_2–phenyl pendant groups on

$$-[CH_2-CH-CH_2-CH-CH_2-CH]_n-$$

(Poly(vinyl benzyl trimethylammonium chloride) polycation)

Strong polyelectrolyte complexes are insoluble in water, acids, bases, and organic solvents. Note that the individual components are generally soluble in these media. Solubility of the complex is possible, however, in mixtures of water, polar organic solvents, and electrolytes. Although not water soluble, these complexes can sorb considerable amounts of water with properties typical of hydrogels.

Polyelectrolyte complexes were first recognized by Kossell [55] with the precipitation of the naturally occurring, oppositely charged polymers of egg albumen and protamine. More recently Michaels and co-workers have studied strong electrolyte complexes in detail, particularly with the example shown above [56–59].

Miscibility resulting from ionic interactions has implications in the area of naturally occurring materials. Shinoda et al. [60–62] investigated the interactions between poly(L-lysine) and synthetic anionic polyelectrolytes [i.e., poly(acrylic acid), poly(methacrylic acid), as well as the modified natural polymer sodium cellulose sulfate]. These studies were conducted because the interactions of poly(L-lysine) and various acidic polyelectrolytes (i.e., DNA, RNA, and mucopolysaccharides) had been previously proposed and investigated as a model complex for biological materials (i.e., nucleoproteins and protein–polysaccharide complexes).

A classic example of polyelectrolyte complexes from naturally occurring materials is the complex from gelatin (polycation) and gum arabic (polyanion) investigated by de Jong at the University of Leiden in the 1930s. Extension of this work led to the development of microencapsulation materials by National Cash Register Co. [63]. Another example of polyelectro-

4.5. Introduction of Interacting Groups

lyte complexes concerns the cationic polyethyleneimine complex with sodium dextran sulfate [64]. Polyethyleneimine adsorption on monocarboxyl cellulose, as demonstrated by Nedelcheva and Stoilkov [65], implied complex formation.

Water-soluble, nonionic polymers often exhibit miscibility with anionic polyelectrolytes. These nonionic polymers include polyethylene oxide, poly(vinyl pyrrolidone), and poly(N,N-dimethylaminoethyl methacrylate). Each of these polymers has been shown to form water-insoluble complexes with poly(acrylic acid), poly(methacrylic acid), and carboxymethyl cellulose [66–71]. Strong acid–base interactions are achieved with these water-soluble polymers in accordance with the proton donor–proton acceptor relationships shown in Table 4.2.

The characterization of polyelectrolyte complexes is considerably different than for the other miscible polymer systems covered in this treatise. Generally, electrical conductivity of solutions, hydrodynamic properties, and precipitation behavior of the complexes are measured. Characterization of the thermomechanical properties of the dry complexes has rarely been reported. T_g versus composition data, however, were determined by Smith et al. [66] for the poly(ethylene oxide)–poly(acrylic acid) complex. The initial addition of poly(ethylene oxide) to poly(acrylic acid) resulted in an increase in T_g to a level of 12% where a monotonic decrease in T_g was observed up tp 50% poly(ethylene oxide). A reasonably constant T_g was observed between 50 and 90% poly(ethylene oxide), due largely to the crystallization of poly(ethylene oxide) from the polymer blend.

Charge-transfer complexation in polymer mixtures was studied by Sulzberg and Cotter [72]. Electron-acceptor polyesters prepared from nitrophthalic acid variations were shown to exhibit miscibility with electron-donor polymers derived from aryliminodiethanols. The resultant blends exhibited single glass transition temperatures, different energy absorption maxima in the visible spectra than either of the components, and a yellow to yellow–orange color (components were colorless).

Hydrogen bonding-type interactions (also described variously as weak acid–base or acceptor–donor interactions) have been proposed as the key to achieving miscibility in many of the blends cited in this treatise. Poly(vinyl chloride) (hydrogen-bond donor) exhibits miscibility with many polymers containing acceptor units, e.g., butadiene/acrylonitrile copolymers [73], α-methylstyrene/methylacrylonitrile copolymers [74], ethylene/vinyl acetate copolymers [75], ethylene/vinyl acetate/sulfur dioxide terpolymers [76], poly(ε-caprolactone) [77], and syndiotactic poly(methyl methacrylate) [14]. Ethylene/ethyl acrylate or ethylene/vinyl acetate copolymers containing carbon monoxide units were shown by Robeson and McGrath [49] to exhibit miscibility with poly(vinyl chloride) over a broad composition range. This was proposed to be due to the specific interaction between the carbonyl of the terpolymer (H-bond acceptor) and the α-hydrogen of poly(vinyl chloride) (H-bond donor). Similar conclusions were also drawn for the blend of poly(vinyl chloride) and copolymers of ethylene and N,N-dimethyl acrylamide [50]. Infrared studies were cited [78] for model systems that suggest that interactions exist between di-substituted amides and compounds similar to poly(vinyl chloride). Poly(butylene terephthalate)–poly(vinyl chloride) miscibility was observed by Robeson [79] and attributed to the specific interaction of the ester carbonyl with the α-hydrogen of PVC. Other systems in which hydrogen bonding was proposed as the key to the observed miscibility include poly(butylene terephthalate)–poly(hydroxyether of bisphenol A) [46] and (nylon-6)–(ethylene/acrylic acid) copolymers [37].

With weak to moderate hydrogen bonding interactions, the dispersive and polar components of the solubility parameter presumably are of important consideration, as pointed out by Shaw [18]. Unless the δ_d and δ_p values of the constituent polymers are in reasonable proximity, the weak to moderate hydrogen bonding interactions may not be sufficient to yield one-phase mixtures. The geometry of the molecules may also be a key factor because of the large angle dependence of hydrogen bonds.

4.6 MISCELLANEOUS

Intensive mixing, as in a mixing section of an extruder, has been found to enhance the physical properties of many polymer blends. No evidence for the influence of mixing, per se, on phase relationships has been found, but graft and block formation may occur under intensive shear, leading to improved miscibility [46].

Mack [80], using a twin-screw extruder with kneading blocks, has demonstrated improvements in properties with intensive mixing for the following systems: (nylon-6)–LDPE, (nylon-6)–(acrylate-modified polyethylene), and PET–LDPE. Analysis of the phases is difficult in these crystalline mixtures and was not attempted.

Another technique of gaining a well-mixed polymer system involves the blending of dispersions of the two components. Bauer et al. [81] demonstrated that poly(ethyl acrylate) and poly(methyl methacrylate), when mixed in the emulsion state, form a system with a mechanical damping peak located between those for the PEA and PMMA at $-5°$ and $110°C$, respectively. Annealing completedly removed the intermediate peak, which was probably an interphase between the two main phases. Its presence could be detected because of the large amount of surface area and, therefore, interphase.

Another method of mixing is by freeze-drying of the system polymer 1–

polymer 2–solvent. If the amount of solvent is adequate to produce a single phase, it might be expected that some interesting polymer–polymer systems would result upon freeze-drying. Using the system poly(vinyl acetate)–poly(methyl methacylate), Miyata and Hata [82] showed that freeze-drying from benzene produced metastable miscible systems with intermediate glass transition temperatures. Ichihara *et al.* [83] also investigated PMMA–PVA blends prepared by freeze-drying dilute solutions in benzene. The glass transition of the resultant powder was determined using dilatometric techniques, and single transitions were observed for all freeze-dried blends. Heating the samples above the glass transition resulted in phase separation, demonstrating the metastable nature of the freeze-dried mixtures. Shultz and Mankin [84] have indirectly shown that freeze-drying can increase the miscibility of the system poly(methyl methacrylate)–polystyrene, a very immiscible system, while Berghmans and Overbergh [85] demonstrated that freeze-drying of the miscible system PPO–isotactic polystyrene from dilute solution could lead to phase separation which was removed by annealing. The latter example presumedly resulted from the tendency of macromolecules to aggregate in dilute solutions.

There are some hints in the literature [86–88] that small amounts of fillers or plasticizers can influence the miscibility of polymer–polymer systems. However, no definitive studies of these subjects have been published. Many useful improvements of physical properties can be achieved in polymer–polymer mixtures through the use of fillers and plasticizers [89, 90], possibly by a mechanism involving enhancement of the fineness of the mixture, or of the properties of one or both of the phases.

Another technique for enhancing compatibility, but not miscibility, is to add an AB block or graft copolymer to a simple mixture of A and B. As discussed in Sections 2.6 and 4.2, a phase-separated block or graft copolymer can do nothing to improve the miscibility of the homopolymer mixture. However, Reiss *et al.* [91] have found that block copolymers can inhibit demixing of the analogous homopolymer mixture dissolved in a common solvent, and several workers [31, 92] have noted a compatibilizing effect in bulk mixtures, particularly if the homopolymers were of lower molecular weight than the blocks in the block copolymer.

REFERENCES

1. A. A. Donatelli, L. H. Sperling, and D. A. Thomas, *J. Appl. Polym. Sci.* **21**, 1189 (1977).
2. R. L. Scott, *J. Polym. Sci.* **9**, 423 (1952).
3. G. E. Molau, *J. Polym. Sci., Polym. Lett.* **3**, 1007 (1965).
4. F. Kollinsky and G. Markert, *Adv. Chem. Ser.* **99**, 175 (1971).
5. R. Koningsveld, L. A. Kleintjens, and G. Markert, *Macromolecules* **10**, 1105 (1977).

6. M. R. Ambler, *J. Polym. Sci., Polym. Chem. Ed.* **11**, 1505 (1973).
7. V. R. Landi, *Appl. Polym. Symp.* **25**, 223 (1974).
8. L. M. Robeson and J. E. McGrath, *Polym. Eng. Sci.* **17**, 300 (1977).
9. W. A. Kruse, R. G. Kirste, J. A. Haas, B. J. Schmitt, and D. J. Stein, *Makromol. Chem.* **177**, 1145 (1976).
10. J. H. G. M. Lohn, G. Kransen, Y. Y. Tan, and G. Challa, *J. Polym. Sci., Polymer Lett.* **13**, 725 (1975).
11. R. G. Bauer and N. C. Bletso, *Polym. Prepr., Am. Chem. Soc., Div. Polym. Chem.* **10**, 632 (1969).
12. A. M. Liquori, M. De Santis Savino, and M. D'Alagani, *J. Polym. Sci., Polymer Lett.* **4**, 943 (1966).
13. H. Z. Liu and K.-J. Liu, *Macromolecules* **1**, 157 (1968).
14. J. W. Schurer, A. deBoer, and G. Challa, *Polymer* **16**, 201 (1975).
15. R. J. Kern, *J. Polym. Sci.* **21**, 19 (1956).
16. R. J. Kern, *J. Polym. Sci.* **33**, 524 (1958).
17. L. J. Hughes and G. E. Britt, *J. Appl. Polym. Sci.* **5**, 449 (1974).
18. M. T. Shaw, *J. Appl. Polym. Sci.* **18**, 449 (1974).
19. L. J. Fetters, *in* "Block and Graft Copolymerization" (R. J. Ceresa, ed.), Chapter 5. Wiley, New York, 1973; *J. Res. Natl. Bur. Stand., Sect. A* **70**, 421 (1966); *J. Polym. Sci., Part C* **26**, 1 (1969).
20. L. M. Robeson, M. Matzner, L. J. Fetters, and J. E. McGrath, *in* "Recent Advances in Polymer Blends, Grafts, and Blocks" (L. H. Sperling, ed.), p. 281. Plenum, New York, 1974.
21. M. Baer, *J. Polym. Sci., Part A* **2**, 417 (1964).
22. D. R. Hansen and M. Shen, *Macromolecules* **8**, 903 (1975).
23. D. J. Dunn and S. Krause, *J. Polym. Sci., Polym. Lett.* **12**, 591 (1974).
24. D. Suong and M. Shen, *Macromolecules* **10**, 357 (1977).
25. S. Krause, D. J. Dunn, A. S. Mozzaffori, and A. M. Biswas, *Macromolecules* **10**, 787 (1977).
26. D. F. Leary and M. C. Williams, *J. Polym. Sci., Polym. Phys. Ed.* **11**, 345 (1973).
27. D. F. Leary and M. C. Williams, *J. Polym. Sci., Phys. Ed.* **12**, 265 (1974).
28. T. Miyamoto, K. Kodama, and K. Shibayama, *J. Polym. Sci., Part A-2* **8**, 2095 (1970).
29. S. L. Aggarwal, R. A. Livigni, L. F. Marker, and T. J. Dudek, *in* "Block and Graft Copolymers" (J. J. Burke and V. Weiss, eds.), p. 157. Syracuse Univ. Press, Syracuse, New York, 1973.
30. J. F. Beecher, L. F. Marker, R. D. Bradford, and S. L. Aggarwal, *J. Polym. Sci., Part C* **26**, 117 (1969).
31. T. Inoue, T. Soen, T. Hasimoto, and H. Kawai, *Macromolecules* **3**, 87 (1970).
32. R. F. Fedors, *J. Polym. Sci., Part C* **26**, 189 (1969).
33. D. J. Meier, *J. Polym. Sci., Part C* **26**, 81 (1969).
34. G. Holden, E. T. Bishop, and N. R. Legge, *Proc. Int. Rubber Conf., 5th, 1967* Paper (1968).
35. J. E. McGrath, L. M. Robeson, M. Matzner, and R. Barclay, Jr., *Pap., Reg. Meet. Am. Chem. Soc., 8th, 1976*.
36. J. E. McGrath, T. C. Ward, E. Shchori, and A. J. Wnuk, *Polym. Eng. Sci.* **17**, 647 (1977).
37. M. Matzner, D. L. Schober, R. N. Johnson, L. M. Robeson, and J. E. McGrath, *in* "Permeability of Plastic Films and Coatings" (H. B. Hopfenberg, ed.), p. 125. Plenum, New York, 1975.
38. J. P. Kennedy, *J. Polym. Sci., Polym. Chem. Ed.* **13**, 2213 (1975).
39. K. C. Frisch, D. Klempner, H. L. Frisch, and H. Ghiradella, *in* "Recent Advances in Polymer Blends, Grafts and Blocks" (L. H. Sperling, ed.), p. 375. Plenum, New York, 1974.
40. V. Huelck, D. A. Thomas, and L. H. Sperling, *Macromolecules* **5**, 348 (1972).

41. L. J. Hughes and G. L. Brown, *J. Appl. Polym. Sci.* **5**, 580 (1961).
42. L. H. Sperling, D. W. Taylor, M. L. Kirkpatrick, H. F. George, and D. R. Bardman, *J. Appl. Polym. Sci.* **14**, 73 (1970).
43. L. H. Sperling, T. W. Chiu, C. P. Hartman, and D. A. Thomas, *Int. J. Polym. Mater.* **1**, 331 (1972).
44. N. Yoshimura and K. Fujimoto, *Rubber Chem. Technol.* **42**, 1009 (1969).
45. P. J. Corish, *Rubber Chem. Technol.* **40**, 324 (1967).
46. L. M. Robeson and A. B. Furtek, *Am. Chem. Soc., Div. Prepr., Org. Coat. Plast. Chem., Pap.* **37**, 136 (1977).
47. B. Schneier, *J. Appl. Polym. Sci.* **17**, 3175 (1973).
48. S. Krause, *J. Macromol. Sci., Rev. Macromol. Chem.* **7**, 251 (1972).
49. L. M. Robeson and J. E. McGrath, *Pap., Natl. Meet. AIChE, 82nd, 1976*.
50. M. Matzner, L. M. Robeson, E. W. Wise, and J. E. McGrath, *Am. Chem. Soc., Div. Org. Coat. Plast. Chem., Pap.* **37**, 123 (1977).
51. O. Olabisi, *Macromolecules* **8**, 316 (1975).
52. H. Burrell, *Interchem. Rev.* **14**, 3 (1955).
53. E. P. Lieberman, *Off. Dig., Fed. Soc. Paint Technol.* **34**, 30 (1962).
54. A. S. Michaels, *Ind. Eng. Chem.* **57**, 32 (1965).
55. A. Kossel, *Z. Physiol. Chem.* **22**, 176 (1896).
56. A. S. Michaels, G. L. Falkenstein, and N. J. Schneider, *J. Phys. Chem.* **69**, 1456 (1965).
57. A. S. Michaels, L. Mir, and N. S. Schneider, *J. Phys. Chem.* **69**, 1447 (1965).
58. A. S. Michaels and R. G. Miekka, *J. Phys. Chem.* **65**, 1765 (1961).
59. A. S. Hoffman, R. W. Lewis, and A. S. Michaels, *Am. Chem. Soc., Div. Org. Coat. Plast. Chem., Pap.* **29**, No. 2, 236 (1969).
60. K. Shinoda, T. Hayashi, T. Yoshida, K. Sakai, and A. Nakajima, *Polym. J.* **8**, 202 (1976).
61. K. Shinoda, K. Sakai, T. Hayashi, and A. Nakajima, *Polym. J.* **8**, 208 (1976).
62. K. Shinoda, T. Hayashi, and A. Nakajima, *Polym. J.* **8**, 216 (1976).
63. L. Holliday, in "Ionic Polymers" (L. Holliday, ed.), p. 24. Wiley, New York, 1975.
64. Y. Kikuchi, Y. Ohishi, and M. Kodama, *J. Appl. Polym. Sci.* **20**, 3205 (1976).
65. M. P. Nedelcheva and G. V. Stoilkov, *J. Appl. Polym. Sci.* **20**, 2131 (1976).
66. K. L. Smith, A. E. Winslow, and D. E. Petersen, *Ind. Eng. Chem.* **51**, 1361 (1959).
67. L. A. Bimendina, V. V. Roganov, and Y. A. Bektarov, *Vysokomol. Soedin., Ser. A* **16**, 2810 (1974).
68. Y. Osada and M. Sato, *J. Polym. Sci., Polym. Lett.* **14**, 129 (1976).
69. F. E. Bailey, Jr., R. D. Lundberg, and R. W. Callard, *J. Polym. Sci., Part A* **2**, 845 (1964).
70. A. D. Antipina, I. M. Papissov, and V. A. Kabanov, *Vysokomol. Soedin., Ser. B* **12**, 329 (1970).
71. J. Ferguson and S. A. O. Shah, *Eur. Polym. J.* **4**, 343 (1968).
72. T. Sulzberg and R. J. Cotter, *J. Polym. Sci., Part A-1* **8**, 2747 (1970).
73. G. A. Zakrzewski, *Polymer* **14**, 348 (1973).
74. J. F. Kenney, in "Recent Advances in Polymer Blends, Grafts, and Blocks" (L. H. Sperling, ed.), p. 137. Plenum, New York, 1974.
75. C. F. Hammer, *Macromolecules* **4**, 69 (1971).
76. J. J. Hickman and R. M. Ikeda, *J. Polym. Sci., Polymer Phys. Ed.* **11**, 1713 (1973).
77. J. V. Koleske and R. D. Lundberg, *J. Polym. Sci., Part A-2* **7**, 795 (1969).
78. M. Baron, J. deVillepin, C. Quiveron, and M. Josien, *J. Chim. Phys.* **67**, 1952 (1970); *Chem. Abstr.* **73**, 56515W (1970).
79. L. M. Robeson, *J. Polym. Sci., Polym. Lett.* **16**, 261 (1978).
80. W. Mack, *Mod. Plast.* **48**, 62 (1971).
81. P. Bauer, J. Hennig, and G. Schreyer, *Angew. Makromol. Chem.* **11**, 145 (1970).

82. S. Miyata and T. Hata, *Proc. Int. Congr. Rheol.*, *5th, 1968* Vol. 3, p. 71 (1970).
83. S. Ichihara, A. Komatsu, and T. Hata, *Polym. J.* **2,** 640 (1971).
84. A. T. Shultz and G. I. Mankin, *J. Polym. Sci., Polym. Symp.* **54,** 341 (1976).
85. H. Berghmans and N. Overbergh, *J. Polym. Sci., Polym. Phys. Ed.* **15,** 1757 (1977).
86. N. G. Gaylord, *Chemtech* **6,** 392 (1976).
87. T. Pazonyi and M. Dimitrov, *Rubber Chem. Technol.* **40,** 1119 (1967).
88. V. N. Kuleznev, Yu. S. Maloshchuk, G. I. Grigoryan, and B. A. Dogadkin, *Polym. Sci. USSR (Engl. Transl.)* **13,** 62 (1971).
89. W. Rovatti and E. G. Bobalek, *J. Appl. Polym. Sci.* **7,** 2269 (1963).
90. C. C. Lee, W. Rovatti, S. M. Skinner, and E. G. Bobalek, *J. Appl. Polym. Sci.* **9,** 2047 (1965).
91. G. Reiss, J. Kohler, C. Tournut, and A. Banderet, *Makromol. Chem.* **101,** 58 (1967).
92. A. Banderet, C. Tournut, and G. Reiss, *J. Polym. Sci., Part C* **16,** 2601 (1967).

Chapter 5

Comprehensive Survey of Miscible Polymer Systems

5.1 CRITERIA FOR SELECTION

A treatise on polymer–polymer miscibility would be far from complete without a detailed listing of the known miscible polymer blends. The lists compiled in this chapter are quite comprehensive, particularly in comparison to previous compilations [1–3]. Previously, the most comprehensive accounting of miscible polymer blends had been reported by Krause [4] in reviews which also included two-phase polymer blends. In this chapter, a discussion of the relevant papers utilizing various techniques to establish miscibility for each specific system is also presented.

The basic methods of establishing certain criteria for polymer blend miscibility have been previously discussed. Most methods cited have utilized the appearance of a single glass transition temperature from various experimental techniques (mechanical, calorimetric, dilatometric, dielectric, etc.), with mechanical methods most prevalent. In the case of the polyelectrolyte complexes, glass transition data are rarely cited. The characteristics of polyelectrolyte complexes indicate miscibility from other well-established techniques. Mechanical compatibility as denoted by a reasonable average of the constituent properties exhibited by the blend is not a sufficient criterion, and papers either showing only mechanical property data or, even more vaguely, stating that mechanical compatibility was achieved were not included in the compilation. One-phase behavior in a solution containing a common solvent is also not considered relevant for discussion in this chapter. In some cases, solvent-free film clarity combined with microscopic methods used to observe phase behavior is cited and prefaced by the comment "not sufficient evidence." Obviously, cases are present which are rather vague or present results which may be subject to interpretations other than those stated by the authors. These are included with the hope that our interpretations fairly delineate the questionable miscibility. Various cases (e.g., interpenetrating networks) are cited where broad transitions exist. Where these transitions exist with sufficient differences between those expected for the polymer components acting as individual phases, the polymer–polymer blend, interpenetrating network, or block or graft copolymer in question has been included in this review.

Sections of this review have been divided to cover separately simple physical blends, block copolymers, graft copolymers, and interpenetrating networks. In categorizing the miscible polymer blends, polymer types have been sectioned into the following classes, chosen primarily due to the prevalence of miscible systems containing the listed component.

- 5.2.1 Poly(vinyl chloride)
- 5.2.2 Polystyrene and Styrene-Containing Polymers
- 5.2.3 Cellulosics
- 5.2.4 Polyacrylates
- 5.2.5 Poly(vinyl acetate) and Vinyl Acetate Copolymers
- 5.2.6 Polyesters and Polycarbonates
- 5.2.7 Polyethers
- 5.2.8 Polyamides
- 5.2.9 Water-Soluble Polymers and Polyelectrolyte Complexes
- 5.2.10 Unsaturated Hydrocarbon-Based Polymers
- 5.2.11 Block Copolymers
- 5.2.12 Graft Copolymers
- 5.2.13 Interpenetrating Networks
- 5.2.14 Isomorphic Polymer Blends
- 5.2.15 Miscellaneous

In this list, numbers 1 through 10 represent physical blends, whereas numbers 11 through 13 represent special modifications of polymer blends. Water-soluble polymers and polyelectrolyte complexes were chosen as a special class because the acidic or basic properties of these systems lead to strong specific interactions. In the case of complexes of strong polyelectrolytes, the term "polysalt" has been applied to describe their behavior. This type of miscible polymer system has been all but overlooked in prior reviews concerning polymer miscibility. Cocrystallinity (i.e., isomorphism) is a rare situation in polymer blends but, nevertheless, an important aspect of polymer miscibility and is therefore included.

One of the basic problems of comparing results of different investigators lies in the utilization of widely differing polymers in the experimental investigation as well as techniques of preparing the polymer blends. Due to incomplete descriptions in many papers (e.g., molecular weight characterization), problems may exist in comparing the results of different investigations. In some cases, commercial polymers have been utilized, and, for the sake of completeness, commercial designations have been entered in lieu of additional data. In certain cases, low molecular weight oligomer blends have been cited depending somewhat on the nature and relevance of the study. However, a lower limit in molecular weight was necessary to prevent inclusion of systems which exhibit miscibility only at very low molecular weights. This molecular weight limit has been chosen to be $M_n = 5000$.

The vast majority of the references cited are from the technical literature. The patent literature generally covers these systems as well as many others not reported in the technical literature. However, most patents do not adequately describe the nature of polymer blends or present sufficient data to ascribe miscibility to specific blends. In several cases, patent literature is cited where it is believed that the information in the patent is of interest and can be used to complement the technical literature.

With each detailed listing of the miscible polymer classifications, a discussion of the pertinent characteristics of these blends will follow. As previously mentioned, specific interactions are quite effective in promoting polymer–polymer miscibility. Where specific interactions occur, or have been at least hypothesized, mention will be made in the relevant discussion sections.

5.2 REFERENCED LISTING

5.2.1 Poly(vinyl chloride)

Poly(vinyl chloride) represents one of the most rigorously investigated components of polymer blends. By virtue of the vast number of blends prepared, it is not surprising that PVC has been found to be miscible with a number of structurally different polymers and copolymers. The capability of weak specific interactions is possible with PVC. The α-hydrogen of PVC is capable of hydrogen bonding, particularly with polymers which have "basic" properties (i.e., amides, carbonyl) [5]. The possibility of charge-transfer interactions with PVC has also been proposed [6], involving the ester oxygen and the pendant chloride for the miscible blend of PVC and poly(ε-caprolactone).

$$\begin{array}{c} \delta^+ \\ H \\ | \\ -CH_2-C- \\ | \\ Cl \\ \delta^- \end{array}$$

Poly(vinyl chloride)

Blends of PVC and butadiene/acrylonitrile copolymer historically represent the initial observation that miscibility with polymer mixtures is possible [7, 8]. The investigations with this particular blend outnumber any other miscible polymer blend. This is also the result of commercial interest in this blend over the past three decades. In the technical literature, this blend has been described as miscible, partially miscible, and even heterogenous based on different experimental techniques. Generally, dynamic mechanical results indicate miscibility, but with some broadening of the glass transition [9]. Using microscopic techniques, separate phase resolution is possible [10, 11]. With the staining procedure potentially promoting phase separation (via chemical reaction on the butadiene, thus promoting structural changes), with PVC crystallization yielding a minor separate crystalline phase, and with the existence of structure even with homopolymers in the amorphous state using similar microscopic techniques [12, 13], it is not surprising that this blend has been described as partially miscible and even heterogeneous. Ambler [14] observed that under certain polymerization conditions compositional variation in butadiene/acrylonitrile copolymers would result in nonhomogeneous materials. Landi [15, 16] observed that two-phase behavior could be obtained in the copolymer if the breadth of the copolymer composition were large enough, as was stated to be the case with commercial compositions of the butadiene/acrylonitrile copolymer. Using calorimetric data, Landi [17] observed that two separate transitions could be isolated in these copolymers. In blends with PVC, only one of these transitions shifted with varying PVC content. These results point out another distinct cause for the observation of two-phase behavior in nitrile rubber–PVC blends using microscopy methods. This blend is, therefore, judged as miscible, although a certain level of heterogeneity may exist due to the reasons discussed above. The range of acrylonitrile content sufficient to yield miscibility in PVC appears to be quite large and equal to 23–45% [18].

Next to butadiene/acrylonitrile blends with PVC, blends of PVC with ethylene/vinyl acetate copolymers have been most widely studied [19–22]. Miscibility appears optimum at vinyl acetate contents of 65–70% [19], although values as low as 45% VA have been experimentally observed to have limited miscibility [20]. Miscibility has also been inferred from diffusion data of gas molecules, which can be used as probes to assess the level of molecular mixing [22, 23]. Nuclear magnetic resonance (nmr) data on blends based on a copolymer of 45% VA content indicated partial miscibility with the level very dependent on sample preparation conditions [24]. Marcincin et al. [25] studied EVA (45% VA)–PVC and chlorinated EVA–PVC blends. While definite phase separation was observed with the EVA–PVC blends, chlorination of up to 38% based on EVA weight yielded blends with PVC having single T_g's.

5.2. Referenced Listing

Terpolymers of ethylene/vinyl acetate/sulfur dioxide have been shown to exhibit miscibility with PVC over the entire composition range by Hickman and Ikeda [26]. Sulfur dioxide incorporation allowed the utilization of much higher concentrations of ethylene in the terpolymer than in EVA copolymers while still maintaining miscibility with PVC. Similar results were noted by Robeson and McGrath [27, 28] with terpolymers of ethylene/ethyl acrylate/carbon monoxide and ethylene/vinyl acetate/carbon monoxide. With ethylene/ethyl acrylate copolymers, it was noted that no copolymer composition exhibited miscibility with PVC; however, as low as 5 wt% CO in the terpolymer yielded miscible blends with PVC. As with the E/VA/SO$_2$ terpolymers, a broad range of terpolymer compositions was observed to be miscible with PVC. These results were believed to be due to the specific interaction of the carbonyl of the terpolymer with the α-hydrogen of poly(vinyl chloride). This interaction was classified as a weak "acid–base" type where PVC represented the proton donor and the terpolymer carbonyl represented the proton acceptor. This interaction, although weak, allowed a large variation in the composition of the terpolymer with retention of miscibility with PVC.

Poly(ε-caprolactone) was initially demonstrated by Koleske and Lundberg [29] to be miscible with PVC over the entire concentration range. In fact, the T_g–composition data were used to determine the T_g of amorphous poly-(ε-caprolactone) (PCL) by extrapolation of the amorphous blend data to 100% PCL. Crystallization kinetics of PCL in PCL–PVC blends were reported by Robeson [30] and were shown to reasonably agree with the kinetics predicted by the spherulitic growth rate equation [31]. A unique method of determining the degree of crystalline content of PCL from data obtained on blends of PCL–PVC was reported by Robeson and Joesten [32]. This method involved the determination of the glass transition shift from the amorphous state of the PCL–PVC blends to the semicrystalline state. The degree of crystallinity of the blend could then be determined by a simple material balance. Olabisi [6] investigated the PCL–PVC blends using solvent probes in the inverse gas chromatography technique. The experimental data allowed for estimation of the interaction parameter for PCL and PVC which predicted miscibility based on its negative value. Detailed studies were reported by Ong for this blend, particularly concerned with the crystallization characteristics of PCL from PCL–PVC blends [33]. Khambatta *et al.* [34] studied in detail the morphology of these blends using small-angle X-ray and light scattering. Hubbell and Cooper [35] investigated the segmental orientation of the components of the PVC–PCL blends using dynamic differential infrared dichroism.

The systems mentioned to this point are all plasticizers for PVC in that their addition results in lowering of the T_g. Several examples are present where the addition of a higher T_g polymer results in an increase in PVC's T_g. A study of the influence of the poly(methyl methacrylate) tacticity on the miscibility with PVC revealed definite effects due to stereo structure [36]. With isotactic PMMA, immiscible blends with PVC were observed over the entire composition range. With syndiotactic PMMA, however, miscible blends resulted up to a blend composition corresponding to a monomer ratio of 1:1 PMMA:PVC. At higher syndiotactic PMMA contents, two-phase behavior was observed with one phase of pure s-PMMA and the other phase corresponding to the 1:1 PMMA:PVC composition.

The patent survey revealed several high T_g polymers based on α-methyl-styrene, which appear from the data to exhibit miscibility with PVC [37–48]. These are the azeotropic compositions of the α-methylstyrene/acrylonitrile copolymer (69/31 by weight) and the α-methylstyrene/acrylonitrile/methyl methacrylate terpolymer (60/20/20 by weight). Note that these patents imply miscibility with PVC but do not present conclusive evidence. Kenney reported miscibility of PVC with α-methylstyrene/methacrylonitrile co-polymers [49] and α-methylstyrene/methacrylonitrile/ethyl acrylate ter-polymers [50].

tert-Butyl styrene/acrylonitrile copolymers and *tert*-butyl styrene/acrylonitrile/styrene terpolymers were claimed to yield homogeneous compositions with PVC by Hall *et al.* [41] with properties reported that indicate the potential of miscibility. Shur and Ranby [51] studied PVC blends with ABS. Using a series of experimental methods for characterization of these blends, they concluded that the styrene/acrylonitrile matrix of ABS was miscible with PVC.

Poly(butylene terephthalate) (PBT) has a melting point (220°C) in excess of the stability limit of PVC (even highly stabilized); thus, common melt mixing techniques cannot be employed. However, Robeson [52] observed that *N*-methyl pyrrolidone used as a mutual solvent at 150°C could be successfully used to prepare PBT–PVC blends. After extensive mixing, coagulation, and proper drying, rapid molding at 220°C followed by quenching yielded transparent blends exhibiting single, sharp glass transition temperatures. Crystallization of PBT occurred when blends were raised above the respective T_g's; however, the amorphous phase still exhibited a single T_g higher than the more amorphous quenched samples. The increase in T_g was attributed to an increase in the PVC content of the amorphous phase and restriction of the segmental motion of the amorphous phase due to crystallization of one of the constituents.

Polyester oligomers in the range of 2000 to 4000 M_n, offering improved permanence in plasticized PVC over the standard low molecular weight plasticizers, are commercially utilized. These structures are lower in molecular weight than the range designated for this survey. Many of these structures

5.2. Referenced Listing

TABLE 5.1

Poly(vinyl chloride)

Miscible polymer	Method of miscibility determination	Comments	References
Butadiene/acrylonitrile copolymers	Dynamic mechanical; calorimetric (DSC); phase-contrast microscopy	Miscible in the range of 23–45% AN	18
Butadiene/acrylonitrile copolymers	Permeability	Miscibility increases with increasing AN from 20 to 40 wt%	59
Butadiene/acrylonitrile copolymers	Dynamic mechanical	Evidence of micro-heterogeneity	60
Butadiene/acrylonitrile copolymers	Calorimetric (DSC)	Some blends exhibit two-phase behavior due to nitrile rubber inhomogeneity	17
Butadiene/acrylonitrile copolymers		Additional references	15, 16, 61–66
Ethylene/vinyl acetate copolymers	Dynamic mechanical	Single-phase with VA content 65–70%	19
Ethylene/vinyl acetate copolymers	Permeability; dynamic mechanical; light scattering	Miscible at 65% VA, two-phase at 45% VA	22, 23
Ethylene/vinyl acetate copolymers	nmr	Limited miscibility at 45% VA	24
Ethylene/vinyl acetate copolymers		Additional references	67–69
Chlorinated ethylene/vinyl acetate copolymers	Dynamic mechanical	Chlorination of a 45% VA content EVA copolymer improved miscibility	25
Poly(ε-caprolactone)	Dynamic mechanical	Miscible over entire composition range	29
Poly(ε-caprolactone)	Crystallization kinetics	Follows behavior expected of a single-phase mixture	30, 33, 70
Poly(ε-caprolactone)	Permeability	Indicated miscibility	71
Poly(ε-caprolactone)	Gas chromatography	Solvent probes indicate specific interactions	6
Poly(ε-caprolactone)	Small-angle X-ray light scattering	Partial miscibility concluded from results	34
Ethylene/vinyl acetate/ sulfur dioxide terpolymers	Dynamic mechanical; calorimetric (DSC); phase-contrast miscroscopy	Allows higher ethylene content than EVA copolymers	26

(Continued)

TABLE 5.1 (*Continued*)

Miscible polymer	Method of miscibility determination	Comments	References
Ethylene/vinyl acetate/carbon monoxide, ethylene/ethyl acrylate/carbon monoxide, and ethylene/2-ethyl hexyl acrylate/carbon monoxide terpolymers	Dynamic mechanical	Carbon monoxide addition allows broad composition range for miscibility	27, 28
Syndiotactic poly(methyl methacrylate)	Dynamic mechanical	Miscible up to 1:1 PMMA:PVC molar ratio	36
Methyl methacrylate/ethyl acrylate (95/5) and methyl methacrylate/butyl acrylate (90/10) copolymers	Dynamic mechanical	Commercial processing aids for rigid PVC compounds	72
α-Methylstyrene/methacrylonitrile copolymer	Dynamic mechanical	(57 mole% MAN)	49
α-Methylstyrene/methacrylonitrile/ethyl acrylate terpolymer	Dynamic mechanical	(58/40/2 by wt)	50
α-Methylstyrene/acrylonitrile (69/31 by wt) copolymer and other compositions		Not sufficient evidence	37
α-Methylstyrene/acrylonitrile/methyl methacrylate terpolymer (60/20/20 by wt) and other compositions		Not sufficient evidence	39
α-Methylstyrene/acrylonitrile/methyl methacrylate (34/8/58)	Dynamic mechanical	T_g versus composition data imply miscibility	73
Styrene/acrylonitrile (72/28 by wt)	Dynamic mechanical; permeability; differential thermal analysis	ABS used in PVC blends; results indicated the SAN matrix was miscible with PVC	51
Poly(butylene terphthalate)	Dynamical mechanical	PBT crystallizes from blend	52
Poly(1,4-cyclohexylene dimethylene tere-/isophthalate)	Clear molded films	T_g's of blend components equal, thus not sufficient evidence	74
Poly(butylene terphthalate)–poly(tetrahydrofuran) $(AB)_n$ block copolymers	nmr; dynamic mechanical; linear thermal expansion	Microheterogeneous	56, 75

5.2. Referenced Listing

TABLE 5.1 (*Continued*)

Miscible polymer	Method of miscibility determination	Comments	References
Thermoplastic polyurethanes based on poly(ε-caprolactone) soft block			57
Ethylene/N,N-dimethyl acrylamide copolymers	Dynamic mechanical	Miscible in the range of 25–30 wt% DMA; higher DMA contents not studied	58
Vinyl chloride/vinyl acetate (10% VA)–nitrile rubber	Dynamic mechanical		9
Ethylene/N-methyl-N-vinyl acetamide (18.4 wt%)	Transparent molded films	Not sufficient evidence	76
Ethylene/n-butyl urethane; ethylene/methyl urethane	Transparent molded films	Not sufficient evidence	76
Ethylene/N,N-dimethylaminoethyl methacrylate (11 wt%)	Transparent molded films	Not sufficient evidence	76
Ethylene/acrylonitrile (13 wt%)	Transparent molded films	Not sufficient evidence	76
Chlorinated PVC	Dynamic mechanical	Chlorinated PVC (65.2% Cl) miscible with PVC; 67.5% Cl not miscible	77

may indeed be miscible with PVC at higher molecular weight but have not been reported in the published literature. These polyester oligomers (OH terminated) have been extended to high molecular weight by reaction with diisocyanates. Their utility as permanent plasticizers cited in the patent literature [53–55] suggest the possibility of miscibility. Typical polyesters cited included butanediol, hexanediol, or ethylene glycol as the dihydroxy reactant and adipic acid, sebacic acid, azelaic acid, or succinic acid as the dicarboxylic acid reactant. Two elastomeric block copolymers that have been reported as permanent plasticizers for PVC are the poly(butylene terephthalate)–poly(tetrahydrofuran) (AB)$_n$ block copolymer [56] and polyester [i.e., poly(ε-caprolactone)] based thermoplastic polyurethanes [57].

Notably, many of the polymers exhibiting miscibility with PVC have as a common entity the carbonyl unit (s-PMMA, E/EA/CO, EVA, E/VA/CO, PBT). With poly(ε-caprolactone) (PCL), spectroscopic measurements showed a frequency shift of 6 cm^{-1} in the carbonyl band for PCL versus

PCL in the blend with PVC. This corresponds to an interaction parameter of -2.9 cal/cc [6].

Copolymers of ethylene and N,N-dimethyl acrylamide (DMA) in blends with PVC yielded single T_g's when the copolymer contained between 17 and 25 wt% DMA [58]. The specific interaction of the "basic" disubstituted amide (proton acceptor) with the α-hydrogen of PVC was proposed to account for the observed miscibility.

In summary, many of the systems miscible with PVC have a common denominator in that they contain structural units (proton acceptors) capable of hydrogen bonding with the α-hydrogen of PVC. Unlike many of the other polymers to be discussed, many of the PVC blends cited here have commercial importance. The detailed listing of the polymers exhibiting miscibility with PVC is given in Table 5.1 [6, 9, 15–19, 22–30, 33, 34, 36, 37, 39, 49–52, 56–77].

5.2.2 Polystyrene and Styrene-Containing Polymers

In the review of polystyrene, most of the references will be concerned with atactic polystyrene, which exists only as an amorphous polymer. Stereospecific catalysts can be employed to prepare isotactic polystyrene. Isotactic polystyrene is a crystallizable polymer which exhibits a glass transition temperature similar to that of atactic polystyrene (100°C). Due to the equivalence of the T_g's, most methods used to determine polymer miscibility cannot be applied to blends of atactic and isotactic polystyrene. One can extrapolate the results of Yeh and Lambert [78] involving crystallization studies of isotactic polystyrene from atactic–isotactic blends and arrive at a conclusion of miscibility. Their results indicate that the crystallization rate of isotactic polystyrene is linearly decreased by addition of atactic polystyrene due to a dilution effect. This implies single-phase behavior of the blend at the crystallization temperatures investigated. Further studies on this blend have been reported by Keith and Padden [79].

The blend of polystyrene and poly(2,6-dimethyl-1,4-phenylene oxide) (PPO) represents the most investigated miscible system containing polystyrene. This blend also represents the commercial products under the trade name Noryl. Initial studies of PPO and polystyrene blends yielded inconclusive results in that two distinct phases (polystyrene-rich and PPO-rich phases) could be resolved from dynamic mechanical loss peak measurements [80]. The same samples, however, exhibited only a single T_g in calorimetric (DSC) measurements. Further studies involving dielectric relaxation data revealed only a single T_g intermediate between that observed for the blend components [81]. The dielectric relaxation data were, however, noticeably broader than the relaxations for the pure polymers. It was concluded that,

5.2. Referenced Listing

while a certain level of miscibility occurred, compositional fluctuations were prevalent enough to yield data indicating definite levels of microheterogeneity. Shultz concluded that PPO and polystyrene were miscible at all concentration levels and attributed the reported instances of nonhomogeneity to inefficient mixing [82, 83]. Rheological studies indicated miscibility of the PPO–polystyrene blend, as the temperature dependence of the viscoelastic response could be predicted from single experimentally determined glass transition temperatures. Zero-shear viscosity was also determined from the composition of the PPO–polystyrene blend [84]. Jacques studied vapor and liquid equilibria in PPO–polystyrene blends and concluded that the data obtained were consistent with the apparent homogeneous nature of the blend [85–87]. Studies at the University of Massachusetts [88–91] with isotactic polystyrene–PPO blends have shown miscibility behavior similar to that with atactic polystyrene. The crystallization of isotactic polystyrene as well as PPO from these blends was investigated in detail. In a study involving copolymers of styrene and p-chlorostyrene, Shultz and Beach [82] observed miscibility of these copolymers with PPO at mole fractions of styrene ≥ 0.35 in the copolymer. Thermo-optical, differential scanning calorimetry, and dynamic mechanical studies were in excellent agreement as to the miscibility of the blends investigated. Tkacik [92] reached similar conclusions in his studies of styrene/p-chlorostyrene copolymer blends with PPO (miscibility at mole fractions of styrene ≥ 0.40). In addition to the experimental techniques utilized by Shultz and Beach, dielectric relaxation and electron microscopy studies were also employed. Miscibility has also been reported for PPO blends with poly(α-methylstyrene)–polystyrene AB block copolymers [93]. This result is not surprising as some level of miscibility has been claimed with PPO–poly(α-methylstyrene) blends [93, 94] although detailed studies of this system have not been reported. Blending of PPO with a styrene–butadiene–styrene ABA block copolymer was investigated by Shultz and Beach [95] using thermo-optical analysis. The polystyrene transition increased monotonically with PPO addition, implying molecular mixing of PPO with the polystyrene phase of the block copolymer.

Single calorimetric T_g's were observed for blends of poly(2,6-dimethyl-1,4-phenylene oxide)–poly(p-chlorostyrene–co–o-chlorostyrene) in the range of 23 to 64% p-chlorostyrene [96]. All of these miscible blends exhibit lcst (lower critical solution temperature) behavior. It is interesting to note that both homopolymers based on p-chlorostyrene and o-chlorostyrene are not miscible with PPO.

Although not of equal commercial importance as the previously discussed PPO–polystyrene blend, the blend of poly(vinyl methyl ether) and polystyrene has been investigated as rigorously in fundamental studies of polymer–polymer miscibility. Bank et al. [97] reported that the phase behavior of these blends was solvent dependent. Two-phase behavior was reported for blends cast from trichloroethylene or chloroform, whereas single-phase behavior was exhibited by films cast from toluene or xylene. Further studies indicated the initially single-phase films cast from the appropriate solvent exhibited phase separation when heated above 125°C [98]. Moreover, the two-phase films cast from the above-listed solvents retained two-phase behavior after thermal treatment consisting of extensive annealing and slow cooling. McMaster [99] investigated this system and observed similar behavior for toluene-cast films. The cloud-point temperature–composition data clearly illustrated this blend to be an example of lcst behavior. Using a modification of Flory's equation-of-state approach to thermodynamics, McMaster found that lcst behavior should be generally anticipated for miscible polymer–polymer blends due to differences in the pure component thermal expansion coefficients. Kwei *et al.* [100] indicated that extensive mixing of the polystyrene–poly(vinyl methyl ether) blends had occurred, based on density measurements. Vapor sorption techniques allowed further evidence in that the interaction parameter χ_{12} was found to be negative in the range of 35–65% polystyrene, thus satisfying the thermodynamic criteria of miscibility. The temperature dependence of χ_{12} suggested the existence of both lcst and ucst (upper critical solution temperature) behavior for this blend. Nishi *et al.* [101] offered evidence of spinodal decomposition of this blend by nmr data which indicated a gradual change in the composition of newly formed phases, whereas phase separation via nucleation and growth leads to discrete phase domains. Robard *et al.* [102] studied the phase separation of polystyrene and poly(vinyl methyl ether) diluted with a series of solvents chosen from those previously reported to give miscible and immiscible blends, after solvent evaporation. Using the gas–liquid chromatography (GLC) method, the solvents yielding immiscible blends were also shown to have high values of $\Delta\chi$ ($\Delta\chi = \chi_{12} - \chi_{13}$) where the magnitude of $\Delta\chi$ represents the difference in interaction of the solvent for the different polymer components of the blend.

Shaw [103] observed that tetramethyl bisphenol A polycarbonate exhibited

Tetramethyl bisphenol A polycarbonate

miscibility with polystyrene with features hinting at lcst behavior. Polystyrene–poly(*o*-methylstyrene) and poly(*o*-methylstyrene)–poly(*m*-methylstyrene) were reported to be miscible by Kern and Slocombe [104] based on

5.2. Referenced Listing

chloroform solution phase behavior. This is not conclusive evidence of polymer–polymer miscibility, and phase separation in solution for polystyrene and poly(o-methylstyrene) was reported by Kern in a later publication [105]. Polystyrene and benzyl cellulose were reported to be miscible in chloroform by Dobry and Boyer-Kawenoki [3]; however, phase separation in cyclohexanone was reported by Slonimskii [106] and in chloroform by Struminskii and Slonimskii [107]. Therefore, these systems, having been reported miscible in previous reviews, are considered either immiscible due to data presented after the initial observations or at least ambiguous due to lack of sufficient characterization.

An important copolymer containing styrene is the matrix component of ABS, namely, styrene/acrylonitrile (SAN) copolymers generally with acrylonitrile contents of 24–32 wt%. Molau [108] studied a series of SAN blends differing in acrylonitrile contents by 3.4, 4.7, 4.9, and 10.2%. Phase microscopy indicated that only the blend differing by 3.4% exhibited miscibility, whereas the other three blends phase-separated.

Poly(ε-caprolactone) was shown to be miscible with styrene/acrylonitrile copolymers of 28 wt% AN by dynamic mechanical loss measurements [109]. McMaster [110] also investigated this blend and observed lcst behavior using equilibrium cloud-point data. McMaster also observed lcst behavior for the styrene/acrylonitrile copolymer (28% AN)–poly(methyl methacrylate) blend with a minimum temperature of 150°C for phase separation observed at $\sim 12\%$ SAN [110]. Transmission electron microscopy clearly indicated that spinodal decompositions could be obtained when the blends were subjected to temperatures above the cloud-point curve. Stein [111] observed miscibility of SAN with PMMA in the range of 9–27 wt% AN using microscopic and mechanical characterization methods.

Slocombe [112] studied blends of a series of co- and terpolymers based on styrene, acrylonitrile, and α-methylstyrene monomers. Many of these blends were transparent when cast from solution. The most noteworthy blend exhibiting clarity was the azeotropic composition of styrene/acrylonitrile (76/24) with the azeotropic composition of α-methylstyrene/acrylonitrile (69/31). This provides the basis of high-heat ABS whereby the α-methylstyrene/acrylonitrile copolymer is added to ABS to raise the heat distortion temperature [113]. The data presented by Slocombe are, however, not conclusive concerning miscibility. Olabisi and Farnham [114] observed that the terpolymer of α-methylstyrene/methyl methacrylate/acrylonitrile (60/20/20 by weight) was miscible at all concentrations in poly(methyl methacrylate) [114]. The flexural and tensile strengths of these blends were found to be higher than expected from averaged properties.

In studies of the miscibility of styrene/acrylonitrile copolymers with polysulfones, miscibility was noted by Shaw [103] with poly[oxy-1,4-phenylene

TABLE 5.2

Polystyrene and Styrene Copolymers

Miscible polymer	Method of miscibility determination	Comments	References
Atactic–isotactic polystyrene blends	Crystallization kinetics	Kinetics indicate single-phase behavior	78
Poly(2,6-dimethyl-1,4-phenylene oxide) (PPO)	Dynamic mechanical; calorimetric	Inconclusive results drawn from data	80
Poly(2,6-dimethyl-1,4-phenylene oxide) (PPO)	Dielectric	Broad transitions	81
Poly(2,6-dimethyl-1,4-phenylene oxide) (PPO)	Solvent equilibria data	Consistent with single-phase behavior	85–87
Poly(2,6-dimethyl-1,4-phenylene oxide) (PPO)	Rheological	Consistent with single-phase behavior	84
PPO–isotactic polystyrene	T_g–composition data; melting point depression	T_g data equal to atactic polystyrene	88–91
PPO–poly(α-methylstyrene)	Dynamic mechanical	PPO–polystyrene/poly-(α-methylstyrene) block copolymers also miscible	93, 94
Polystyrene–poly(2,6-diethyl-1,4-phenylene oxide); poly(2-methyl,6-ethyl-1,4-phenylene oxide); poly(2-methyl,6-propyl-1,4-phenylene oxide); poly(2,6-dipropyl-1,4-phenylene oxide); poly(2-ethyl,6-propyl-1,4-phenylene oxide)	T_g	Single T_g in blends claimed; data not presented	94
PPO–styrene/p-chlorostyrene copolymers	Thermo-optical; calorimetric; dynamic mechanical; dielectric; electron microscopy	Miscible with styrene content > 0.35 mole fraction	82, 92
Polystyrene–poly(vinyl methyl ether)	Calorimetric; dielectric	Miscibility solvent dependent	97
Polystyrene–poly(vinyl methyl ether)	Calorimetric	lcst behavior observed	98
Polystyrene–poly(vinyl methyl ether)	Light transmission; nmr	Spinodal decomposition observed	101
Polystyrene–poly(vinyl methyl ether)	Density; vapor sorption	Negative interaction parameter determined	100
Polystyrene–poly(vinyl methyl ether)	Cloud-point data	lcst behavior observed	99

5.2. Referenced Listing

Table 5.2 (*Continued*)

Miscible polymer	Method of miscibility determination	Comments	References
Polystyrene–tetramethyl bisphenol A polycarbonate	Modulus–temperature data	lcst behavior possible	73, 103
Styrene/acrylonitrile copolymers of different compositions	Phase microscopy	Blends differing in AN content of more than 3.4 wt% were immiscible	108
Styrene/acrylonitrile (28% AN)–poly(ε-caprolactone)	Dynamic mechanical		109
Styrene/acrylonitrile (28% AN)–poly(ε-caprolactone)	Cloud-point curves	lcst behavior	99
Styrene/acrylonitrile (28% AN)–poly(methyl methacrylate)	Cloud-point curves	lcst behavior	99
Styrene/acrylonitrile (28% AN)–poly(methyl methacrylate)	Dynamic mechanical; microscopy	Miscibility with AN range of 9–27 wt%	111
α-Methylstyrene/methyl methacrylate/acrylonitrile terpolymer–poly(methyl methacrylate)	Modulus–temperature data	Miscible at all concentrations	114
Poly[oxy-1,4-phenylene sulfonyl-1,4-phenyleneoxy(2,6-diisopropyl-1,4-phenylene)isopropylidene(3,5-diisopropyl-1,4-phenylene)] styrene/acrylonitrile copolymer	Modulus–temperature data	Miscible at 13–16% acrylonitrile	103
Poly[oxy(2,5-dimethyl-1,4-phenylene)sulfonyl-(2,5-dimethyl-1,4-phenylene)oxy-1,4-phenylene isopropylidene-1,4-phenylene]–styrene/acrylonitrile copolymer	Modulus–temperature data	Miscible at 13 wt% acrylonitrile	103
Styrene/acrylonitrile copolymer (76/24)–α-methylstyrene/acrylonitrile copolymer (69/31)	Clear cast film	Not sufficient evidence	112

(*Continued*)

TABLE 5.2 (*Continued*)

Miscible polymer	Method of miscibility determination	Comments	References
Nitrocellulose–styrene/acrylonitrile copolymer	Clear cast film	Not sufficient evidence	115
Poly(*p*-chlorostyrene-co-*o*-chlorostyrene)–poly(2,6-dimethyl-1,4-phenylene oxide)	Calorimetric	Miscible in range of 23–64% *p*-chlorostyrene; all miscible blends exhibit lcst behavior	96
Styrene/methyl methacrylate (Zerlon 150, Dow)–nitrocellulose	Cast film clarity	Not sufficient evidence	115
Styrene/methyl methacrylate (Zerlon, Dow)–methyl methacrylate)	Cast film clarity	Not sufficient evidence	115
Styrene and α-methylstyrene containing co- and terpolymers miscible with PVC	Refer to Table 5.1		
Additional references			116–121

sulfonyl-1,4-phenyleneoxy(2,6-diisopropyl-1,4-phenylene)isopropylidene-(3,5-diisopropyl-1,4-phenylene)] blended with styrene/acrylonitrile copolymers (13 to 16 wt% AN). Miscibility of another sulfone-based poly(aryl ether), was noted with a styrene/acrylonitrile copolymer containing 13 wt% AN.

Poly[oxy(2,5-dimethyl-1,4-phenylene) sulfonyl (2,5-dimethyl-1,4-phenylene) oxy-1,4-phenyleneisopropylidene-1,4-phenylene]

Styrene/methyl methacrylate copolymers were investigated by Petersen et al. [115], in a series of polymer blends. Miscibility of the copolymer (Zerlon 150, Dow Chemical Co.) was claimed with nitrocellulose and poly(methyl methacrylate). However, it must be noted that these conclusions were based on solution homogeneity and film clarity and, therefore, must be judged as inconclusive evidence. Styrene and α-methylstyrene-based copolymers exhibiting miscibility with PVC have been discussed in Section 5.2.1.

The detailed listing of the polymers exhibiting miscibility with polystyrene and styrene copolymers is given in Table 5.2.

5.2.3 Cellulosics

The basic cellulose structure

as well as modified cellulosics contain high concentrations of hydroxyls quite amenable for specific interactions (i.e., hydrogen bonding). It is therefore not surprising that polymers containing the cellulose unit exhibit miscibility with various polymers. Petersen et al. [115] reported seven miscible polymers with nitrocellulose: polyester-based urethane, styrene/methyl methacrylate copolymer, styrene/acrylonitrile copolymer, poly(methyl methacrylate), poly(vinyl acetate), ethyl cellulose, and cellulose acetate propionate. These results were based on cast film clarity and microscopy and are judged as insufficient evidence of miscibility. While there is a reasonable probability that these blends are miscible, it should be noted that the same procedures showed poly(vinyl acetate) and poly(methyl methacrylate) blends to be miscible, although other investigators have found evidence to the contrary [122]. Dobry and Boyer-Kawenoki [3] listed nitrocellulose–poly(vinyl acetate) and nitrocellulose–poly(methyl methacrylate) blends as miscible, based on solution studies, thus agreeing with the results of Petersen et al. [115].

Brode and Koleske reported miscibility for the blends of nitrocellulose (12% nitrogen) and poly(ε-caprolactone) (PCL) over the entire composition range [123]. At higher levels of poly(ε-caprolactone), crystallization of PCL resulted in nonplasticized systems similar to results previously reported for PCL blends with PVC. Hubbell and Cooper noted miscibility of poly-(ε-caprolactone)–nitrocellulose above 50% PCL but noted phase separation at lower PCL concentrations [35]. Cellulose butyrate and cellulose propionate were also found to exhibit some level of miscibility with PCL by virtue of a broad transition observed in the dynamic mechanical testing [124, 125]. The miscibility of the polyester poly(ε-caprolactone) with several modified cellulosics may be the result of a specific interaction between the hydroxyl of the cellulosic (proton donor) and the carbonyl of poly(ε-caprolactone) (proton acceptor).

Nitrocellulose–poly(methyl acrylate) physical blends and graft copolymers exhibited similar single glass transition behavior as determined by thermomechanical testing [126]. Miscibility of nitrocellulose with poly-(methyl acrylate) as well as polymers based on ethyl acrylate, n-butyl

TABLE 5.3

Cellulosics

Miscible polymer	Method of miscibility determination	Comments	References
Nitrocellulose–polyester urethane	Cast film clarity; microscopy	Not sufficient evidence	115
Nitrocellulose–styrene/methyl methacrylate copolymer	See Table 5.2		
Nitrocellulose–styrene/acrylonitrile copolymer	Cast film clarity; microscopy	Not sufficient evidence	115
Nitrocellulose–poly(methyl methacrylate)	Cast film clarity; microscopy	Not sufficient evidence	3, 115
Nitrocellulose–poly(vinyl acetate)	Cast film clarity; microscopy	Not sufficient evidence	3, 115
Nitrocellulose–ethyl cellulose	Cast film clarity; microscopy	Not sufficient evidence	115
Nitrocellulose–cellulose acetate proprionate	Cast film clarity; microscopy	Not sufficient evidence	115
Nitrocellulose–poly(methyl acrylate)	Thermomechanical	Physical blends and graft exhibited similar T_g's	126
Nitrocellulose–poly(ε-caprolactone)	Dynamic mechanical	Miscible over entire composition range	123
Cellulose butyrate–poly(ε-caprolactone)	Dynamic mechanical	Blend glass transition indicates microheterogeneity	124, 125
Cellulose proprionate–poly(ε-caprolactone)	Dynamic mechanical	Blend glass transition indicates microheterogeneity	124, 125
Cellulose acetate butyrate–poly(ε-caprolactone)	Dynamic mechanical; calorimetric; small-angle light scattering; microscopy	Data indicate only limited mixing at intermediate concentrations; partially miscible	132
Nitrocellulose–poly(vinyl acetate)	Sorption isotherms	Blend isotherm below pure-component isotherms	128
Benzyl cellulose (46.5% benzyl)–polystyrene	Cast film clarity	Not sufficient evidence	114
Cellulose acetate butyrate–ethylene/vinyl acetate copolymer	Dynamic mechanical	Miscible at 80 wt% vinyl acetate in copolymer; immiscible at 70 wt%	73
Phosphorylated polystyrene or phosphorylated poly(2,6-dimethyl-1,4-phenylene oxide)–(cellulose acetate, cellulose acetate butyrate, cellulose butyrate, and nitrocellulose)	Solution and cast film behavior	Not sufficient evidence	129
			129
Additional references			133–136

acrylate, ethyl methacrylate, and *n*-butyl methacrylate was claimed by van Eijnsbergen, but without adequate documentation [1, 127].

Cellulose acetate butyrate miscibility with an ethylene/vinyl acetate copolymer (80% vinyl acetate) was claimed by Casper and Morbitzer [73] based on dynamic mechanical data. Two-phase blends were, however, observed with a copolymer containing 70% vinyl acetate.

Tager [128] investigated the sorption isotherms for two systems [nitrocellulose and poly(vinyl acetate), stated to be miscible, and cellulose acetate and nitrocellulose, stated to be immiscible]. The sorption isotherms for the immiscible blend were between the pure-component polymer isotherms. With the miscible blend, the sorption isotherms were lower than both purecomponent isotherms. This was interpreted as the prevention of the solvent probe interaction with the miscible blend due to polymer–polymer interactions (thus lower sorption isotherms).

Phosphorylated polystyrene and phosphorylated poly(2,6-dimethyl-1,4-phenylene oxide) were claimed to be miscible with a series of cellulose derivatives including cellulose acetate, cellulose acetate butyrate, cellulose butyrate, and nitrocellulose [129]. Further data would be required to verify these observations, which were based on solution studies and cast film clarity.

Polymer modification of nitrocellulose [130] and cellulose acetate butyrate [131] to improve film properties has been a commercial practice for years. Little definitive data relating to miscibility have been reported. Typical polymeric (or oligometic) modifications include alkyl resins, phenol/formaldehyde, urea/formaldehyde, estergum, arylsulfonamide/formaldehyde condensate, and natural resins including dammar, elemi, mastic, pontianak, and shellac.

Table 5.3 gives the detailed listing of polymers exhibiting miscibility with cellulosics.

5.2.4 Polyacrylates

The most significant commercial polyacrylate is poly(methyl methacrylate) (PMMA), and therefore the literature references concerning polymer blends are more frequent with this member of the family of polyacrylates. In previous sections, PMMA–styrene/acrylonitrile copolymers, syndiotactic PMMA–poly(vinyl chloride) and PMMA–nitrocellulose blends have been cited as miscible and will not be discussed here.

Several workers have investigated the mutual miscibility of members of the polyacrylate family [137, 138]. Hughes and Britt [137] noted general immiscibility of members of this family, which included PMMA–poly(methyl acrylate) and PMMA–poly(ethyl acrylate). This illustrates the expectation of polymer-polymer blend immiscibility in the absence of specific interactions, even when the structural variations are slight.

The blend of PMMA and poly(vinylidene fluoride) (PVF_2) has received the most technical interest of the miscible polymer blends containing PMMA. Many authors have noted that blends containing more than 50% PVF_2 were observed to exhibit PVF_2 crystallinity [139–142]. Noland et al. [140] showed that the T_g data for the blends were consistent with extrapolation to a T_g of $-46°C$ for pure PVF_2. Paul et al. [141] pointed out that uncertainty exists as to the T_g of PVF_2 and indicated that the blend data were not conclusive regarding extrapolation to pure PVF_2. Nishi and Wang [139] observed a melting point depression and depressed crystallization temperatures with the addition of PMMA to PVF_2. Their T_g data extrapolation was similar to that reported by Noland, yielding a value of $-50°C$ for pure PVF_2. Patterson et al. [142] employed Brillouin scattering to study the thermal behavior of the PMMA–PVF_2 blend and concluded that the blend was miscible due to the existence of a single pair of fully polarized Brillouin peaks with splitting intermediate between the constituents. Poly(ethyl methacrylate) (PEMA) also exhibits miscibility with PVF_2 [140, 143, 144]. Kwei et al. [143] noted that crystallization of PVF_2 occurs even with rapid quenching at low PEMA content, with two resultant amorphous phases containing 100 and 45% PVF_2. Extrapolation of high PEMA content blend T_g's yielded a value of $-50°C$ for the T_g of amorphous PVF_2. Imken et al. [144], noting the complex behavior of the transition data of this blend, postulated PVF_2 crystallization as the contributing factor. They extrapolated the T_g data on blends (using DTA data) to $38°C$ in the limit of pure PVF_2. They noted that the dynamic mechanical (110 Hz) tan δ peak (α-transition of the blend) yielded a uniform curve extrapolating to $60°C$ in the limit of pure PVF_2.

Kwei et al. [145, 146] investigated ternary blends of PMMA–PEMA–PVF_2 and observed miscibility, whereas the binary blends of PMMA–PEMA were phase-separated. Crystallization kinetics of PVF_2 in PMMA–PVF_2 blends were reported by Wang and Nishi [147]. The rapid decrease in crystallization rate with increasing PMMA content was attributed to the T_g increase, melting point depression, and presence of depletion layers at spherulitic growth fronts. Poly(methyl acrylate) and poly(ethyl acrylate) have been shown to exhibit miscibility with PVF_2, and both blends exhibit lcst behavior [148–150]. While all authors agree on the miscibility of PVF_2 with PMMA and PEMA, certain differences exist in the interpretation of the results.

Isotactic PMMA exhibits a noticeably lower T_g ($45°C$) than syndiotactic PMMA ($120°C$). Krause and Roman [151] observed that the dilatometric T_g determination of mixtures of these stereo-isomers yielded T_g's intermediate

5.2. Referenced Listing

between T_g's of the pure isomers. The data, however, did not adequately fit the predictions of the Fox, Gordon and Taylor, or Kanig equations. Bauer and Bletso [152] investigated blends of isotactic PMMA and atactic PMMA and observed two T_g's from dilatometric data, indicating two-phase behavior.

Mixtures of PMMA and poly(vinyl acetate) prepared using common techniques (i.e., solution casting or melt blending) yield two-phase systems [122, 153]. Freeze-drying of benzene solutions of the above blend gave single glass transition temperatures as determined by differential scanning calorimetry [153]. The T_g–composition data fit the Gordon–Taylor equation perfectly. Heating the miscible freeze-dried blends to above the T_g's of both blend components resulted in two-phase behavior when specific heat determinations were repeated.

Miscibility of poly(methyl acrylate) with nitrocellulose has been cited previously in the section on cellulosics. A review by Krause [1] illustrated the immiscibility of poly(methyl acrylate) with various members of the acrylate family, including the polyacrylates based on ethyl acrylate, n-butyl acrylate, n-butyl methacrylate, and methyl methacrylate.

Poly(methyl acrylate) has been reported to be miscible with poly(vinyl acetate) by Kern and Slocombe [104]. Hughes and Britt [137] studied this blend prepared via three different mixing techniques: (1) mixing preformed polymer emulsions, (2) solution of polymers in a mutual solvent, and (3) emulsion polymerization of one polymer in the presence of the other preformed polymer. Only the solution cast film yielded behavior expected of a single-phase blend.

Miscible blends of poly(isopropyl acrylate) and poly(isopropyl methacrylate) were determined based on dilatometric data [151]. Blend T_g's were intermediate between the constituents [$-5°$C, poly(isopropyl acrylate); 78°C, poly(isopropyl methacrylate)] and were equivalent to those determined on copolymers of the same composition.

Stereoassociation complexes have been proposed for mixtures of isotactic PMMA and syndiotactic PMMA, based on the observation that dilute solutions of these polymers in various solvents resulted in immediate gelation [154]. Liquori [155] proposed that this complex was formed by the positioning of extended syndiotactic PMMA chains into the grooves of helices of the isotactic molecules.

Using high-resolution nmr spectra of solution mixtures of s-PMMA–i-PMMA, the minimum length of associated syndiotactic sequences was found to be about 10 in aromatic solvents and about 3 in CCl_4 [156]. Viscosity studies of the interaction between PMMA and poly(methacrylic acid) indicated that complexation occurs between isotactic PMMA and syndiotactic poly(methacrylic acid) but not between syndiotactic PMMA and isotactic poly(methacrylic acid) [157].

TABLE 5.4

Polyacrylates

Miscible polymer	Method of miscibility determination	Comments	References
Syndiotactic poly(methyl methacrylate)–isotactic poly(methyl methacrylate)	Dilatometric		151
Syndiotactic poly(methyl methacrylate)–isotactic poly(methyl methacrylate	Precipitation from solution	Complex formed	154, 155
Syndiotactic poly(methyl methacrylate)–isotactic poly(methyl methacrylate)	High resolution nmr	Sequence length determined	156, 159
Isotactic poly(methyl methacrylate)–syndiotactic poly(methacrylic acid)	Viscosity	Maximum in solution viscosity at 50/50 molar ratio; time dependent	157
Poly(methyl methacrylate)–poly(vinylidene fluoride)	Dynamic mechanical	PVF_2 crystallization above 50 wt% PVF_2; specific volumes not additive	141
Poly(methyl methacrylate)–poly(vinylidene fluoride)	Dilatometric; differential thermal analysis	T_g behavior agreed with modified Kelly–Bueche equation	140
Poly(methyl methacrylate)–poly(vinylidene fluoride)	Differential scanning calorimeter; melting point depression	Negative interaction parameter determined	139
Poly(methyl methacrylate)–poly(vinylidene fluoride)	Brillouin scattering	Results agreed with one-phase system model	142
Poly(methyl methacrylate)–poly(vinylidene fluoride)	Calorimetric	Atactic, isotactic, and syndiotactic PMMA all miscible with PVF_2	160
Poly(ethyl methacrylate)–poly(vinylidene fluoride)	Differential thermal analysis; Thermomechanical analysis	T_g behavior agreed with modified Kelly–Bueche equation	140
Poly(ethyl methacrylate)–poly(vinylidene fluoride)	Dynamic mechanical; calorimetric	Complex T_g behavior observed	144
Poly(ethyl methacrylate)–poly(vinylidene fluoride)	Calorimetric	Proposed a two-phase system at high PVF_2 content	143
Poly(methyl methacrylate)–poly(ethyl methacrylate)–poly(vinylidene fluoride)	Calorimetric	Single-phase behavior in ternary blends; PEMA–PMMA blends are two phase	145, 146

5.2. Referenced Listing

TABLE 5.4 (*Continued*)

Miscible polymer	Method of miscibility determination	Comments	References
Poly(methyl methacrylate)–poly(vinyl acetate) (freeze-dried)	Calorimetric	Two-phase behavior when samples were heated above T_g's of both components	122, 153
Poly(methyl methacrylate)–methyl vinyl ether/maleic anhydride copolymer	Cast film clarity	Not sufficient evidence	115
Poly(methyl acrylate)–poly(vinyl acetate)	Solution miscibility	Not sufficient evidence	104
Poly(methyl acrylate)–poly(vinyl acetate)	Torsional modulus data	Single transitions for solution cast films	137
Poly(methyl acrylate)–poly(vinylidene fluoride)	Dynamic mechanical	lcst behavior noted	148, 149, 150
Poly(ethyl acrylate)–poly(vinylidene fluoride)	Dynamic mechanical	lcst behavior noted	148, 149, 150
Poly(isopropyl acrylate)–poly(isopropyl methacrylate)	Dilatometric	Blends gave same T_g as random copolymers	151
Poly(*n*-propyl acrylate)–poly(vinyl butyrate)	Clear films	Not sufficient evidence	158
Poly(methyl methacrylate)–styrene/ acrylonitrile copolymers	See Section 5.2.2		
Syndiotactic poly(methyl methacrylate)–poly(vinyl chloride)	See Section 5.2.1		
Various polyacrylates–nitrocellulose	See Section 5.2.3		
Additional references			161–165

Poly(methyl methacrylate) and methyl vinyl ether/maleic anhydride copolymer blends were claimed to be miscible, based on solution behavior and cast film clarity [115]. Miscible blends of poly(vinyl butyrate) and poly(*n*-propyl acrylate) were claimed in a U.S. patent of Kern [158]. Miscible blends of polyacrylates with PVC or styrene copolymers have been previously discussed in Sections 5.2.1 and 5.2.2.

Table 5.4 gives the detailed listing of polymers exhibiting miscibility with polyacrylates.

TABLE 5.5

Poly(vinyl acetate) and Vinyl Acetate Copolymers

Miscible polymer	Method of miscibility determination	Comments	References
Poly(vinyl acetate)–poly(vinyl nitrate)	Thermo-dielectric loss		166
Ethylene/vinyl acetate copolymer (86% vinyl acetate)–poly(vinyl nitrate)	Thermo-dielectric loss	Not as homogeneous as poly(vinyl acetate)–(vinyl nitrate) blend	166
Poly(vinyl acetate)–methyl vinyl ether/maleic anhydride copolymer	Cast film clarity	Not sufficient evidence	115
Poly(vinyl acetate)–vinyl chloride/vinyl acetate (90/10) copolymer	Mechanical loss	Only specific compositions miscible	170
Poly(vinyl acetate)–poly(vinylidene fluoride)	Dynamic mechanical; thermal analysis	No lcst behavior up to 350°C observed	171, 172
Poly(vinyl acetate)–nitrocellulose	See Section 5.2.3		
Poly(vinyl acetate)–poly(methyl acrylate)	See Section 5.2.4		
Poly(vinyl acetate)–poly(methyl methacrylate) freeze-dried blends	See Section 5.2.4		
Ethylene/vinyl acetate copolymer–poly(vinyl chloride)	See Section 5.2.1		
Ethylene/vinyl acetate/carbon monoxide terpolymer–poly(vinyl chloride)	See Section 5.2.1		
Ethylene/vinyl acetate/sulfur dioxide terpolymer–poly(vinyl chloride)	See Section 5.2.1		
Ethylene/vinyl acetate copolymer–cellulose acetate butyrate	See Section 5.2.3		

5.2.5 Poly(vinyl acetate) and Vinyl Acetate Copolymers

Poly(vinyl acetate)–poly(vinyl nitrate) and ethylene/vinyl acetate (86% vinyl acetate)–poly(vinyl nitrate) blends were characterized using the thermodielectric loss method and found to be miscible [166]. This technique was

5.2. Referenced Listing

capable of resolving differences in homogeneity with these blends, with the conclusion that the poly(vinyl acetate)–poly(vinyl nitrate) blend was more homogeneous. Other techniques have also been used to establish miscibility for these two blends [167–169].

Poly(vinyl acetate) and vinyl acetate/vinyl chloride copolymer (10/90) blends were claimed to be miscible with certain compositions [20/80 and 60/40 poly(vinyl acetate)–vinyl acetate/vinyl chloride copolymer] as determined by mechanical loss peaks, whereas 50/50 and 40/60 mixtures exhibited resolution of separate peaks [170]. The separate peaks were, however, shifted in the direction of the T_g of the other component, indicating limited phase mixing.

Petersen [115] noted cast film clarity with a blend of poly(vinyl acetate) and methyl vinyl ether/maleic anhydride copolymer. Poly(vinyl acetate) miscibility with nitrocellulose, poly(methyl acrylate), and freeze-dried blends with poly(methyl methacrylate) has been cited in previous sections. Ethylene/vinyl acetate copolymers and terpolymers with carbon monoxide or sulfur dioxide were also previously cited as miscible with poly(vinyl chloride).

Paul et al. observed miscibility of poly(vinyl acetate) with poly(vinylidene fluoride) based on differential thermal analysis and dynamic mechanical testing [171, 172]. Heating these blends up to 350°C did not reveal any evidence of lcst behavior.

Table 5.5 gives the list of polymers that exhibit miscibility with poly(vinyl acetate) and vinyl acetate copolymers.

5.2.6 Polyesters and Polycarbonates

In the family of polyesters, poly(ε-caprolactone) exhibits a unique situation among the many polymers reviewed in this chapter in that so many diverse miscible systems containing this polymer have been observed. These include poly(vinyl chloride), styrene/acrylonitrile copolymers, nitrocellulose, cellulose propionate, and cellulose butyrate, which have been discussed in earlier sections of this chapter.

The polyhydroxyether from the condensation of bisphenol A and epichlorohydrin (Phenoxy, Union Carbide), was found to be miscible with

Phenoxy

poly(ε-caprolactone) by Brode and Koleske [123]. Dynamic mechanical measurements yielded T_g data for this blend which gave an excellent fit with the Fox equation. Extrapolation of the T_g data to 100% amorphous poly(ε-caprolactone) yielded $-71°C$, in excellent agreement with the extrapolation of poly(vinyl chloride)–poly(ε-caprolactone) T_g data. Blends of Phenoxy–poly(ε-caprolactone) containing >50% PCL were reported to crystallize at room temperature due to the blend T_g being lower than this temperature.

Blends of polyepichlorohydrin were also shown to miscible with poly(ε-caprolactone) [123]. As polyepichlorohydrin has a T_g below room temperature, crystallization of poly(ε-caprolactone) from this blend is much more rapid at the 50% level than observed with similar concentrations of poly(ε-caprolactone) in blends with either poly(vinyl chloride) or Phenoxy. Miscibility of poly(ε-caprolactone) with a chlorinated polyether (Penton, Hercules, Inc.) has also been demonstrated [135].

Calorimetric and dynamic mechanical data indicated single glass transition values for blends of poly(ε-caprolactone) and the polycarbonate of bisphenol A [173]. Lower critical solution temperature behavior was noted, and polycarbonate crystallization at high PCL levels occurred due to the increased mobility of polycarbonate resulting from the greatly reduced T_g.

Poly(butylene terephthalate), poly(ethylene terephthalate), and poly(1,4-cyclohexylene dimethylene tere/isophthalate) have also been shown to be miscible with Phenoxy [74, 174]. The crystallization rate of PBT was shown to be noticeably depressed in the case of the Phenoxy–PBT blends; however, the degree of crystallinity of PBT was found to be higher in the blends than for unblended PBT. The melting point depression was slight ($1°–2°C$), indicative of a very small negative interaction parameter. Blends of Phenoxy with either poly(butylene terephthalate) or poly(ethylene terephthalate) were shown to cross-link due to transesterification reactions when held at elevated temperatures for moderate time periods (i.e., 250°C for 30 min). The miscibility of Phenoxy with poly(ε-caprolactone), poly(butylene terephthalate), poly(ethylene terephthalate), or poly(1,4-cyclohexylene dimethylene tere-/isophthalate) is believed to be the result of specific interactions between the hydroxyl of Phenoxy (proton donor) and the carbonyl of the polyesters (proton acceptor).

Blends of poly(ethylene terephthalate) (PET) and the polycarbonate of bisphenol A exhibited single T_g's up to 50% polycarbonate [173]. Above 50% polycarbonate, phase separation occurred with T_g data indicating a polycarbonate-rich phase and a PET-rich phase. Poly(1,4-cyclohexylene dimethylene tere-/isophthalate) blends with polycarbonate were shown to be miscible over the entire composition range [173]. Only limited mixing of poly(butylene terephthalate) and polycarbonate was observed based on minimum shifting of the T_g's from the respective unblended values [173]. Two crystallization exotherms for PBT were interpreted as PBT crystallization from separate phases [175].

5.2. Referenced Listing

TABLE 5.6

Polyesters and Polycarbonates

Miscible polymer	Method of miscibility determination	Comments	References
Poly(ε-caprolactone)–poly(vinyl chloride)	See Section 5.2.1		
Poly(ε-caprolactone)–styrene/acrylonitrile (28% AN)	See Section 5.2.2		
Poly(ε-caprolactone)–nitrocellulose	See Section 5.2.3		
Poly(ε-caprolactone)–cellulose proprionate	See Section 5.2.3		
Poly(ε-caprolactone)–cellulose butyrate	See Section 5.2.3		
Poly(ε-caprolactone)–polyhydroxy ether of bisphenol A (phenoxy)	Dynamic mechanical	T_g data satisfied the Fox equation	123
Poly(ε-caprolactone)–polyepichlorohydrin	Dynamic mechanical		123
Poly(ε-caprolactone)–chlorinated polyether (Penton, Hercules, Inc.)	Dynamic mechanical	Single T_g over entire composition range	135
Poly(ε-caprolactone)–polycarbonate of bisphenol A	Dynamic mechanical	Polycarbonate crystallizes at high PCL levels	173
Poly(butylene terephthalate)–Phenoxy	Dynamic mechanical	Depressed crystallization rate; blend cross-linked at elevated temperatures	74, 174
Poly(ethylene terephthalate)–Phenoxy	Dynamic mechanical	Blend cross-linked at elevated temperatures	74
Poly(1,4-cyclohexylene dimethylene tere-/isophthalate)–Phenoxy	Modulus and resilience–temperature data		74
Poly(ethylene terephthalate)–polycarbonate of bisphenol A	Dynamic mechanical	Phase separation above 50% PET	173, 179
Poly(1,4-cyclohexylene dimethylene tere/isophthalate)–polycarbonate of bisphenol A	Dynamic mechanical	Miscible at all concentrations	173, 180
Poly(phenyliminodiethanol-isophthalate)–poly(1,6-hexane diol-5-nitroisophthalate)	Modulus and resilience–temperature data	Pure components were colorless; mixture was yellow	176

TABLE 5.6 (*Continued*)

Miscible polymer	Method of miscibility determination	Comments	References
Poly(*p*-anisyliminodiethanol-isophthalate)–poly(bisphenol A-5-nitroisophthalate)	Modulus and resilience–temperature data	Pure components were colorless; mixture was yellow–orange	176
Poly(*p*-anisyliminodiethanol-bisphenol A carbonate)–poly[bis-(2-hydroxyethyl)-5-nitro-isophthalate-bisphenol A carbonate]	Modulus and resilience–temperature data	Pure components were colorless; mixture was yellow–orange	176
Poly(ethylene adipate)–poly(vinyl acetate)	Melting point depression	More data needed for conclusive evidence	178
Poly(ethylene adipate)–poly(methyl methacrylate)	Melting point depression	More data needed for conclusive evidence	178
Poly(ethylene sebacinate)–poly(methyl methacrylate)	Melting point depression	More data needed for conclusive evidence	178
Poly(butylene terephthalate)–poly(vinyl chloride)	See Section 5.2.1		
Poly(butylene terephthalate)–polycarbonate of bisphenol A	Dynamic mechanical; calorimetric	Only limited miscibility based on shifted component T_g's	175

Mixtures of electron acceptor polyesters prepared from 5-nitroisophthalic, nitroterephthalic, and 4,6-dinitroisophthalic acids were blended by Sulzberg and Cotter [176] with electron donor polymers based on aryliminodiethanols where the aryl group was phenyl, *p*-anisyl, 2,5-dimethoxyphenyl, or 3,4,5-trimethoxyphenyl. Several of these blends were shown to exhibit miscibility due to charge-transfer complexation. These blends were

(1) poly(phenyliminodiethanol isophthalate)–poly(1,6-hexane diol-5-nitroisophthalate)
(2) poly(*p*-anisyliminodiethanol isophthalate)–poly(bisphenol A-5-nitroisophthalate)
(3) poly(*p*-anisyliminodiethanol-bisphenol A carbonate)–poly[bis(2-hydroxyethyl)-5-nitroisophthalate-bisphenol A carbonate]

Several interesting aspects of these blends were noted, particularly the situation whereby the individual unblended components were colorless and the blends were yellow to yellow–orange. Unique absorption maxima in the

5.2. Referenced Listing

visible spectra significantly different than component values were observed for the blends. Blends also exhibited higher electrical conductivity than the components, although a polymer containing both donor and acceptor groups had the highest conductivity.

Poly(ethylene terephthalate)–poly(butylene terephthalate) blends yielded single T_g values (calorimetric determination) between the component values [177]. Melting point depression of each component with increasing concentration of the second component was observed.

Miscible mixtures of poly(ethylene adipate)–poly(vinyl acetate), poly(ethylene adipate)–poly(methyl methacrylate), and poly(ethylene sebacinate)–poly(methyl methacrylate) were claimed by Natov *et al.* [178], based on melting point depression data and microscopy studies. Further data would be necessary to unambiguously conclude miscibility for these blends.

Table 5.6 lists the polymers exhibiting miscibility with polyesters and polycarbonates.

5.2.7 Polyethers

Poly(ethylene oxide) is an interesting member of the poly(alkylene oxides) in that it is the only one exhibiting water solubility. The most prominent case of polymer miscibility with poly(ethylene oxide) involves poly(acrylic acid). Although both polymers exhibit complete water solubility, when solutions of each in water are mixed, a precipitate immediately appears due to the complexation of these two polymers, thus providing an excellent example of a strong specific interaction. It is also an excellent example of how solution studies can be misleading in cases where polymer–polymer or polymer–solvent interactions are strong, because the precipitation of this complex might be easily interpreted as immiscibility. Water-free mixtures of poly(ethylene oxide) and poly(acrylic acid) have single glass temperatures intermediate between the constituents, although the T_g-composition data are quite complex [181]. In fact, low concentrations of poly(ethylene oxide) in poly(acrylic acid) actually result in T_g's higher than that observed with unblended poly(acrylic acid). This is additional evidence of the high level of specific interaction. This blend is a primary example of water-soluble polymers with acidic properties miscible with water-soluble polymers with basic properties. There are many other examples, which will be covered in a later section dealing exclusively with water-soluble polymers.

Poly(ethylene oxide) has also been claimed to exhibit complexation with polyureas, poly(vinyl methyl ether/maleic anhydride) copolymer, carboxymethyl cellulose (sodium salt), carboxymethyl dextran, and phenolic resins [182, 183]. In the cases cited, poly(ethylene oxide) is the proton acceptor and the other miscible polymers listed are the proton donors for the expected

TABLE 5.7

Polyethers

Miscible polymer	Method of miscibility determination	Comments	References
Poly(ethylene oxide)–poly(acrylic acid)	Modulus–temperature data	T_g of blend at low poly(ethylene oxide) level higher than either component	181
Poly(ethylene oxide)–poly(methacrylic acid)	Hydrodynamic data	Complexation observed	185a
Poly(ethylene oxide)–polyureas	Lowered water solubility; clear films	Property data imply complexation	182
Poly(ethylene oxide)–poly(vinyl methyl ether/maleic anhydride)	Complexation observed	Microencapsulation utility suggested	186
Poly(ethylene oxide)–carboxymethyl cellulose sodium salt	Complexation observed	Water solutions precipitate at low pH; resultant films are water insoluble	182
Poly(ethylene oxide)–carboxymethyl dextran	Complexation observed	Water solutions precipitate at low pH; resultant films are water insoluble	182
Poly(ethylene oxide)–phenolic resins	Clear flexible films at high phenolic content	More data needed to clearly establish miscibility	182
Poly(vinyl methyl ether)–polystyrene	See Section 5.2.2		
Poly(2,6-dimethyl-1,4-phenylene oxide)–polystyrene	See Section 5.2.2		
Poly(hydroxy ether) of bisphenol A (Phenoxy)–polyesters	See Section 5.2.7		
Poly(2,6-dimethyl-1,4-phenylene oxide)–poly(2-methyl-6-benzyl-1,4-phenylene oxide) or poly(2-methyl-6-phenyl-1,4-phenylene oxide)	Thermal optical analysis		184, 185

specific interactions that result. The addition of the above-cited polymers to poly(ethylene oxide) yields clear, homogeneous films with greatly reduced water solubility. The water insolubility of these blends is quite sensitive to pH and, generally, a neutral to acidic pH is required to prevent water solubility.

5.2. Referenced Listing

Poly(2,6-dimethyl-1,4-phenylene oxide) miscibility with poly(2-methyl-6-phenyl-1,4-phenylene oxide) [184] and poly(2-methyl-6-benzyl-1,4-phenylene oxide) [185] has been noted using thermo-optical analysis.

Poly(vinyl methyl ether) and polystyrene miscibility has been cited in a previous section (5.2.2). The polyhydroxyether of bisphenol A (Phenoxy) miscibility with a series of polyesters was also discussed in a previous section (5.2.6). Poly(2,6-dimethyl-1,4-phenylene oxide) and variations of that structure have been shown to be miscible with polystyrene and styrene copolymers, as discussed in Section 5.2.2.

Table 5.7 lists the polymers that exhibit miscibility with polyethers.

5.2.8 Polyamides

Polyamides, such as the nylon series, contain the amide group, which is capable of specific interactions. The basic properties of the amide group (proton acceptor) suggests its capability of bonding with polymers containing acidic (proton donor) segments. In spite of this potential, miscible polymer systems involving polyamides cited in the literature are quite rare.

Ethylene/acrylic acid copolymers ($\sim 14\%$ acrylic acid) and nylon 6 were found to exhibit an intermediate, broad T_g by Matzner *et al.* [187]. Cationic polymerization of nylon 6 onto ethylene/acrylic acid copolymers was also investigated, and transition data determined by dynamic mechanical methods yielded a single, sharp glass transition temperature. Permeability data of the graft copolymer as a function of composition also indicated single-phase behavior.

Blends of poly(vinyl pyrrolidone) and a macrocyclic polyether–polyamide (crown ether), exhibited single-phase behavior in thermomechanical,

Macrocyclic polyether–polyamide (crown ether)

calorimetry, and microscopy studies [188]. The water-soluble poly(vinyl pyrrolidone) was not leached out of the blends after prolonged water storage. The thermal stability of the blends was found to be higher than the unblended constituents.

Blends of nylon 6,6 with the polyamide from *m*-phenylene diamine and adipic acid or isophthalamide were found to exhibit clarity and single T_g behavior [189]. The experimental T_g values gave a positive deviation over that expected from a linear extrapolation of pure component values. Positive deviation of the modulus of the blend was also observed. It was noted that a certain degree of amide interchange was possible, thus yielding block copolymer formation. This reaction would greatly improve the miscibility; therefore, attempts were made to minimize the amide interchange rate.

Ethylene/N,N-dimethyl acrylamide copolymer miscibility with PVC has been discussed in Section 5.2.1. Other ethylene copolymers containing amide groups have been shown by McGrath and Matzner to yield transparent blends, indicating potential miscibility [76].

Table 5.8 lists the polymers that exhibit miscibility with polyamides.

TABLE 5.8

Polyamides

Miscible polymer	Method of miscibility determination	Comments	References
Nylon 6-ethylene/acrylic acid (14% acrylic acid) copolymers	Dynamic mechanical	Graft copolymer yielded sharper transition than simple mixture	187
Macrocyclic polyether–polyamide (see text)–poly(vinyl pyrrolidone)	Thermomechanical; microscopy; calorimetry	Blends were more thermally stable than pure components	188
Nylon 6,6–polyamide from m-phenylene diamine and adipic acid	Dynamic mechanical	Blend T_g gave a positive deviation over linear extrapolation	189
Nylon 6,6–polyamide from isophthalamide	Dynamic mechanical	Blend T_g gave a positive deviation over linear extrapolation	189
Ethylene/N,N-dimethyl acrylamide–poly(vinyl chloride)	See Section 5.2.1		
Additional references			190

5.2.9 Water-Soluble Polymers and Polyelectrolyte Complexes

Polymers which exhibit water solubility generally have functional groups with high polarity and sites for specific interactions with other molecular species with complementary structures. These characteristics also lead to many miscible polymer pairs from members of this special class of polymers and, therefore, a specific section of this review has been reserved for these polymers. The primary example of this behavior was discussed in the previous section on polyethers (5.2.7) concerning poly(ethylene oxide) and poly(acrylic acid). Upon mixing the water solutions of these two polymers, a gelatinous material precipitates that exhibits miscibility in the dry state. Many other polymer–polymer complexes exist where both polymers in the unmixed state are water soluble but mixtures are water insoluble. These examples are discussed in the remainder of this section.

Poly(vinyl pyrrolidone), a highly water-soluble polymer, exhibits base-like properties due to the substituted amide group. Complexation with tannic acid, poly(acrylic acid), and methyl vinyl ether/maleic anhydride copolymers is claimed [191]. As is typical with the other polymers discussed in this section, insoluble complexes are formed when water solutions of poly(vinyl pyrrolidone) are mixed with water solutions of the complexing polymers. Solubilization of these complexes is only achieved at high pH. This ability to complex with other substances has led to applications including an iodine complex (increases solubility of iodine in water), calamine lotion (complex with phenolic-type irritants such as the constituents of poison ivy), a retardant vehicle and suspending agent for specific pharmaceutical products, and dye complexation in fibers. Poly(vinyl pyrrolidone)–poly(methacrylic acid) complexes yield minimum electrical conductivity in solution when the polymeric components have a ratio of 1 (i.e., equimolar complex) [192].

Poly(ethylene imine) $[(CH_2CH_2NH)_n]$ has properties of a weak base in aqueous solutions. Due to extensive branching, the polymer has primary, secondary, and tertiary amine groups. The polymer is more basic than ammonia, with the primary amine groups exhibiting the highest level of basicity. This is the reverse of typical monomeric amines and is ascribed to steric hindrance of the secondary and tertiary amines [191]. Because of the cationic nature of poly(ethylene imine), a high affinity for anionic substances is observed. Sato and Nakajima studied the complex of poly(ethylene imine) and carboxymethyl cellulose as a function of pH. The experimental data from conductometric and potentiometric titrations and turbidity led to the conclusions that the complex did not obey stoichiometry as various cases of polyelectrolyte complexes cited by the authors [193]. Physical adsorption of poly(ethylene imine) on monocarboxy cellulose was observed to yield an adsorption isotherm complying with Langmuir's equation [194].

Zezin et al. [195] studied the formation of amide links in polyelectrolyte complexes of poly(ethylene imine) and poly(acrylic acid) or poly(methacrylic acid). Heating above 150°C produced amide linkages yielding a "ladder" polymer. The high degree of amidation occurring after heat treatment was attributed to the good steric conformation of the poly(acrylic acid)–poly(ethylene imine) complex.

Aleksina et al. [196] showed that poly(methacrylic acid)–poly(L-lysine) and poly(methacrylic acid)–poly(N,N-dimethylaminoethyl methacrylate) polyelectrolyte complexes could be obtained either by mixing water solution of the components or by polymerization of one of the components in the presence of the other polymer. An effective method for the separation of these complexes by the addition of polyethylene oxide was demonstrated. The poly(methacrylic acid)–poly(ethylene oxide) complex is formed in quantitative yield and is substantially free of the base polymers cited above.

Complex formation has also been noted for poly(acrylic acid)–poly(N,N-dimethylaminoethyl methacrylate), poly(acrylic acid)–poly(1,2-ethylene piperidine), poly(acrylic acid)–poly(ethylene piperazine) [197], poly(ethylene imine)–poly(L-glutanic acid) [198], and poly(methacrylic acid)–poly-(N,N,N',N'-tetraethyl-N-p-xylylene ethylene diammonium chloride) [199]. Zezin, in a review of polyelectrolyte complexes [200], cited these additional examples:

(1) poly(acrylic acid)–poly(L-lysine)
(2) poly(acrylic acid)–poly(4-vinyl-N-ethyl pyridinium bromide)
(3) poly(acrylic acid)–poly[4-vinyl-N-cetyl(n) ethyl($1-n$) pyridinium bromide]
(4) poly(acrylic acid)–poly(N,N-dimethyl-N-ethylaminoethyl methacrylate)
(5) poly(acrylic acid)–polybetaine
(6) poly(L-glutamic acid)–poly(N,N-dimethylaminoethyl methacrylate)
(7) poly(L-glutamic acid)–poly(vinyl amine)
(8) poly(methacrylic acid)–poly(L-lysine)
(9) poly(methacrylic acid)–poly(4-vinyl-N-ethyl pyridinium bromide)
(10) poly(vinyl pyridine)–poly(sodium styrene sulfonate)
(11) poly(N,N-dimethylaminoethyl methacrylate)–poly(sodium styrene sulfonate)

The complex formation of polyelectrolytes, of opposite charge in many cases, is similar to that of salt formation of microelectrolytes in that stoichiometry is observed. Cases where this is observed include partially sulfated poly(vinyl alcohol)–acetylated poly(vinyl alcohol) [201], sodium polystyrene sulfonate–poly(vinyl benzyltrimethylammonium chloride) [202], poly(4-vinyl-n-N-butyl pyridonium bromide)–sodium polyacrylate [203], and sulfated poly(vinyl alcohol)–aminoacetalyzed poly(vinyl alcohol) [204]. Sato [205] pointed out that stoichiometry need not always be the case with polyelectrolyte complexes, particularly if the chain flexibilities of the polyanion and polycation are widely different.

Michaels and co-workers studied the properties of poly(sodium styrene sulfonate)–poly(vinyl benzyltrimethylammonium chloride) complexes in detail [206–209]. These studies also included application studies for these complexes, and the commercial exploitation of their work was carried out at Amicon Corporation. In their study of the mechanical properties of this

5.2. Referenced Listing

TABLE 5.9

Water-Soluble Polymers and Polyelectrolyte Complexes

Miscible polymer	Method of miscibility determination	Comments	References
Poly(ethylene oxide)–poly(acrylic acid)	See Section 5.2.7		
Poly(ethylene oxide)–vinyl methyl ether/maleic anhydride copolymer	See Section 5.2.7		
Poly(ethylene oxide)–carboxymethyl cellulose sodium salt	See Section 5.2.7		
Poly(ethylene oxide)–carboxymethyl dextran	See Section 5.2.7		
Poly(ethylene oxide)–poly(methacrylic acid)	See Section 5.2.7		
Poly(vinyl pyrrolidone)–macrocyclic polyether–polyamide	See Section 5.2.8		
Poly(vinyl pyrrolidone)–poly(acrylic acid)	Complex formation		191
Poly(vinyl pyrrolidone)–methyl vinyl ether/maleic anhydride copolymer	Complex formation		191
Poly(vinyl pyrrolidone)–poly(methacrylic acid)	Complex formation	Equimolar complex	192
Poly(ethylene imine)–carboxymethyl cellulose	Complex formation; conductometric titrations	Nonstoichiometric complex	193
Poly(ethylene imine)–monocarboxycellulose	Physical adsorption	Data imply complex formation behavior	194
Poly(methacrylic acid)–poly(L-lysine)	Complex formation	Could be separated by adding poly(ethylene oxide)	196, 200
Poly(methacrylic acid)–poly(N,N-dimethylaminoethyl methacrylate)	Complex formation	Could be separated by adding poly(ethylene oxide)	196
Sulfated poly(vinyl alcohol)–acetylated poly(vinyl alcohol)	Complex formation	Equimolar complex	201
Poly(4-vinyl-n,N-butyl pyridonium bromide)–sodium polyacrylate	Complex formation	Equimolar complex	203

(*Continued*)

TABLE 5.9 (*Continued*)

Miscible polymer	Method of miscibility determination	Comments	References
Sulfated poly(vinyl alcohol)–aminoacetylized poly(vinyl alcohol)	Complex formation	Equimolar complex	204
Poly(sodium styrene sulfonate)–poly(vinyl benzyl trimethylammonium chloride)	Complex formation	Studied in detail by Michaels et al.	206–208
Poly(sodium styrene sulfonate)–poly(vinyl pyridinium bromide)	Complex formation	Only soluble in specific ternary solvent mixtures	210, 216
Poly(L-lysine)–poly(acrylic acid)	Complex formation	Complex structure studied using induced dichroism spectra	200, 201
Poly(L-lysine)–poly(methacrylic acid)	Complex formation	Dichroism spectroscopy used to study structure	202
Poly(L-lysine)–sodium cellulose sulfate	Complex formation	Dichroism spectroscopy used to study structure	213
Poly(riboadenylic acid)–poly(ribouridylic acid)	Complex formation; calorimetry	Heat of mixing determined; equimolar mixture yields double-stranded helix structure	214, 217
Poly(ethylene imine)–poly(L-glutamic acid)	Complex formation; potentiometric, turbidimetric, circular dichroism spectroscopy	Studied conformational changes of polypeptides in the complex	198, 218
Carboxymethyl cellulose–poly(L-lysine)	Complex formation; potentiometric, turbidimetric, circular dichroism spectroscopy	Studied conformational changes of polypeptides in the complex	218
Poly(ethylene imine)–sodium dextran sulfate	Complex formation	No solvents were found	219
Basic proteins(ribonuclease, γ-globulin, or lysozyme)–potassium hyaluronate	Conductometric titrations; turbidometry		215, 220
Poly(ethylene imine)–poly(acrylic acid)	Complex formation	Amide formation upon heating	195
Poly(ethylene imine)–poly(methacrylic acid)	Complex formation	Amide formation upon heating	195
Poly(acrylic acid)–poly(N,N-dimethylaminoethyl methacrylate)	Complex formation		197
Poly(acrylic acid)–poly(1,2-ethylene piperadine)	Complex formation		197

5.2. Referenced Listing

TABLE 5.9 (*Continued*)

Miscible polymer	Method of miscibility determination	Comments	References
Poly(acrylic acid)–poly(ethylene piperazine)	Complex formation		197
Poly(methacrylic acid)–poly(N,N,N',N'-tetraethyl-N-p-xylylene ethylene diammonium chloride)	Complex formation		199
Poly(acrylic acid)–polyvinyl-N-ethyl pyridinium bromide)	Complex formation		200
Poly(acrylic acid)–poly[4-vinyl-N-acetyl(n)-ethyl(1 − n) pyridinium bromide]	Complex formation		200
Poly(acrylic acid)–poly(N,N-dimethyl-N-ethylaminoethyl methacrylate)	Complex formation		200
Poly(acrylic acid)–polybetaine	Complex formation		200
Poly(L-glutamic acid)–poly(N,N-dimethylaminoethyl methacrylate)	Complex formation		200
Poly(L-glutamic acid)–poly(vinyl amine)	Complex formation		200
Poly(methacrylic acid)–poly(4-vinyl-N-ethyl pyridinium bromide)	Complex formation		200
Poly(vinyl pyridine)–poly(sodium styrene sulfonate)	Complex formation		200
Poly(N,N-dimethylaminoethyl methacrylate)–poly(sodium styrene sulfonate)	Complex formation		200
Poly(ethylene imine)–pectic acid	Complex formation		221
Poly(methylacrylic acid)–poly(vinyl alcohol)	Complex formation		222, 223
Egg albumen–protamine	Complex formation	First observation of polyelectrolyte complexes	224
Additional references			225–232

complex [209], stress relaxation experiments demonstrated the validity of time–temperature superposition; however, time–concentration superposition was not possible. They concluded from their data that a different structure exists in salt-free films versus that in films containing salt. They proposed a three-region model for this complex: (1) a region containing strong polyanion–polycation interactions with low water concentration, (2) a region containing weak polyanion–polycation interactions with higher water concentrations, and (3) a region with negligible ionic interactions containing water and imbibed salt.

Lysaght [210] listed typical commercial polyanions and polycations in his review of the technology of polyelectrolyte complexes. These were

    Polyanion: sodium poly(styrene sulfonate)
                   poly(vinyl sulfonate)
                   polytak RMA
    Polycation: poly(vinyl benzyltrimethylammonium chloride)
                   poly(vinyl pyridinium bromide)
                   poly(diallyldimethylammonium chloride)

The interactions between poly(L-lysine) and acidic polyelectrolytes have been studied by Shinoda et al. [211–213] as a model for biological systems such as nucleoproteins and protein–polysaccharide complexes. Complex formation of poly(L-lysine) with poly(acrylic acid), poly(methacrylic acid), or sodium cellulose sulfate was observed. Poly(riboadenylic acid) and poly(ribouridylic acid) form a complex with a double-stranded helical structure [214]. The interactions of a series of basic proteins (polycations) with the polyanion potassium hyaluronate provide additional evidence of the interest in complex formation with naturally occurring polyelectrolytes [215]. Various other polyelectrolyte complexes are cited in Table 5.9.

5.2.10 Unsaturated Hydrocarbon-Based Polymers

Elastomer–elastomer blends have been covered in several detailed review articles [233–235]. In the characterization of these blends, as with polymer–polymer blends, two-phase behavior has been cited in most cases. Several cases of miscibility have been observed, as will be discussed.

Polybutadiene can be polymerized into various structures (i.e., cis-1,4; trans-1,4; as well as 1,2 and all combinations of these variations). The T_g can vary from $-110°C$ (cis-1,4) to $-25°C$ (1,2). The phase behavior in polymer mixtures will likewise be affected by the polymer microstructure. Bartenev and Kongarev investigated natural rubber–polybutadiene ($T_g = -48°C$, high 1,2 content) and observed single sharp T_g's determined dilatometrically [236]. Fujimoto and Yoshimura [237] studied cis-1,4-poly-

5.2. Referenced Listing

TABLE 5.10

Unsaturated Hydrocarbon-Based Polymers

Miscible polymer	Method of miscibility determination	Comments	References
Natural rubber–polybutadiene (high 1,2 content)	Dilatometric	Blend T_g linear with volume fraction	236
cis-1,4-Polybutadiene–styrene/butadiene copolymer (SBR 1500)	Electron microscopy	Nine other elastomer blends were shown to be two phase	238, 242
cis-1,4-Polybutadiene–styrene/butadiene copolymer (JSR 1500)	Dielectric loss; dynamic mechanical	Transitions of blends not as sharp as pure components	237
Styrene/butadiene copolymers with different styrene contents	Modulus–temperature data	Blends of 16 and 23.5% styrene miscible; 37.5 and 50% show broadening of T_g; 16 and 50% two phase	240
Butadiene/acrylonitrile copolymers with 18 and 40% acrylonitrile	Dilatometric	Blend T_g linear with volume fraction	236
Chloroprene (Neoprene W)–styrene/butadiene copolymer (SBR 1500)	Dilatometric	Crystallization rate of chloroprene retarded by SBR addition	241
Polychloroprene–nitrile rubber (18 wt% acrylonitrile)	Dynamic mechanical	Two phase if acrylonitrile content is 39 wt%	73
Butadiene/acrylonitrile (65/35) copolymer–styrene/acrylonitrile (80/20) copolymer	Stated to be homogeneous	>70% SAN	243
Styrene/butadiene copolymer (SBR 1000)–polysulfide	Modulus–temperature data	SBR vulcanized with 22% sulfur (hard rubber)	244

butadiene–natural rubber blends by dielectric loss, mechanical loss, and modulus–temperature data and observed heterogeneous behavior. In spite of the heterogeneous character, property advantages for these blends, such as improved elasticity, abrasion resistance, and heat stability, have been cited.

Marsh et al. [238] studied eleven elastomer–elastomer blends by electron microscopy and rated only two blends miscible: high cis-1,4-polybutadiene mixed with a styrene/butadiene copolymer ($T_g = -56°C$) and an emulsion cis-1,4-polybutadiene with the same styrene/butadiene copolymer. Fujimoto and Yoshimura [239] studied a low cis content polybutadiene blend with styrene–butadiene copolymers of varying styrene contents using dielectric loss, dynamic mechanical methods, and phase-contrast microscopy. Above 30% styrene in the copolymer, phase separation of the mixtures occurred. In another study by the same authors, a high cis-1,4-polybutadiene blend with a styrene/butadiene (characterized only by the designation JSR 1500) copolymer yielded homogeneous blends [237]. Livingston and Brown [240] studied the phase behavior of styrene/butadiene copolymers of varying styrene contents. Mixtures with copolymer contents of 16 and 23.5% styrene gave homogeneous behavior judged from modulus–temperature data. In particular, the slope of the modulus–temperature curve in the region of the T_g was similar for the pure components and the blends. With mixtures based on copolymers containing 37.5 and 50% styrene, the slopes of the mixture modulus–temperature curves were less steep than the corresponding pure-component data. When the styrene contents were 16 and 50% in the copolymers, two-phase behavior was apparent.

Bartenev and Kongarov [236] studied butadiene/acrylonitrile copolymer blends at different acrylonitrile contents (18 and 40%) and concluded that the mixtures were miscible based on dilatometric measurements.

Polychloroprene [poly(2-chloro-1,3-butadiene)] and styrene/butadiene copolymer blends investigated by Kell et al. [241] exhibited single T_g's as determined by dilatometric data. Crystallization rates of polychloroprene were observed to decrease, and the level of crystallinity of polychloroprene (Neoprene W) was also found to be slightly reduced. Polychloroprene and nitrile rubber (18% acrylonitrile) gave miscible blends, whereas nitrile rubber with 39% acrylonitrile exhibited two-phase behavior with polychloroprene [73].

Table 5.10 lists the polymers exhibiting miscibility with unsaturated hydrocarbon-based polymers.

5.2.11 Block Copolymers

Single-phase behavior in block copolymers has been shown to be thermodynamically more favorable than homopolymer mixtures composed of components with molecular weights equal to that of the individual block molecular weights. For this reason, it is not surprising that many single-phase block copolymers have been observed. One of the most studied systems is the polystyrene–poly(α-methylstyrene) block copolymer. Baer [245] first observed single-phase behavior for an ABA polystyrene–poly(α-methylstyrene)–polystyrene block copolymer. More recent studies by Robeson et al. [93] involving the diblock styrene–α-methylstyrene copolymers found single-phase behavior for block molecular weights in the range of 200,000

5.2. Referenced Listing

M_n ($M_w/M_n = 1.1$). Blends with this high M_w block copolymer with polystyrene or poly(α-methylstyrene) showed different, unexpected behavior. Poly(α-methylstyrene) was found to be more miscible with the equal block content copolymer than polystyrene. Hansen and Shen [246] studied ABA block copolymers of styrene and α-methylstyrene and also observed single-phase behavior. Blends of a high styrene content block copolymer, however, were found to exhibit better miscibility for polystyrene than poly(α-methylstyrene) as judged by the loss modulus, E'', and peak broadening. It was pointed out that this was not in agreement with the previously reported results by Robeson et al. [93]. However, it must also be pointed out that the choice of a high styrene content block copolymer would definitely favor miscibility with polystyrene over poly(α-methylstyrene); thus, the quoted observations are as expected and are not comparable to the results of Robeson et al., where equal composition block copolymers were investigated.

Theoretically predicted phase separation in AB block copolymers composed of polystyrene and poly(α-methylstyrene), using the theory developed by Krause [247], corresponded to a block copolymer molecular weight of 800,000. Triblock copolymers would require a total molecular weight of 1,200,000. Dunn and Krause [248] observed the onset of phase separation in simple mixtures of polystyrene and poly(α-methylstyrene) at $M_n = 160,000$ and 147,000, respectively, using reasonably monodisperse samples (i.e., M_w/M_n close to 1.0). With diblock copolymers, phase separation, determined by calorimetric and dilatometric data, occurred with total block copolymer $M_n = 500,000$. Recent studies have provided further confirmation of the earlier observations for the styrene–α-methylstyrene block copolymers [249–251].

McGrath et al. [252] studied a series of condensation block copolymers based on rigid block constituents (i.e., $T_g > 23°C$). Polysulfone (condensation polymer from bisphenol A and 4,4'-dichlorodiphenyl sulfone) and bisphenol A polycarbonate blends exhibit two-phase behavior. (AB)$_n$ block copolymers of these components (5000 M_n blocks) show single-phase behavior, as determined by modulus–temperature data. Block copolymers of the polyhydroxyether of bisphenol A and polysulfone (also 5000 M_n blocks) exhibited an intermediate single T_g between the component polymers, whereas the blends of the homopolymer constituents were two-phase materials. Phase separation in polystyrene–polysulfone block copolymers was observed above 4000 M_n for the constituent blocks. The calculated solubility parameter difference for the above-stated miscible blocks was less than for the two-phase polystyrene–polysulfone block copolymer.

Polysulfone–poly(butylene terephthalate) block copolymers were capable of being obtained in a wholly amorphous state with single-phase behavior ($T_g = 139°C$; 30% PBT content). Annealing above 140°C resulted in a slight increase in the T_g of the block copolymer with a modulus–temperature behavior suggesting the onset of poly(butylene terephthalate) crystallization. Recent work by McGrath et al. [253] extended the polysulfone–polycarbonate block copolymer to significantly higher block molecular weights. Single-phase behavior was noted with 16,000 M_n polycarbonate/17,000 M_n polysulfone block molecular weights, whereas unreacted blocks of even lower molecular weight were clearly immiscible.

Compositions of nylon 6,6 and the polyamide from *m*-phenylene diamine and adipic acid were melt blended and allowed to amide-interchange to a low level ($<3\%$), leading to block copolymer formation. The amide-interchanged block copolymers exhibited single T_g's higher than that expected from linear extrapolation [189]. Other compositions also exhibiting single-phase behavior with nylon 6,6 after amide interchange were the polyamide of 2-methylhexamethylenediamine-terephthalic acid and that of adipic acid and isophthalamide. These block copolymers were observed to exhibit improved tire flat-spotting resistance over that of unmodified nylon 6,6. This result was cited in Section 5.2.8 as the level of block formation was considered to be quite low.

Epoxy–polyester and epoxy–polyether thermosetting block copolymers were investigated by Noshay and Robeson [254]. Variations in the molecular weight of hydroxyl-terminated poly(ε-caprolactone) incorporated into the two epoxy structures as shown below were studied:

> ECMC   3,4-epoxycyclohexylmethyl-3,4-
> epoxycyclohexane carboxylate
> ESD   2(3,4-epoxycyclohexyl)-5,5-spiro-(3,4-
> epoxycyclohexane-*m*-dioxane)

The ECMC and ESD cycloaliphatic epoxies were cross-linked with hexahydrophthalic anhydride, which also reacts with the –OH-terminated polyesters or polyethers employed in this study. At poly(ε-caprolactone) $M_n = 1300$, single-phase behavior was observed. Increasing the molecular weight to 2000 resulted in a broader transition with the appearance of a definite shoulder in the mechanical loss curve indicative of microheterogeneous behavior. At polyester molecular weights of 5000 and 10,000, phase separation was clearly depicted by two distinct glass transition temperatures. Similar behavior was observed with ECMC modified with hydroxyl-terminated poly(propylene oxide).

A block copolymer composed of 65 wt% poly(ethylene adipate) and 35 wt% cyclohexylidene bisphenol polycarbonate was shown to exhibit a single damping peak in dynamic mechanical testing [255]. Bisphenol A polycarbonate–poly(ethylene oxide) $(AB)_n$ block copolymers have been studied by several investigators [256–258]. Single glass transitions were cited for

5.2. Referenced Listing

compositions based on block segment M_n in the range of 2000 to 20,000. Goldberg [258] demonstrated a linear decrease in the glass transition temperature with increasing poly(ethylene oxide) content. The crystallization rate of bisphenol A polycarbonate was found to increase significantly with increasing poly(ethylene oxide) content, presumably due to the favorable kinetics resulting from a lowered T_g. At high block M_n for the poly(ethylene oxide) component, a low temperature crystalline melting point ascribed to poly(ethylene oxide) was observed.

Bisphenol A polycarbonate–poly(ε-caprolactone) [259] and poly(2,2,4,4-tetramethyl cyclobutane diol carbonate) (50% trans isomer)–poly(ε-caprolactone) (AB)$_n$ block copolymers were also observed to have intermediate

Poly(2,2,4,4-tetramethyl cyclobutane diol carbonate)

glass transitions between the constituent values [260]. Crystallization rates of the higher T_g blocks were found to increase with the incorporation of increasing amounts of poly(ε-caprolactone) in the block copolymers. At levels of >40 wt% poly(ε-caprolactone), elastomeric properties were observed for these block copolymers.

Thermoplastic polyurethanes based on 4,4′-diphenylmethane diisocyanate (MDI)–butane diol hard blocks and polyester or polyether soft blocks comprise a family of important commercial materials. Although considerable characterization of the morphology of thermoplastic polyurethanes has been reported [261–265], a universally accepted explanation of the phase behavior has not been forthcoming. Generally, the MDI–butane diol hard block is believed to exhibit crystallinity, whereas the amorphous soft blocks are phase separated, yielding a low temperature T_g phase. Phase mixing of the amorphous fraction of the hard block with the soft blocks is not clearly defined and probably varies with each specific thermoplastic polyurethane. It is interesting to note that thermoplastic polyurethanes based on poly-(ε-caprolactone) soft blocks have been cited to be miscible with poly(vinyl chloride) [57] and the polyhydroxy ether of bisphenol A (Phenoxy) [266]. The modulus–temperature data on these blends indicate that the crystallinity of the hard block is noticeably suppressed. Note that poly(ε-caprolactone) has been previously cited as miscible with poly(vinyl chloride) and Phenoxy.

The block copolymer of poly(butylene terephthalate) and poly(tetrahydrofuran) has a property profile remarkably similar to thermoplastic polyurethanes [267, 268]. Detailed analysis of the phase behavior of this block copolymer has revealed miscibility of the amorphous segments of the block copolymer with partial crystallization of poly(butylene terephthalate) from the single-phase amorphous structure [269–271].

TABLE 5.11

Block Copolymers

Miscible polymer	Method of miscibility determination	Comments	References
Polystyrene–poly(α-methylstyrene) AB and ABA block copolymers	Dynamic mechanical	Miscible in AB block copolymer with block $M_n = 200,000$; phase separation at 250,000 M_n	93, 245 246, 247 248–251
Polysulfone–bisphenol A polycarbonate $(AB)_n$ block copolymers	Modulus–temperature data	Miscible at block $M_n = 5000$ and up to $M_n = 16,000$	252, 253
Polysulfone–poly(hydroxy ether) of bisphenol A (Phenoxy) $(AB)_n$ block copolymers	Modulus–temperature data	Miscible at block $M_n = 5000$	252
Polysulfone–poly(butylene terephthalate) $(AB)_n$ block copolymers	Modulus–temperature data	Annealing yields higher T_g values due to development of PBT crystallinity	252
Nylon 6,6–polyamide of m-phenylene diamine and adipic acid block copolymer	Dynamic mechanical	T_g of block copolymer higher than linear extrapolation of components	189
Nylon 6,6–polyamide of 2-methylhexamethylene diamine and terephthalic acid	Dynamic mechanical	T_g of block copolymer higher than linear extrapolation of components	189
Nylon 6,6–polyamide of adipic acid and isophthalamide	Dynamic mechanical	T_g of block copolymer higher than linear extrapolation of components	189
Poly(ethylene adipate)–cyclohexylidene bisphenol polycarbonate $(AB)_n$ block copolymer	Dynamic mechanical		255
Cycloaliphatic epoxy–poly(ε-caprolactone) thermoset block copolymers	Dynamic mechanical	Miscible at 2000 block M_n; two phase at 5000 M_n	254
Cycloaliphatic epoxy–poly(propylene oxide) thermoset block copolymers	Dynamic mechanical	Miscible at 2000 block M_n; two phase at 5000 M_n	254
Bisphenol A polycarbonate–poly(ethylene oxide) $(AB)_n$ block copolymers	Thermomechanical; differential thermal analysis	Bisphenol A polycarbonate crystallization rate increased with increasing poly(ethylene oxide) content	256–258

5.2. Referenced Listing

TABLE 5.11 (*Continued*)

Miscible polymer	Method of miscibility determination	Comments	References
Bisphenol A polycarbonate–poly(ε-caprolactone)	Modulus–temperature data	Polycarbonate crystallinity increased with poly(ε-caprolactone) content	259
Poly[2,2,4,4-tetramethylcyclobutane diol carbonate (50% trans isomer)]–poly(ε-caprolactone)	Modulus–temperature data	Elastomeric properties at poly(ε-caprolactone) content > 40 wt%	260
Poly(butylene terephthalate)–poly(tetrahydrofuran) $(AB)_n$ block copolymers	Dynamic mechanical; calorimetry	PBT crystallizes from single-phase amorphous state	269–271
Thermoplastic polyurethane [poly(ε-caprolactone) soft blocks]–poly(vinyl chloride)	Dynamic mechanical	Permanent plasticizer for PVC	57
Thermoplastic polyurethane [poly(ε-caprolactone) soft blocks]–poly(hydroxy ether) of bisphenol A (Phenoxy)	Dynamic mechanical	Urethane hard block crystallinity is suppressed	266
Poly(butylene terephthalate)–poly(tetrahydrofuran) $(AB)_n$ block copolymers–poly(vinyl chloride)	Dynamic mechanical; nmr	Poly(butylene terephthalate) crystallinity suppressed	56
Additional references			272–277

Nishi *et al.* [56] observed that the poly(butylene terephthalate)–poly(tetrahydrofuran) block copolymer exhibited a level of miscibility with poly(vinyl chloride). Annealing the blends at 130°C resulted in phase separation. Crystallization of the poly(butylene terephthalate) could have occurred; however, the transitions ascribed to the amorphous phase showed definite broadening with a low temperature shoulder. This was interpreted as the existence of an upper critical solution temperature for this blend. It was concluded that the hard segments of the block copolymer [poly(butylene terephthalate)] remained segregated in the blend with poly(vinyl chloride). As miscibility of poly(butylene terephthalate) with PVC has been demonstrated, the hypothesis of Nishi *et al.* has been questioned [52].

Table 5.11 lists the polymers exhibiting miscibility with block copolymers.

5.2.12 Graft Copolymers

The thermodynamic criteria developed for block copolymers should be applicable to graft copolymers in that the potential for phase separation will require higher individual component molecular weights than with simple polymer mixtures. Kargin [126], in a detailed study of the solid-state properties of graft copolymers, observed that graft copolymers of nitrocellulose and poly(methyl acrylate) yielded single-phase behavior. Simple polymer mixtures of the same components also were single phase, with the same T_g-composition function as the graft copolymers. Poly(ethylene oxide) grafted onto a nylon 6 backbone was also observed to be a single-phase system. A noticeable decrease in the melting point of nylon 6 was observed and no poly(ethylene oxide) crystallinity could be resolved. The thermomechanical data indicated T_g's between published values for nylon 6 and poly(ethylene oxide). A simple mixture of the two pure components exhibited spherulitic structures of both polymers without a reduction in their respective melting points.

Matzner et al. [187] synthesized graft copolymers of nylon 6 grafted onto an ethylene/acrylic acid copolymer backbone. These copolymers exhibited single-phase behavior for the amorphous phase. A noticeable reduction in the melting point of nylon 6 was observed, due, at least in part, to the low molecular weight of the nylon 6 grafts. The mechanical loss peak of graft copolymer at the T_g was significantly sharper than with simple mixtures. Simple mixtures were partially miscible at acrylic acid levels of 14 wt%. Miscibility of the graft copolymer was obtained over a wider range of acrylic acid concentrations than with simple mixtures. The permeability as a function of composition showed a positive deviation over that expected from two-component miscible blends. This was believed to be the result of a decreased level of both residual polyethylene crystallinity as well as nylon 6 crystallinity compared to the unblended or ungrafted polymer values.

A bigraft copolymer of poly(ethylene/propylene/1,4-hexadiene) grafted with polystyrene as well as polyisobutylene was subjected to dynamic mechanical and calorimetric characterization [278]. Only a single low temperature transition was observed, indicative of miscibility of the ethylene/propylene/hexadiene terpolymer with polyisobutylene. The polystyrene branches, however, exhibited a separate phase. Similar behavior in this bigraft copolymer was observed when polystyrene was replaced by poly-(α-methylstyrene) [279]. When both poly(α-methylstyrene) and polystyrene were grafted onto the ethylene/propylene/hexadiene terpolymer, only one T_g was observed [280] for the rigid graft, thus yielding results similar to those previously quoted for polystyrene–poly(α-methylstyrene) block copolymers. Similar phase behavior was also cited for PVC–g–polystyrene copolymers [195].

5.2. Referenced Listing

Graft copolymers of polystyrene and poly(ε-caprolactone) were prepared by incorporation of hydroxyethyl methacrylate into the polystyrene backbone, followed by the *in situ* polymerization of ε-caprolactone [281]. The hydroxyl group of hydroxyethyl methacrylate in the backbone of polystyrene serves as an initiation site for poly(ε-caprolactone) polymerization.

TABLE 5.12

Graft Copolymers

Miscible polymer	Method of miscibility determination	Comments	References
Nitrocellulose–poly-(methyl acrylate) graft copolymer	Thermomechanical	T_g–composition data equal to simple mixtures	126
Nylon 6–poly(ethylene oxide) graft copolymer	Thermomechanical; crystallization behavior	Single T_g observed; melting point depression of Nylon 6; no observed poly(ethylene oxide) crystallinity	126
Nylon 6–ethylene/acrylic acid graft copolymers	Dynamic mechanical	T_g sharper than simple mixtures; permeability data agreed with single-phase analysis	187
Poly(ethylene/propylene/1,4-hexadiene) grafted with polystyrene and polyisobutylene	Dynamic mechanical; calorimetric	Ethylene/propylene/1,4-hexadiene terpolymer miscible with polyisobutylene graft but not with polystyrene graft	278, 279
Poly(ethylene/propylene/1,4-hexadiene) grafted with poly(α-methylstyrene) and polyisobutylene	Dynamic mechanical	Ethylene/propylene/1,4-hexadiene terpolymer miscible with polyisobutylene graft but not with poly(α-methylstyrene)	279
Poly(ethylene/propylene/1,4-hexadiene) grafted with polystyrene and poly(α-methylstyrene)	Dynamic mechanical	Polystyrene graft miscible with poly(α-methylstyrene) grafts; grafts not miscible with terpolymer backbone	280
Poly(vinyl chloride)–polystyrene graft copolymer		Stated to be miscible; data to be published later	279
Polystyrene–poly(ε-caprolactone) graft copolymers–poly(vinyl chloride)	Dynamic mechanical	Graft copolymer components not miscible; blend with PVC yields miscible PVC–poly(ε-caprolactone) phase	281

These graft copolymers exhibit two-phase behavior. Studies involving blends of this graft copolymer with poly(vinyl chloride) revealed that the poly-(ε-caprolactone) phase is miscible with poly(vinyl chloride), whereas the polystyrene phase remains a separate entity. The crystallization kinetics of poly(ε-caprolactone) in the single-phase amorphous blend with poly(vinyl chloride) was studied as a function of composition, polystyrene content of the graft copolymer, and number of grafted poly(ε-caprolactone) chains per graft copolymer molecule. Increasing polystyrene content or number of grafts per graft copolymer molecule significantly reduced the crystallization rate, primarily due to restriction of the end of the crystallizable chain, which is firmly held in position at the polystyrene interface with the poly(vinyl chloride)–poly(ε-caprolactone) phase.

Table 5.12 lists the polymers that exhibit miscibility with graft copolymers.

5.2.13 Interpenetrating Networks

Since the introduction of the concept of polymeric catenane structures termed interpenetrating networks (IPN), the transitional characteristics of IPNs versus the simple polymer mixtures have been used to qualitatively determine the extent of interpenetration. In fact, unless a measurable change in the glass transitions of the components of the mixture has occurred, the degree of interpenetration is judged insignificant.

Frisch *et al.* [282–285] prepared a series of IPNs by mixing the linear prepolymers together with cross-linking agents and appropriate catalysts. The series of IPNs consisted of a polyurethane, with the other component in the mixture being a polyester, epoxy, or a polyacrylate. As a large number of different structural variables were investigated, only the generalized polymer class will be discussed here. The glass transition determinations were determined calorimetrically, and electron microscopy was conducted on osmium tetroxide-stained samples. Specific cases involving IPNs of polyurethane–epoxy, polyurethane–polyester, and polyurethane–polyacrylate gave single glass transitions slightly lower than calculated from composition weighted averages of component T_g's. In the case of polyurethane–polyacrylate [282], the electron micrographs showed no evidence of phase separation. A maximum in tensile strength at intermediate concentrations for these IPNs was believed to be the result of a high level of interpenetration of the respective networks. Increased densities (relative to that expected for noninteracting, phase-separated mixtures) were observed and ascribed to the mixing of the component networks [285]. Latex blends of polyurethanes and polyacrylates with admixed cross-linking agents to yield IPNs were also prepared and characterized. A single broad transition resulted, indi-

5.2. Referenced Listing

cating a degree of interpenetration lower than that of the IPNs of polyurethane and polyacrylate with similar compositions as discussed above.

IPNs of poly(n-butyl acrylate)-poly(ethyl methacrylate) were prepared by polymerizing (via emulsion polymerization) one of the component polymers followed by the addition of the second monomer and polymerization of it with the appropriate cross-linking agents added to yield a catenane-type structure [286]. The resultant dynamic mechanical spectra revealed broad transitions, which indicated some miscibility. Reversal of the order of polymerization changed the dynamic mechanical results, indicating that the continuous phase was predominately the second polymer that was polymerized.

Poly(ethyl acrylate)-poly(styrene/methyl methacrylate) IPNs were studied

TABLE 5.13

Interpenetrating Networks

Miscible polymer	Method of miscibility determination	Comments	References
Polyurethane–polyester	Calorimetric	Unsaturated polyesters cross-linked with styrene monomer	282, 283
Polyurethane–epoxy	Calorimetric	Maximum in tensile strength at intermediate concentration; bisphenol A and novolac-based expoxies	282, 283
Polyurethane–polyacrylate	Calorimetric; electron microscopy	Maximum in tensile strength at intermediate concentrations	282–284
Polyurethane–poly(methyl methacrylate)	Calorimetric; dynamic mechanical	Two glass transitions observed, shifted toward the other component; density higher than expected	285
Poly(n-butyl acrylate)–poly(ethyl methacrylate)	Dynamic mechanical	Broad transition	286
Poly(ethyl acrylate)–poly(methyl methacrylate/styrene)	Dynamic mechanical	Broad transitions show-increased miscibility with lower styrene concentrations	287
Castor oil/urethane–polystyrene	Dynamic mechanical	Single broad transition at high cross-link density	288
Additional references			292–295

as a function of methyl methacrylate content in the copolymer [287]. Dynamic mechanical spectroscopy measurements indicated that an improved level of miscibility was obtained as the methyl methacrylate concentration was increased. The extent of mixing was, however, incomplete even with poly(ethyl acrylate)–poly(methyl methacrylate) IPNs.

Castor oil, urethane, and polystyrene IPNs have been extensively investigated by Sperling and co-workers [288–290]. When 2,4-toluene diisocyanate was used to cross-link castor oil, an intermediate, broad T_g was observed at high cross-link densities, whereas distinct two-phase behavior was typical for low cross-link densities.

In the majority of the IPNs investigated, only minor changes in the glass transitions of the unblended constituents have been reported; therefore, these IPNs have not been included in Table 5.13. In many of these cases, the minor change in T_g is significant in that it represents more variation than observed with simple blends of the same constituents. Many of these examples are cited in a recent book by Manson and Sperling [291].

5.2.14 Isomorphic Polymer Blends

Miscibility, up to this point, has only been considered for the amorphous state. Cocrystallinity, or isomorphism, is a rare situation for which only several cases involving macromolecules have been observed. Two different types of macromolecular isomorphisms have been experimentally observed: isomorphism of repeating chain units (e.g., hexamethylene adipamide–hexamethylene terephthalamide copolymers [296]) and isomorphism of polymeric chains. Only isomorphism of uniquely different polymer chains will be reviewed here. Isomorphism is believed to be possible when the polymer chains (blend) or monomer units (copolymer) are similar in conformation and size in the crystalline state. A trivial situation in which isomorphism obviously occurs is with chemically identical polymers containing different isotopes. Substitution of deuterium for hydrogen in polyolefins (i.e., polyethylene, polypropylene) results in isomorphic behavior.

Poly(vinyl fluoride)–poly(vinylidene fluoride) exhibits isomorphic characteristics at all compositions. The crystalline phase structure of the blend is identical with that of poly(vinyl fluoride) and one of the crystalline structures of poly(vinylidene fluoride) [297]. The melting point curve as a function of composition shows no minimum.

Isomorphism of poly(isopropyl vinyl ether) with poly(sec-butyl vinyl ether) was observed at all concentrations by Natta *et al.* [298]. Mixtures of isotactic poly(4-methylpentene) with isotactic poly(4-methylhexene) were isomorphic only in the range of 0–25 wt% of either component [298]. The unit cells of isotactic poly(4-methyl-1-pentene) and isotactic poly(4-methyl-

5.2. Referenced Listing

1-hexene) are tetragonal, and the chains have a helix conformation with 3.5 monomer units per pitch. The poly(vinyl alkyl ethers) cited above likewise have tetragonal unit cells with identical pitch of the helices (3.4 monomer units). Therefore, for both of these isomorphic systems, the chain axes and chain symmetries are identical, as expected for isomorphic behavior.

Tordella and Dunion [299] studied blends of ethylene/vinyl acetate copolymers and paraffin wax. They concluded from the low-angle X-ray diffraction data that cocrystallization of paraffin wax with the ethylene sequences of the copolymer occurred. For a review of isomorphism behavior in macromolecules, consult Wunderlich [300] and Allegra and Bassi [301].

Table 5.14 gives the polymers exhibiting miscibility with isomorphic polymer blends.

TABLE 5.14

Isomorphic Polymer Blends

Miscible polymer	Method of miscibility determination	Comments	References
Poly(vinyl fluoride)–polyvinylidene fluoride)	Crystal phase structure	Melting point curve shows no minimum	297
Poly(isopropyl vinyl ether)–poly(sec-butyl vinyl ether)	Crystal phase structure	Isomorphic at all concentrations	298
Poly(4-methylpentene) – poly(4-methylhexene)	Crystal phase structure	Isomorphic only at 0–25 wt% of each component	298
Ethylene/vinyl acetate copolymer–paraffin wax	X-Ray diffraction		299

5.2.15 Miscellaneous

A study of the miscibility of poly(butadiene/acrylonitrile) and polychloroprene rubbers with phenolic resins established miscibility by utilizing calorimetric measurements [302]. The phenolic resins investigated in the rubber blends were poly(*tert*-butyl phenol/formaldehyde) and a rosin-modified phenol/formaldehyde novolac resin. A slight negative heat of mixing for all blends was observed and all cast films were transparent. Ebonite–polysulfide blends were considered miscible by Meltzer *et al.* [303] at ebonite levels of 80–95 wt%.

The structure of the racemic mixture of poly(γ-benzyl glutamate) is different from its d and l forms [304]. Several structural studies on the racemic mixture have led to the conclusion that regular side-chain conformations are formed between the benzyl groups. This stacking of the benzyl groups formed between neighboring d and l polymeric forms was found to yield a first-order transition at 100°C. This study and miscible polymer pairs are somewhat outside the scope of this study in that the d and l forms would be expected to be miscible; however, the regular side-chain conformations that occur are an interesting result that deserves mention.

Poly(butene-1)–isotactic polypropylene blends were characterized by Piloz et al. [305] using dilatometric, thermal analysis, and dynamic mechanical methods. Single T_g values for the blends, intermediate between the unblended homopolymers, were observed, and therefore miscibility in the amorphous phase of the blend was concluded.

Fluorocarbon polymer mixtures have rarely been cited. In one case [306], poly(vinylidene fluoride)–poly(vinyl fluoride) and poly(vinylidene fluoride)–tetrafluorethylene/vinylidene fluoride copolymer blends were stated to be molecularly miscible. However, in reviewing this work, the results had to be rejected based on their interpretation of the data. It was stated that for the molecularly miscible compositions three transitions were observed, with two transitions corresponding to the original polymers and the third relating to the molecularly miscible composition. Their choice of $-120°C$ as the T_g for poly(vinylidene fluoride) is not universally accepted (most investigators argue whether the T_g is near $-40°$ or $40°C$). Isomorphism of poly(vinyl fluoride) and poly(vinylidene fluoride) (Section 5.2.14) indicates molecular mixing at least in the crystalline state.

Blends of poly(vinyl chloride) and butadiene/α-methylstyrene elastomer were stated to be miscible at greater than 60% elastomer content [307, 308]. The conclusions were based on dielectric and dynamic mechanical properties as well as ultimate property data. We draw a different conclusion than these authors, as no change in the T_g of the elastomeric component is observed and the disappearance of the PVC transition at lower PVC content appears to be due to the resolution of the experimental techniques employed. More definitive data are needed before this system can be classified as miscible.

Chlorinated rubber (65% chlorine) and ethylene/vinyl acetate copolymers (28 and 40% vinyl acetate) were found to exhibit single-phase behavior when cast from xylene or perchloroethylene, but they phase-separated when cast from chloroform or methylene chloride [309, 310]. The miscible mixtures phase-separated at elevated temperatures (188° to 225°C for 40% vinyl acetate copolymer and 120° to 190°C for 28% vinyl acetate copolymer). This system is, therefore, another case of lower critical solution temperature behavior.

Table 5.15 lists the miscellaneous miscible polymers.

5.3. Discussion

TABLE 5.15

Miscellaneous

Miscible polymer	Method of miscibility determination	Comments	References
Poly(butadiene/acrylonitrile)–poly(*tert*-butyl phenol/formaldehyde) or a rosin-modified phenol/formaldehyde) resin	Heat of mixing	Cast films were transparent	302
Polychloroprene–poly(*tert*-butyl phenol/formaldehyde) or a rosin-modified phenol/formaldehyde resin	Heat of mixing	Cast films were transparent	302
Ebonite–polysulfide		80–95% ebonite	303
Poly(γ-benzyl-*d*-glutamate)–poly(γ-benzyl-*l*-glutamate)	Dilatometric; thermal analysis	Regular side-chain conformations between d and l forms are observed; first-order transition at 100°C results	304
Poly(butene-1)–polypropylene	Dilatometric; thermal analysis; dynamic mechanical	Single T_g behavior observed with each experimental technique	305
Chlorinated rubber (65% chlorine)–ethylene/vinyl acetate copolymers (28 or 40% vinyl acetate)	Calorimetric; cloud point	Miscibility dependent upon solvent choice; exhibits lcst behavior	308, 309
Poly(vinylidene fluoride)–poly(vinyl methyl ketone)	Dynamic mechanical; thermal analysis	lcst behavior observed	171, 311

5.3 DISCUSSION

The multitude of miscible systems described in this chapter clearly demonstrates that miscibility in polymer blends is not as rare an occurrence as was once believed. It is of interest to point out several conclusions reached in one of the initial reviews [2]. That is, miscibility is especially improved by raising the temperature of a blend, and a miscible mixture of a crystalline polymer with another polymer is unlikely except in the case of isomorphic behavior. This chapter clearly shows that lower critical solution temperature behavior is definitely more common than upper critical solution temperature behavior in polymer blends; therefore, increasing the temperature decreases the miscibility. A considerable number of polymer blends composed of crystalline components have been shown here to exhibit miscibility, thus providing experimental verification to contradict the earlier conclusion of Bohn [2]. The review of Bohn cited 13 cases of polymer–polymer miscibility, of which 5 cases were variations in composition of a copolymer structure (e.g., styrene/acrylonitrile copolymers varying in composition by 3.5 wt%). This chapter clearly demonstrates the experimental progress made since the late 1960s in defining new miscible polymer blends.

Specific interactions (defined by the proton donor–proton acceptor concept) are clearly important for achieving miscibility in polymer blends, as a large number of the miscible blends in this chapter clearly demonstrate. This is particularly obvious in the section reviewing water-soluble polymers and polyelectrolytes. The multiplicity of miscible systems containing poly(vinyl chloride) or poly(ε-caprolactone) is clearly the result of specific interactions.

To predict polymer–polymer miscibility, a useful approach must be capable of assessing the relative strength of specific interactions. The use of the one-component solubility parameter approach (while reasonable for low molecular weight materials) is virtually useless unless only dispersive interactions are present. The two-component solubility parameter has been shown to work for those cases where no specific interactions are present but where dispersive and polar interactions are present [103]. An extension of the two-component solubility parameter approach to include corrections for specific interactions superficially appears to be the best empirical approach for predicting polymer–polymer miscibility. An adequate development of this approach is yet to be demonstrated.

REFERENCES

1. S. Krause, *J. Macromol. Sci., Rev. Macromol. Chem.* **7**, 251 (1972).
2. L. Bohn, *Rubber Chem. Technol.* **41**, 495 (1968).
3. A. Dobry and F. Boyer-Kawenoki, *J. Polym. Sci.* **2**, 90 (1947).
4. S. Krause, in "Polymer Blends" (D. R. Paul and S. Newman, eds.), Chapter 2. Academic Press, New York, 1978.
5. E. P. Lieberman, *Off. Dig., Fed. Soc. Paint Technol.* **34**, 30 (1962).
6. O. Olabisi, *Macromolecules* **9**, 316 (1975).
7. R. A. Emmett, *Ind. Eng. Chem.* **36**, 730 (1944).
8. J. E. Pittenger and G. F. Cohan, *Mod. Plast.* **25**, 81 (1947).
9. S. Manabe, P. Murakami, and M. Takayanagi, *Mem. Fac. Eng., Kyushu Univ.* **28** (4), 295 (1969).

References

10. J. Horvath, W. Wilson, H. Lundström, and J. Purdun, *Appl. Polym. Symp.* **7,** 95 (1968).
11. M. Matsuo, *Polym. Eng. Sci.* **9,** 197 (1969).
12. P. H. Geil, *Ind. Eng. Chem., Prod. Res. Dev.* **14,** 59 (1975).
13. P. H. Geil, in "Polymeric Materials," p. 119. Am. Soc. Metals, Metals Park, Ohio, 1975.
14. M. R. Ambler, *J. Polym. Sci., Polym. Chem. Ed.* **11,** 1505 (1973).
15. V. Landi, *Rubber Chem. Technol.* **45,** 222 (1972).
16. V. Landi, *Rubber Chem. Technol.* **45,** 1684 (1972).
17. V. Landi, *Appl. Polym. Symp.* **25,** 223 (1974).
18. G. A. Zakrzewski, *Polymer* **14,** 348 (1973).
19. C. F. Hammer, *Macromolecules* **4,** 69 (1971).
20. C. Elmqvist and S. E. Svanson, *Eur. Polym. J.* **12,** 559 (1976).
21. D. Feldman and M. Rusu, *Eur. Polym. J.* **10,** 41 (1974).
22. Y. J. Shur and B. G. Ranby, *J. Appl. Polym. Sci.* **19,** 1337 (1975).
23. B. G. Ranby, *J. Polym. Sci., Polym. Symp.* **51,** 89 (1975).
24. C. Elmqvist and S. E. Svanson, *Eur. Polym. J.* **11,** 789 (1975).
25. K. Marcincin, A. Romanov, and V. Pollack, *J. Appl. Polym. Sci.* **16,** 2239 (1972).
26. J. J. Hickman and R. M. Ikeda, *J. Polym. Sci., Polym. Phys. Ed.* **11,** 1713 (1973).
27. L. M. Robeson and J. E. McGrath, *Pap., 82nd Nat. Meet. AIChE, 1976.*
28. L. M. Robeson and J. E. McGrath, *Polym. Eng. Sci.* **17,** 300 (1977).
29. J. V. Koleske and R. D. Lundberg, *J. Polym. Sci., Part A-2* **7,** 795 (1969).
30. L. M. Robeson, *J. Appl. Polym. Sci.* **17,** 3607 (1973).
31. F. Gornich and J. D. Hoffman, in "Nucleation Phenomena" (A. S. Michaels, ed.), Am. Chem. Soc., Washington, D. C., 1966.
32. L. M. Robeson and B. L. Joesten, *Pap., N.Y. Acad. Sci., 1975.*
33. C. Ong, Ph.D. Thesis, University of Massachusetts, Amherst (1974).
34. F. B. Khambatta, F. Warner, T. Russell, and R. S. Stein, *J. Polym. Sci., Polym. Phys. Ed.* **14,** 1391 (1976).
35. D. S. Hubbell and S. L. Cooper, *J. Polym. Sci., Polym. Phys. Ed.* **15,** 1143 (1977).
36. J. W. Schurer, A. deBoer, and G. Challa, *Polymer* **16,** 201 (1975).
37. C. W. Roberts and D. H. Haigh, U. S. Patent 3,043,795 (1962) (assigned to Dow Chem. Co.).
38. Great Britain Patent 1,400,848 (1975) (assigned to BASF AG).
39. F. Ide, K. Kishida, and S. Deguchi, Japan. Kokai 76/26,953 (assigned to Mitsubishi Rayon Co., Ltd.)
40. F. Ide, K. Kishida, and N. Osaka, German Offen. 2,621,522 (1976) (assigned to Mitsubishi Rayon Co., Ltd.).
41. R. H. Hall, J. J. Lamson, A. J. Sikkema, and C. W. Roberts, U. S. Patent 3,424,823 (1969) (assigned to Dow Chem. Co.).
42. A. J. Urbanic and F. J. Mauver, U. S. Patent 3,283,034 (1966) (assigned to the General Tire & Rubber Co.).
43. K. Saito, M. Yoshino, and S. Yoshioka, U. S. Patent 3,287,443 (1966) (assigned to Kanegafuchi Chemical Industry Co.).
44. K. Sugimoto, S. Tanaka, and H. Fujita, U. S. Patent 3,520,953 (1970) (assigned to the Japanese Geon Co., Ltd.).
45. Y. C. Lee and Q. A. Trementozzi, U. S. Patent 3,644,577 (1972) (assigned to Monsanto Co.).
46. S. Yonezu, T. Tanaka, M. Tsuzuki, and T. Kobayaski, U. S. Patent 3,652,727 (1972) (assigned to Kanegafuchi Kagaku—Kogyo Kabushiki—Kaisha).
47. K. Saito, T. Tanaka, and I. Saito, U. S. Patent 3,670,052 (1972) (assigned to Kanegafuchi Chemical Industry Co.).

48. L. Scarso, E. Cerri, and G. Pezzin, U. S. Patent 3,772,409 (1973) (assigned to Montecatini Edison Sp. A.).
49. J. F. Kenney, in "Recent Advances in Polymer Blends, Grafts and Blocks" (L. H. Sperling, ed.), p. 117. Plenum, New York, 1974.
50. J. F. Kenney, J. Polym. Sci., Polym. Chem. Ed. 14, 123 (1976).
51. Y. J. Shur and B. G. Ranby, J. Appl. Polym. Sci. 20, 3121 (1976).
52. L. M. Robeson, J. Polym. Sci., Polym. Lett. Ed. 16, 261 (1978).
53. W. Keberle and W. Gobel, U. S. Patent 3,637,553 (1972) (assigned to Farbenfabriken Bayer Aktiengesellschaft).
54. A. Reischl, W. Gobel, and K. Schmidt, U. S. Patent 3,444,266 (1969) (assigned to Farbenfakriken Bayer Aktiengesellschaft).
55. R. P. Carter, Jr., U. S. Patent 3,487,126 (1969) (assigned to Goodyear Tire and Rubber Co.).
56. T. Nishi, T. K. Kwei, and T. T. Wang, J. Appl. Phys. 46, 4157 (1975).
57. H. W. Bonk, A. A. Sardanopoli, H. Ulrich, and A. A. R. Sayigh, J. Elastoplast. 3, 157 (1971).
58. M. Matzner, L. M. Robeson, E. W. Wise, and J. E. McGrath, Amer. Chem. Soc., Div. Org. Coat. Plast. Chem., Pap. 37(1), 123 (1977).
59. Y. J. Shur and B. G. Ranby, J. Appl. Polym. Sci. 19, 2143 (1975).
60. M. Takayanagi, H. Harima, and Y. Iwatu, Mem. Fac. Eng., Kyushu Univ. 23, 1 (1963).
61. D. Feldman and M. Rusu, Eur. Polym. J. 6, 627 (1970).
62. L. Chandler and E. Collins, J. Appl. Polym. Sci. 13, 1385 (1969).
63. H. F. Schwarz and W. S. Edwards, Appl. Polym. Symp. 25, 243 (1974).
64. V. I. Aleksieyenko and I. U. Mishustin, Vysokomol. Soedin. 1(11), 1593 (1959).
65. E. F. Jordan, Jr., B. Artymyshyn, G. R. Riser, and A. N. Wrigley, J. Appl. Polym. Sci. 21, 2715 (1976).
66. L. E. Nielsen, J. Am. Chem. Soc. 75, 1435 (1953).
67. D. Hardt, Br. Polym. J. 1, 225 (1969).
68. C. Elmqvist, Eur. Polym. J. 13, 95 (1977).
69. S. E. Svanson, C. Elmqvist, Y. J. Shur, and B. Ranby, J. Appl. Polym. Sci. 21, 943 (1977).
70. R. S. Stein, A. Misra, T. Yuasa, and F. B. Khambatta, Pure Appl. Chem. 49, 915 (1977).
71. Y. J. Shur and B. Ranby, J. Macromol. Sci., Phys. 14(4), 565 (1977).
72. J. Zelinger, E. Volfova, H. Zahradnikova, and Z. Pelzbauer, Int. J. Polym. Mater. 5, 99 (1976).
73. R. Casper and L. Morbitzer, Angew. Markromol. Chem. 58/59, 1 (1977).
74. L. M. Robeson and A. B. Furtek, J. Appl. Polym. Sci. 23, 645 (1979).
75. T. Nishi and T. K. Kwei, J. Appl. Polym. Sci. 20, 1331 (1976).
76. J. E. McGrath and M. Matzner, U. S. Patent 3,798,289 (1974) (assigned to Union Carbide Corp.).
77. B. Carmoin, G. Villoutreix, and R. Berlot, J. Macromol. Sci, Phys. 14 (2), 307 (1977).
78. G. S. Y. Yeh and S. L. Lambert, J. Polym. Sci., Part A-2 10, 1183 (1972).
79. H. D. Keith and F. J. Padden, Jr., J. Appl. Phys. 35, 1286 (1964).
80. J. Stoelting, F. E. Karasz, and W. J. MacKnight, Polym. Eng. Sci. 10, 133 (1970).
81. W. J. MacKnight, J. Stoelting, and F. E. Karasz, Adv. Chem. Ser. 99, 29 (1971).
82. A. R. Shultz and B. M. Beach, Macromolecules 7, 902 (1974).
83. A. R. Shultz and B. M. Gendron, J. Appl. Polym. Sci. 16, 461 (1972).
84. W. M. Prest, Jr. and R. S. Porter, J. Polym. Sci., Part A-2 10, 1639 (1972).
85. C. H. M. Jacques, H. B. Hopfenberg, and V. Stannett, Polym. Eng. Sci. 13, 82 (1973).
86. C. H. M. Jacques and H. B. Hopfenberg, Polym. Eng. Sci. 14, 441 (1974).
87. C. H. M. Jacques, Ph.D. Thesis, North Carolina State University, Raleigh (1973).
88. R. Hammel, W. J. MacKnight, and F. E. Karasz, J. Appl. Phys. 46, 4199 (1975).

References

89. W. Wenig, F. E. Karasz, and W. J. MacKnight, *J. Appl. Phys.* **46**, 4194 (1975).
90. R. N. Lemos, Ph.D. Thesis, University of Massachusetts, Amherst (1975).
91. E. Wilusz, Ph.D. Thesis, University of Massachusetts, Amherst (1976).
92. J. Tkacik, Ph.D. Thesis, University of Massachusetts, Amherst (1975).
93. L. M. Robeson, M. Matzner, L. J. Fetters, and J. E. McGrath, *in* "Recent Advances in Polymer Blends, Grafts, and Blocks" (L. H. Sperling, ed.), p. 281. Plenum, New York, 1974.
94. E. P. Cizek, U. S. Patent 3,383,435 (1968) (assigned to General Electric Co.).
95. A. R. Shultz and B. M. Beach, *J. Appl. Polym. Sci.* **21**, 2305 (1977).
96. P. Alexandrovich, F. E. Karasz, and W. J. MacKnight, *Polymer* **18**, 1022 (1977).
97. M. Bank, J. Leffingwell, and C. Thies, *Macromolecules* **4**, 43 (1971).
98. M. Bank, J. Leffingwell, and C. Thies, *J. Polym. Sci., Part A-2* **10**, 1097 (1972).
99. L. P. McMaster, *Macromolecules* **6**, 760 (1973).
100. T. K. Kwei, T. Nishi, and R. F. Roberts, *Macromolecules* **7**, 667 (1974).
101. T. Nishi, T. T. Wang, and T. K. Kwei, *Macromolecules* **8**, 227 (1975).
102. A. Robard, D. Patterson, and G. Delmas, *Macromolecules* **10**, 706 (1977).
103. M. T. Shaw, *J. Appl. Polym. Sci.* **18**, 449 (1974).
104. R. J. Kern and R. J. Slocombe, *J. Polym. Sci.* **25**, 183 (1955).
105. R. J. Kern, *J. Polym. Sci.* **21**, 19 (1956).
106. G. L. Slonimskii, *J. Polym. Sci.* **30**, 625 (1958).
107. G. V. Struminskii and G. L. Slonimskii, *Zh. Fiz. Khim.* **30**, 1941 (1956) (cited in Krause [1]).
108. G. E. Molau, *J. Polym. Sci., Polym. Lett. Ed.* **3**, 1007 (1965).
109. C. G. Seefried, Jr. and J. V. Koleske, *J. Test. Eval.* **4**, 220 (1976).
110. L. P. McMaster, *Adv. Chem. Ser.* **142**, 43 (1975).
111. V. D. J. Stein, R. H. Jung, K. H. Illers, and H. Hendus, *Angew. Makromol. Chem.* **36**, 89 (1974).
112. R. J. Slocombe, *J. Polym. Sci.* **26**, 9 (1957).
113. H. H. Irvin, U. S. Patent 3,010,936 (1961) (assigned to Borg-Warner Corp.).
114. O. Olabisi and A. G. Farnham, *Adv. Chem. Ser.* **176**, 559 (1979).
115. R. J. Petersen, R. D. Corneliussen, and L. T. Rozelle, *Polym. Prepr., Am. Chem. Soc., Div. Polym. Chem.* **10**, 385 (1969).
116. F. P. Warner, W. J. MacKnight, and R. S. Stein, *J. Polym. Sci., Polym. Phys. Ed.* **15**, 2113 (1977).
117. A. Robard and D. Patterson, *Macromolecules* **10**, 1021 (1977).
118. S. T. Wellinghoff, J. L. Koenig, and E. Baer, *J. Polym. Sci., Polym. Sci. Ed.* **15**, 1913 (1977).
119. Z. Slama and J. Majer, *Plaste Kautsch.* **24**, 423 (1977).
120. F. E. Karasz and W. J. MacKnight, *Contemp. Top. Polym. Sci.* **2**, 143 (1977).
121. N. E. Weeks, F. E. Karasz, and W. J. MacKnight, *J. Appl. Phys.* **48**, 4068 (1977).
122. S. Miyata and T. Hata, *Proc. Int. Congr. Rheol., 5th, 1968* Vol. 3, p. 71 (1970).
123. G. L. Brode and J. V. Koleske, *J. Macromol. Sci., Chem.* **6** (6), 1109 (1972).
124. J. V. Koleske, C. J. Whitworth, Jr., and R. D. Lundberg, U. S. Patent 3,922,239 (1975) (assigned to Union Carbide Corporation).
125. G. V. Olhoft, N. R. Eldred, and J. V. Koleske, U. S. Patent 3,642,507 (1972) (assigned to Union Carbide Corp.).
126. V. A. Kargin, *J. Polym. Sci., Part C* **4**, 1601 (1963).
127. J. F. H. van Eijnsbergen, *Chim. Peint.* **4**, 253 (1941) (cited in Krause [1]).
128. A. A. Tager, *Vysokomol. Soedin., A. Ser.* **14** (12), 2690 (1972).
129. I. Cabasso, *Am. Chem. Soc., Div. Org. Coat. Plast. Chem., Pap.* **37** (1), 110 (1977).

130. "Nitrocellulose: Properties and Uses," Publication of Hercules Powder Co., 1955.
131. C. J. Malm and H. L. Smith, Jr., *Ind. Eng. Chem.* **41,** 2325 (1949).
132. D. S. Hubbell and S. L. Cooper, *J. Appl. Polym. Sci.* **21,** 3035 (1977).
133. K. Friese, *Plaste Kautsch.* **13,** 65 (1966).
134. K. Friese, *Plaste Kautsch.* **15,** 646 (1968).
135. J. V. Koleske, *in* "Polymer Blends" (D. R. Paul and S. Newman, eds.), Chapter 22. Academic Press, New York, 1978.
136. I. Cabasso, J. Jagur-Grodzinski, and D. Vofsi, *in* "Polymer Alloys: Blends, Blocks, Grafts, and Interpenetrating Networks" (D. Klempner and K. C. Frissch, eds.), p. 1. Plenum, New York, 1977.
137. L. J. Hughes and G. E. Britt, *J. Appl. Polym. Sci.* **5,** 337 (1961).
138. L. J. Hughes and G. L. Brown, *J. Appl. Polym. Sci.* **5,** 580 (1961).
139. T. Nishi and T. T. Wang, *Macromolecules* **8,** 909 (1975).
140. J. S. Noland, N. N. -C. Hsu, R. Saxon, and J. M. Schmitt, *Adv. Chem. Ser.* **99,** 15 (1971).
141. D. R. Paul and J. O. Altamirano, *Adv. Chem. Ser.* **142,** 354 (1975).
142. G. D. Patterson, T. Nishi, and T. T. Wang, *Macromolecules* **9,** 603 (1976).
143. T. K. Kwei, G. D. Patterson, and T. T. Wang, *Macromolecules* **9,** 780 (1976).
144. R. L. Imken, D. R. Paul, and J. W. Barlow, *Polym. Eng. Sci.* **16,** 593 (1976).
145. T. K. Kwei, H. L. Frisch, W. Radigan, and S. Vogel, *Macromolecules* **10,** 157 (1977).
146. T. K. Kwei, H. L. Frisch, W. Radigan, and S. Vogel, *Am. Chem. Soc. Div. Org. Coat. Plast. Chem., Pap.* **37** (1), 116 (1977).
147. T. T. Wang and T. Nishi, *Macromolecules* **10,** 421 (1977).
148. R. E. Bernstein, C. A. Cruz, D. R. Paul, and J. W. Barlow, *Polym. Prepr., Am. Chem. Soc., Div. Polym. Chem.* **37** (2), 574 (1977).
149. D. C. Wahrmund, M.S. Thesis, University of Texas, Austin (1975).
150. D. C. Wahrmund, R. E. Bernstein, J. W. Barlow, and D. R. Paul, *Polym. Eng. Sci.* **18,** 677 (1978).
151. S. Krause and N. Roman, *J. Polym. Sci., Part A* **3,** 1631 (1965).
152. R. G. Bauer and N. C. Bletso, *Polym. Prepr., Am. Chem. Soc., Div. Polym. Chem.* **10** (2), 632 (1969).
153. S. Ichihara, A. Komatsu, and T. Hata, *Polym. J.* **2,** 640 (1971).
154. H. Z. Liu and K. J. Liu, *Macromolecules* **1,** 157 (1968).
155. A. M. Liquori, G. Anzuino, V. M. Coiro, M. D'Alagni, P. DeSantis, and M. Savino, *Nature (London)* **206,** 358 (1965).
156. J. Spevacek and B. Schneider, *Makromol. Chem.* **175,** 2939 (1974).
157. J. H. G. M. Lo, G. Kransen, Y. Y. Tan, and G. Challa, *J. Polym. Sci., Polym. Lett. Ed.* **13,** 725 (1975).
158. R. J. Kern, U. S. Patent 2,806,015 (1957) (assigned to Monsanto Chem. Co.).
159. J. Spevacek, *J. Polym. Sci., Polym. Phys. Ed.* **16,** 523 (1978).
160. E. Roerdink and G. Challa, *Polymer* **19,** 173 (1978).
161. D. J. Hourston and I. D. Hughes, *Polymer* **18,** 1175 (1977).
162. L. P. Bubnova, I. N. Rasinskaya, L. L. Burshtein, T. I. Borisova, and B. P. Shtarkman, *Vysokomol. Soedin., Ser. A* **18** (11), 2535 (1976); *Polym. Sci. USSR (Engl. Transl.)* **18,** 2897 (1976).
163. K. Friese, *Plaste Kautsch.* **12,** 90 (1965).
164. I. Cho, K. D. Ahn, *J. Polym. Sci., Polym. Lett. Ed.* **15,** 745 (1977).
165. T. K. Kwei, *Contemp. Top. Polym. Sci.* **2,** 157 (1977).
166. S. Akiyama, Y. Komatsu, and R. Kaneko, *Polym. J.* **7,** 172 (1975).
167. S. Akiyama, N. Inaka, and R. Kaneko, *Kobunshi Kagaku* **26,** 529 (1969).
168. S. Akiyama, *Bull. Chem. Soc.* **45,** 1381 (1972).

References

169. S. Akiyama and R. Kaneko, *Kobunshi Ronbunshu* **31**, 12 (1974).
170. H. Wolff, *Plaste Kautsch.* **4**, 244 (1957).
171. D. R. Paul, J. W. Barlow, R. E. Bernstein, and D. C. Wahrmund, *Polym. Eng. Sci.* **18**, 1225 (1978).
172. R. E. Bernstein, D. R. Paul, and J. W. Barlow, *Polym. Eng. Sci.* **18**, 683 (1978).
173. D. R. Paul, J. W. Barlow, C. A. Cruz, R. N. Mohn, T. R. Wassar, and D. C. Wahrmund, *Am. Chem. Soc., Div. Org. Coat. Plast. Chem., Pap.* **37** (1), 130 (1977).
174. L. M. Robeson and A. B. Furtek, *Am. Chem. Soc., Div. Org. Coat. Plast. Chem., Pap.* **37** (1), 136 (1977).
175. D. C. Wahrmund, D. R. Paul, and J. W. Barlow, *J. Appl. Polym. Sci.* **22**, 2155 (1978).
176. T. Sulzberg and R. J. Cotter, *J. Polym. Sci., Part A-1* **8**, 2747 (1970).
177. A. Escala, E. Balizer, and R. S. Stein, *Polym. Prepr., Am. Chem. Soc., Div. Polym. Chem.* **19** (1), 152 (1978).
178. M. Natov, L. Peeva, and E. Djagarova, *J. Polym. Sci., Part C* **16**, 4197 (1968).
179. T. R. Nassar, D. R. Paul, and J. W. Barlow, *J. Appl. Polym. Sci.* **23**, 85 (1979).
180. R. N. Mohn, D. R. Paul, J. W. Barlow, and C. A. Cruz, *J. Appl. Polym. Sci.* **23**, 575 (1979).
181. K. L. Smith, A. E. Winslow, and D. E. Petersen, *Ind. Eng. Chem.* **51**, 1361 (1959).
182. F. E. Bailey and J. V. Koleske, "Polyethylene Oxide," Academic Press, New York, 1976.
183. "Forming Association Compounds," F-43272. Publication of Union Carbide Corp.
184. A. R. Shultz and B. M. Gendron, *in* "Polymer Characterization by Thermal Methods of Analysis" (J. Chin, ed.), p. 175. Dekker, New York, 1974.
185. A. R. Shultz and B. M. Gendron, *J. Polym. Sci., Polym. Symp.* **43**, 89 (1973).
185a. Y. Osada and M. Sato, *J. Polym. Sci., Polym. Lett. Ed.* **14**, 129 (1976).
186. Stanford Research Institute, Report S-2915 (1960) (quoted in reference [183]).
187. M. Matzner, D. L. Schober, R. N. Johnson, L. M. Robeson, and J. E. McGrath, *in* "Permeability of Plastic Films and Coatings" (H. B. Hofenberg, ed.), p. 125. Plenum, New York, 1975.
188. E. Schchori and J. J. Grodzinski, *J. Appl. Polym. Sci.* **20**, 1665 (1976).
189. J. Zimmerman, E. M. Pearce, I. K. Miller, J. A. Muzzio, I. G. Epstein, and E. A. Hosegood, *J. Appl. Polym. Sci.* **17**, 849 (1973).
190. F. Wingler, L. Liebig, R. V. Meyer, and G. Wassmuth, Ger. Offen. 2,616,092 (1977) (assigned to Bayer).
191. R. L. Davidson and M. Sittig, eds., "Water Soluble Resins," Van Nostrand-Reinhold, Princeton, New Jersey, 1968.
192. L. A. Bimendina, V. V. Roganov, and Y. A. Bekturov, *Vysokomol. Soedin., Ser. A* **16** (12), 2810 (1974); *Polym. Sci. USSR (Engl. Transl.)* **16**, 3274 (1974).
193. H. Sato and A. Nakajima, *Polym. J.* **7**, 241 (1975).
194. M. P. Nedelcheva and G. V. Stoilkov, *J. Appl. Polym. Sci.* **20**, 2131 (1976).
195. A. B. Zezin, V. B. Rogacheva, V. S. Komarov, and Y. F. Razvodovskii, *Vysokomol. Soedin., Ser A* **17** (12), 2637 (1975); *Polym. Sci. USSR (Engl. Transl.)* **17**, 3032 (1975).
196. O. A. Aleksina, I. M. Papisov, and A. B. Zezin, *Vysokomol. Soedin. Ser. A* **13** (5), 1199 (1971); *Polym. Sci. USSR (Engl. Transl.)* **13**, 1350 (1971).
197. R. I. Kalyuzhnaya, A. L. Volynskii, A. R. Rudman, N. A. Vengerova, Y. F. Razvodovskii, B. S. El'tsefon, and A. B. Zezin, *Vysokomol. Soedin., Ser. A* **18** (1), 71 (1976); *Polym. Sci. (USSR) (Engl. Transl.)* **18**, 83 (1976).
198. K. Abe, M. Koide, and E. Tsuchida, *Polym. J.* **9**, 73 (1977).
199. K. Abe, H. Ohno, and E. Tsuchida, *Makromol. Chem.* **178**, 2285 (1977).
200. A. B. Zezin and V. B. Rogacheva, *in* "Advances in Polymer Science" (Z. A. Rogovin, ed.), p. 1. Wiley, New York (Russian translation).
201. M. Hosono, S. Sagii, O. Kusudo, and W. Tsuji, *Rep. Poval Comm., Kyoto* **61**, 79 (1972).

202. A. S. Michaels and R. G. Miekka, *J. Phys. Chem.* **65,** 176 (1961).
203. R. M. Fuoss and H. Sadek, *Science* **110,** 552 (1949).
204. M. Hosono, O. Kusudo, and W. Tsuji, *Rep. Poval Comm., Kyoto* **57,** 99 (1970).
205. H. Sato and A. Nakajima, *Polym. J.* **7,** 241 (1975).
206. C. N. Gray, M. I. T., Ph.D. Thesis (1965).
207. A. S. Michaels, L. Mir, and N. S. Schneider, *J. Phys. Chem.* **69,** 1447 (1965).
208. A. S. Michaels, G. L. Falkenstein, and N. S. Schneider, *J. Phys. Chem.* **69,** 1456 (1965).
209. A. S. Hoffman, R. W. Lewis, and A. S. Michaels, *Am. Chem. Soc., Div. Org. Coat. Plast. Chem. Pap.* **29** (2), 236 (1969).
210. M. J. Lysaght, *in* "Ionic Polymers" (L. Holliday, ed.), p. 281. Wiley, New York, 1975.
211. K. Shinoda, T. Hayashi, T. Yoshida, K. Sakai, and A. Nakajima, *Polym. J.* **8,** 202 (1976).
212. K. Shinoda, K. Sakai, T. Hayashi, and A. Nakajima, *Polym. J.* **8,** 208 (1976).
213. K. Shinoda, T. Hayashi and A. Nakajima, *Polym. J.* **8,** 216 (1976).
214. S. Tanaka, Y. Baba, and A. Kagemoto, *Polym. J.* **8,** 325 (1976).
215. A. K. Ghosh, *Makromol. Chem.* **176,** 2005 (1975).
216. A. Najajima and H. Sato, *Bull. Inst. Chem. Res., Kyoto Univ.* **47** (3), 177 (1969).
217. R. Warner, *J. Biol. Chem.* **229,** 711 (1957).
218. H. Sato, T. Hayashi, and A. Nakajima, *Polym. J.* **8,** 517 (1976).
219. Y. Kikuchi, Y. Onishi, M. Kodama, *J. Appl. Polym. Sci.* **20,** 3205 (1976).
220. M. K. Pal and A. K. Ghosh, *Macromol. Chem.* **169,** 273 (1973).
221. H. Deuel, J. Solms, and A. Dengler, *Helv. Chim. Acta* **36,** 1671 (1953).
222. I. S. Okhrimenko and E. B. D'yakonova, *Vysokomol. Soedin.* **6** 10, 1891 (1964); *Polym. Sci. USSR (Engl. Transl.)* **6,** 2095 (1964).
223. G. I. Distler, E. B. D'yakonova, I. F. Yefremov, Y. E. Kurtukova, I. S. Okhrimenko, and P. S. Sotnikov, *Vysokomol. Soedin.* **8** 10, 1737 (1966); *Polym. Sci. (USSR) (Engl. Transl.)* **8,** 1917 (1966).
224. A. Kossel, *J. Physiol. Chem.* **22,** 176 (1896).
225. M. Vogel, R. A. Cross, H. J. Bixler, and R. J. Guzman, *J. Macromol. Sci., Chem.* **4** (3), 675 (1970).
226. R. Miekka, Ph.D. Thesis, Massachusetts Institute of Technology, Cambridge (1961).
227. H. J. Bixler, L. M. Markley, and R. A. Cross, *J. Biomed. Res.* **2,** 145 (1968).
228. K. Abe, M. Koible, and E. Tsuchida, *Macromolecules* **10,** 1259 (1977).
229. L. A. Bimendina, G. S. Tleubayeva, and Y. A. Bekturov, *Vysokomol. Soedin., Ser. A* **19** (1), 71 (1977).
230. M. Iara and A. Nakajima, *Polym. J.* **10,** 37 (1978).
231. T. Mori, K. Imada, R. Tanaka, and T. Tanaka, *Polym. J.* **10,** 45 (1978).
232. Y. Kikuchi and A. Noda, *J. Appl. Polym. Sci.* **20,** 2561 (1976).
233. J. R. Dunn, *Rubber Chem. Technol.* **49,** 978 (1976).
234. P. J. Corish and B. D. W. Powell, *Rubber Chem. Technol.* **47,** 481 (1974).
235. P. J. Corish, *Rubber Chem. Technol.* **40,** 324 (1967).
236. G. M. Bartenev and G. S. Kongarov, *Rubber Chem. Technol.* **36,** 668 (1963).
237. K. Fujimoto and N. Yoshimura, *Rubber Chem. Technol.* **41,** 1109 (1968).
238. P. A. Marsh, A. Voet, L. D. Price, and T. J. Mullens, *Rubber Chem. Technol.* **41,** 344 (1968).
239. K. Fujimoto and N. Yoshimura, *Rubber Chem. Technol.* **41,** 1109 (1968).
240. D. I. Livingston and J. E. Brown, Jr., *Proc. Int. Congr. Rheol., 5th, 1968* Vol. 4, p. 25, (1970).
241. R. M. Kell, B. Bennett, and P. B. Stickney, *J. Appl. Polym. Sci.* **2,** 8 (1959).
242. P. A. Marsh, A. Voet, and L. D. Price, *Rubber Chem. Technol.* **40,** 359 (1967).
243. B. D. Gesner, *Encyc. Polym. Sci. Technol.* **10,** 694 (1969).
244. T. H. Meltzner, W. J. Dermody, and A. V. Tobolsky, *J. Appl. Polym. Sci.* **8,** 765 (1964).
245. M. Baer, *J. Polym. Sci., Part A* **2,** 417 (1964).

References

246. D. R. Hansen and M. Shen, *Macromolecules* **8,** 903 (1975).
247. S. Krause, *in* "Sagamore Conference on Block and Graft Copolymers" (J. J. Burke and V. Weiss, eds.), p. 143. Syracuse Univ. Press, Syracuse, New York, 1973.
248. D. J. Dunn and S. Krause, *J. Polym. Sci., Polym. Lett. Ed.* **12,** 591 (1974).
249. M. Shen, D. Soong, and D. R. Hansen, *Polym. Eng. Sci.* **17,** 560 (1977).
250. D. Soong and M. Shen, *Macromolecules* **10,** 357 (1957).
251. S. Krause, D. J. Dunn, A. Seyed-Mozzaffari, and A. M. Biswas, *Macromolecules* **10,** 786 (1977).
252. J. E. McGrath, L. M. Robeson, M. Matzner, and R. Barclay, Jr., *8th Reg. Am. Chem. Soc. Meet. 1976*, *J. Polym. Sci., Polym. Symp.* **60,** 29 (1977).
253. J. E. McGrath, T. C. Ward, E. Shchori, and A. J. Wnuk, *Polym. Eng. Sci.* **17,** 647 (1977).
254. A. Noshay and L. M. Robeson, *J. Polym. Sci., Polym. Chem. Ed.* **12,** 689 (1974).
255. M. D. Hartley and H. L. Williams, *J. Appl. Polym. Sci.* **19,** 2431 (1975).
256. S. H. Merrill and S. E. Petrie, *J. Polym. Sci., Part A* **3,** 2189 (1965).
257. S. H. Merrill, *J. Polym. Sci.* **55,** 343 (1961).
258. E. P. Golberg, *J. Polym. Sci., Part C* **4,** 707 (1964).
259. M. Matzner, U. S. Patent 3,641,200 (1972) (assigned to Union Carbide Corp.).
260. M. Matzner, U. S. Patent 3,639,503 (1972) (assigned to Union Carbide Corp.).
261. L. L. Harrell, Jr., *Macromolecules* **2,** 607 (1969).
262. S. L. Samuels and G. L. Wilkes, *J. Polym. Sci., Polym. Phys. Ed.* **11,** 807 (1973).
263. S. B. Clough and N. S. Schneider, *J. Macromol. Sci., Phys.* **2** (4), 641 (1968).
264. G. M. Estes, S. L. Cooper, and A. V. Tobolsky, *J. Macromol. Sci., Rev. Macromol. Chem.* **4** (2), 313 (1970).
265. S. B. Clough and N. S. Schneider, *J. Macromol. Sci., Phys.* **2** (4), 553 (1968).
266. C. G. Seefried, Jr., J. V. Koleske, and F. E. Critchfield, *Polym. Eng. Sci.* **16,** 771 (1976).
267. G. K. Hoeschele, *Polym. Eng. Sci.* **14,** 848 (1974).
268. J. R. Wolfe, Jr., *Rubber Chem. Technol.* **50,** 688 (1977).
269. A. Lilaonitkul, J. C. West, and S. L. Cooper, *J. Macromol. Sci., Phys.* **12** (4), 563 (1976).
270. N. K. Kalfoglou, *J. Appl. Polym. Sci.* **21,** 543 (1977).
271. A. Lilaonitkul and S. L. Cooper, *Rubber Chem. Technol.* **50,** 1 (1977).
272. D. J. Hourston and I. D. Hughes, *J. Appl. Polym. Sci.* **21,** 3099 (1977).
273. A. Noshay and J. E. McGrath, "Block Copolymers: Overview and Critical Survey." Academic Press, New York, 1977.
274. M. Brown and W. K. Witsiepe, *Rubber Age* **104,** 35 (1972).
275. S. Krause and A. M. Biswas, *J. Polym. Sci., Polym. Phys. Ed.* **15,** 2033 (1977).
276. R. P. Kane, *J. Elast. Plast.* **9,** 416 (1977).
277. T. C. Ward, A. J. Wnuk, A. R. Henn, S. Tang, and J. E. McGrath, *Polym. Prepr., Am. Chem. Soc., Div. Polym. Chem.* **19** (1), 155 (1978).
278. J. P. Kennedy, *J. Polym. Sci., Polym. Chem. Ed.* **13,** 2213 (1975).
279. A. Vidal and J. P. Kennedy, *J. Polym. Sci., Polym. Lett. Ed.* **14,** 489 (1976).
280. J. P. Kennedy and A. Vidal, *J. Polym. Sci., Polym. Chem. Ed.* **13,** 1765 (1975).
281. L. A. Pilato, J. V. Koleske, B. L. Joesten, and L. M. Robeson, *Polym. Prepr. Am. Chem. Soc., Div. Polym. Chem.* **17** (2), 824 (1976).
282. K. C. Frisch, D. Klempner, H. L. Frisch, and H. Ghiradella, *in* "Recent Advances in Polymer Blends, Grafts, and Blocks" (L. H. Sperling, ed.), p. 395. Plenum, New York, 1974.
283. H. L. Frisch, K. C. Frisch, and D. Klempner, *Polym. Eng. Sci.* **14,** 646 (1974).
284. K. C. Frisch, D. Klempner, S. Migdal, H. L. Frisch, and H. Ghiradella, *Polym. Eng. Sci.* **14,** 76 (1974).
285. S. C. Kim, D. Klempner, K. C. Frisch, and H. L. Frisch, *Macromolecules* **9,** 263 (1976).

286. L. H. Sperling, T. W. Chiu, C. P. Hartman, and D. A. Thomas, *Int. J. Polym. Mater.* **1**, 331 (1972).
287. V. Huelck, D. A. Thomas, and L. H. Sperling, *Macromolecules* **5**, 348 (1972).
288. G. M. Yenwo, L. H. Sperling, J. Pulido, and J. A. Manson, *Polym. Eng. Sci.* **17**, 251 (1977).
289. G. M. Yenwo, L. H. Sperling, and J. A. Manson, *in* "Chemistry and Properties of Crosslinked Polymers" (S. S. Labana, ed.), p. 257. Academic Press, New York, 1977.
290. J. E. Pulido, J. A. Manson, G. M. Yenwo, L. H. Sperling, and A. Conde, *Polym. Prepr., Amer. Chem. Soc., Div. Polym. Chem.* **18** (1), 841 (1977).
291. J. A. Manson and L. H. Sperling, "Polymer Blends and Composites." Plenum, New York, 1976.
292. L. H. Sperling, *J. Polym. Sci.; Macromol. Rev.* **12**, 141 (1977).
293. L. H. Sperling, T. W. Chiu, C. P. Hartman, and D. A. Thomas, *J. Appl. Polym. Sci.* **17**, 2443 (1973).
294. L. H. Sperling, D. A. Thomas, J. E. Lorenz, and E. J. Nagel, *J. Appl. Polym. Sci.* **19**, 2225 (1975).
295. K. C. Frisch, D. Klemper, S. K. Mukherjee, and H. L. Frisch, *J. Appl. Polym. Sci.* **18**, 689 (1974).
296. O. B. Edgar and R. Hill, *J. Polym. Sci.* **8**, 1 (1952).
297. G. Natta, G. Allegra, I. W. Bassi, D. Sianesi, G. Caporicco, and E. Torti, *J. Polym. Sci., Part A* **3**, 4263 (1965).
298. G. Natta, G. Allegra, I. W. Bassi, C. Carlino, E. Chiellini, and G. Montagnoli, *Macromolecules* **2**, 311 (1969).
299. J. P. Tordella and P. F. Dunion, *J. Polym. Sci., Part A-2* **8**, 81 (1970).
300. B. Wunderlich, "Macromolecular Physics," Vol. 1. Academic Press, New York, 1973.
301. G. Allegra and I. W. Bassi, *Adv. Polym. Sci.* **6**, 549 (1969).
302. P. Novakov, C. Konstantinov, and P. Mitanov, *J. Appl. Polym. Sci.* **16**, 1827 (1972).
303. T. H. Meltzer, W. J. Dermody, and A. V. Tobolsky, *J. Appl. Polym. Sci.* **8**, 765 (1964).
304. M. Yoshikawa, Y. Tsujitar, and I. Uematsu, *Polym. J.* **1**, 96 (1973).
305. A. Piloz, J. Y. Decroix, and J. F. May, *Angew. Makromol. Chem.* **54**, 77 (1976).
306. M. P. Zverev, L. A. Polovikhina, A. N. Barash, L. P. Mal'kova, and G. D. Litovchenko, *Vysokomol. Soedin., Ser. A* **16** (8), 1813 (1974).
307. D. Feldman and M. Rusu, *Eur. Polym. J.* **6**, 627 (1970).
308. D. Feldman and M. Rusu, *J. Polym. Sci., Polym. Symp.* **42**, 639 (1973).
309. J. Leffingwell, C. Thies, and H. Gertzman, *Polym. Prepr., Am. Chem. Soc., Div. Polym. Chem.* **14** (1), 596 (1973).
310. A. Purcell and C. Thies, *Polym. Prepr., Am. Chem. Soc., Div. Polym. Chem.* **9** (1), 115 (1968).
311. R. E. Bernstein, D. C. Wahrmund, J. W. Barlow, and D. R. Paul, *Polym. Eng. Sci.* **18**, 1220 (1978).

Chapter 6

Properties of Miscible Polymer Systems

6.1 THERMAL AND THERMOMECHANICAL

6.1.1 Glass Transition

This chapter will discuss the properties of miscible polymer blends. Some repetition will occur with Chapter 3 in that methods for the determination of polymer miscibility rely on the basic physical properties to establish single-phase behavior. More detail on the property relationships, specific data, and contrast of homogeneous versus heterogeneous blends (as well as the intermediate cases) will be covered in this chapter.

The most important property of a polymer is the glass transition. The position of the glass transition temperature determines such properties as the transition from solid to rubber behavior, the creep rate, the rheological characteristics, the crystallization rate, and the toughness (particularly relevant for crystalline or cross-linked polymers). The viscoelastic and rheological properties as well as the crystallization kinetics of polymers have been shown to be empirically related to the T_g by the Williams, Landel, and Ferry equation [1]:

$$\log a_T = 17.44(T - T_g)/[51.6 + (T - T_g)] \qquad (6.1)$$

where a_T represents the temperature variation of the segmental frictional coefficient for a molecular relaxation process. The value of a_T determines the amount of horizontal shift of the time scale for data determined isothermally in short time intervals. This method, using the T_g as a reference temperature, allows one to extrapolate data to cover a time span much greater than is experimentally possible. Commonly referred to as time–temperature superposition, it is quite applicable for the construction of relaxation–modulus master curves or extending creep data covering many decades of time.

Specific aspects of the T_g appear similar to a second-order thermodynamic transition since a change in slope occurs in specific heat or volume versus temperature when passing through the glass transition. Similar changes in slope occur with permeability, dielectric constant, and electrical or thermal conductivity; thus, activation energies (e.g., diffusion) exhibit different values below and above the T_g. Dramatic changes in modulus also occur at the T_g, reflecting the importance of the various thermomechanical techniques in the determination of the glass transition behavior in polymer blends.

A brief review of the nature of the glass transition and the various theories proposed for its existence were covered in Chapter 3. For purposes of discussion, a simplistic view of the glass transition will be presented. The T_g represents the temperature at which the polymeric chains have a combination of energy (i.e., vibrational, translational, and rotational forces) equal to the forces of attraction (i.e., dispersive forces, polar interactions, and specific interactions). Below this temperature, the polymer chains are locked into a random network with motion restricted to vibrational, rotational, and short-order translational movement of small units of the polymer chain. Above this point, translational movement of the entire chain is possible, and diffusion therefore occurs as might be required, for example, by a crystallization process which approaches a negligible rate at the T_g. The observed T_g is affected by the frequency of the observed test method, pressure, crystallinity, and cross-linking.

Heterogeneous blends will exhibit distinctly different T_g's as opposed to

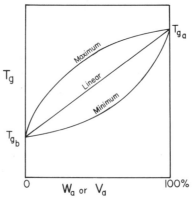

Fig. 6.1. Generalized behavior of T_g relationships for miscible polymer blends.

6.1. Thermal and Thermomechanical

a homogeneous blend with a single T_g. In this section, the predictions capable of being made will be outlined along with the generalized expectations of the properties of a heterogeneous blend versus a homogeneous blend.

For miscible polymer blends, T_g versus composition is not a universally similar relationship, but has many variations, similar to those observed with random copolymers. For polymer blends exhibiting miscibility over the entire composition range, three generalized curves (illustrated in Fig. 6.1) are possible: the linear relationship and the minimum and maximum deviations from linearity.

Examples of a linear variation of T_g versus composition include nitrocellulose–poly(methyl acrylate) [2] and natural rubber–polybutadiene [3].

Examples of a minimum variation from linearity are quite common. In many of these cases, data fit relationships commonly used for random copolymers or for polymer–diluent blends (e.g., plasticized PVC), namely, the Fox equation [4],

$$1/T_g = (W_a/T_{ga}) + (W_b/T_{gb}) \tag{6.2}$$

and the Gordon–Taylor equation [5],

$$T_g = [W_a T_{ga} + k(1 - W_a)T_{gb}]/[W_a + k(1 - W_a)] \tag{6.3}$$

where T_{ga} and T_{gb} represent the glass transitions of the undiluted polymer components, W_a and W_b are the weight fractions of the blend, and k is the ratio of the thermal expansion coefficients between the rubber and glass states of the component polymers, $(\alpha_{1b} - \alpha_{gb})/(\alpha_{1a} - \alpha_{ga})$. The Kelley–Bueche equation [6] is similar to the Gordon–Taylor equation except that the volume fraction, ϕ_i, is used instead of the weight fraction:

$$T_g = [\phi_a T_{ga} + k(1 - \phi_a)T_{gb}]/[\phi_a + k(1 - \phi_a)] \tag{6.4}$$

As $\alpha_1 - \alpha_g$ has been proposed to be constant for all polymers [7], $k = 1.0$ and the Gordon–Taylor and Kelley–Bueche equations reduce to the linear form:

$$T_g = W_a T_{ga} + W_b T_{gb} \tag{6.5}$$

$$T_g = \phi_a T_{ga} + \phi_b T_{gb} \tag{6.6}$$

The above equations, commonly used to express the T_g–composition relationships for copolymers and plasticizer–polymer compositions, are quite useful for miscible polymer blends.

Examples of the successful applications of the Fox equation to miscible blends include poly(vinyl chloride)–butadiene/acrylonitrile copolymers [8], poly(hydroxy ether) of bisphenol A–poly(ε-caprolactone) [9], and poly(vinyl chloride)–poly(ethylene/vinyl acetate/sulfur dioxide) [10]. Miscible

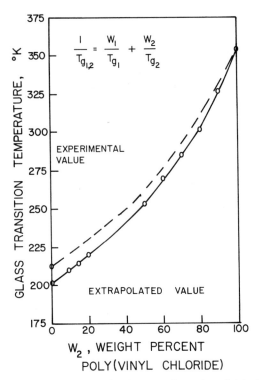

Fig. 6.2. Glass transition behavior of poly(ε-caprolactone)–poly(vinyl chloride) blends. The solid line was calculated from the Fox equation with $T_{g1} = 202°K$ and $T_{g2} = 365°K$. The dashed line was calculated from the Fox equation with $T_{g1} = 213°K$. [Reprinted by permission from J. V. Koleske and R. D. Lundberg, *J. Polym. Sci., Part A-2* **7**, 795 (1969). Copyright by John Wiley and Sons, Inc.]

blends in which the T_g–composition data are satisfied by the Gordon–Taylor expression include freeze-dried poly(methyl methacrylate)–poly(vinyl acetate) [11] and styrene/butadiene copolymer with polybutadiene [12].

In an interesting example of the application of these equations, the miscible polymer approach was used to determine the T_g of amorphous poly(ε-caprolactone) from extrapolation of data from poly(vinyl chloride)–poly(ε-caprolactone) blends using the Fox equation [13]. A direct determination of the T_g of poly(ε-caprolactone) was not possible due to the rapid crystallization of poly(ε-caprolactone), which made the preparation of amorphous specimens impossible even while using extreme quenching methods (immersion in liquid nitrogen). As poly(vinyl chloride) addition dramatically lowered the crystallization rate, amorphous blend specimens were possible, thus allowing the extrapolation illustrated in Fig. 6.2.

A similar extrapolation of T_g data for poly(methyl methacrylate)–poly-

6.1. Thermal and Thermomechanical

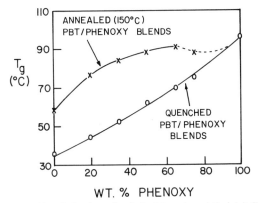

Fig. 6.3. Glass transition behavior of poly(butylene terephthalate) (PBT)–poly(hydroxy ether) of bisphenol A (Phenoxy) blends. [Reprinted by permission from L. M. Robeson and A. B. Furtek, *J. Appl. Polym. Sci.* **23**, 645 (1979). Copyright by John Wiley and Sons, Inc.]

(vinylidene fluoride) and poly(ethyl methacrylate)–poly(vinylidene fluoride) blends has been used to ascertain the T_g of poly(vinylidene fluoride) [14–16]. Disagreement exists in this value extrapolated from miscible blends in that $-45°C$ [14, 15] and $38°C$ [16] have both been claimed.

Maximum deviations of the T_g–composition data from linearity strongly imply the existence of strong intermolecular interactions between the constituents of the blend. This is the case with poly(vinyl nitrate)–poly(vinyl acetate) and poly(vinyl nitrate)–ethylene/vinyl acetate copolymers, where the heat of mixing was determined to be exothermic [17]. Poly(acrylic acid) and poly(ethylene oxide) blends exhibited a complex behavior for the T_g–composition data but, nevertheless, exhibited a maximum at 10% poly(ethylene oxide). This was suggested to be the result of strong intermolecular bonding due to the hydrogen bonding between the ether and carboxyl groups [18]. Poly(methyl methacrylate)–poly(vinylidene fluoride) blends, which exhibited a negative interaction parameter, had a maximum deviation from linearity in the T_g–composition data reported by Nishi and Wang [14]. In the case where the lower T_g component can crystallize out of solution, leaving an amorphous phase richer in the high-T_g component than the bulk mixture, maximum deviation is quite possible and probably explains the results of Nishi and Wang [14]. An example of this behavior is shown in Fig. 6.3 for quenched and annealed blends of the poly(hydroxy ether) of bisphenol A and poly(butylene terephthalate) [19].

6.1.2 Modulus–Temperature Behavior

The thermomechanical response of polymer blends is a sensitive key to the phase behavior of the blend. As discussed in Chapter 3, the difference between homogeneous and heterogeneous blends is easily detected in the mechanical loss and modulus–temperature data (refer to Figs. 3.4 and 3.3 in Chapter 3 for the generalized relation of single-phase versus two-phase behavior). Additional details will be given in this chapter on the thermomechanical aspects of miscible polymer systems.

Dynamic mechanical testing methods (e.g., torsion pendulum, viscoelastomer, resilience) have been the most common techniques to determine polymer–polymer phase behavior. The mechanical damping, tan δ, data parallel the modulus–temperature data in that the slope of the modulus–temperature curve is roughly proportional to the value of tan δ. This is illustrated in Fig. 6.4, which shows generalized relationships for heterogeneous, partially miscible, and miscible polymer blends. Note that, for partially miscible blends, two possibilities are represented. A broad transition region can exist, indicating microheterogeneity in that an infinite number of phases of differing compositions exist. However, the distribution of compositions is such that a dispersion maximum occurs between those of the constituents (case 3). Another possibility (case 2) exists which is expected by consideration of the phase diagram for a polymer blend. Each phase contains a certain concentration of the other component of the blend, although, for heterogeneous blends, this is negligible relative to any property change (i.e., T_g shift) which can be observed. However, in cases of marginal miscibility, the concentration of the minor component in the phase of the major component may have a finite and measurable value, thus shifting the respective transitions in the directions shown in Fig. 6.4. Of course, there are other possibilities than those stated above, and these will be pointed out in the discussion that follows related to experimentally investigated blend phase behavior.

Blends which exhibit sharp transitions similar to that expected of single-phase materials (case 1) include blends of poly(ε-caprolactone) with nitrocellulose [20], poly(hydroxy ether) of bisphenol A [20], or poly(vinyl chloride) [13], poly(vinylidene fluoride)–poly(methyl methacrylate) [14], ethylene/vinyl acetate/sulfur dioxide terpolymer–poly(vinyl chloride) [10], and poly(butylene terephthalate)–poly(hydroxy ether) of bisphenol A [19]. Case 2 blends have been cited for a 1:4 mixture of chlorinated (32% Cl) ethylene–vinyl acetate copolymer and poly(vinyl chloride) [21], an interpenetrating network of poly(ethyl acrylate) and polystyrene [22], and a 1:1 mixture of poly(vinyl acetate) with vinyl chloride/vinyl acetate copolymer of 10% vinyl acetate [23]. Case 3 microheterogeneous blends include butadiene/acrylonitrile rubber–poly(vinyl chloride) [24], interpenetrating networks of poly(ethyl acrylate) and poly(methyl methacrylate) [25], and styrene/butadiene copolymer (37.5% styrene)–styrene/butadiene copolymer (50% styrene) [26]. Case 4 blends exhibit the typical immiscible behavior observed

6.1. Thermal and Thermomechanical

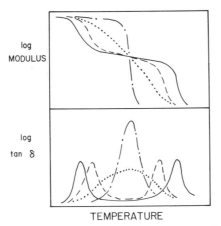

Fig. 6.4. Generalized mechanical loss (tan δ) and modulus behavior for different types of polymer blends. Case 1 (dashed–dotted line), miscible; case 2 (dashed line), limited miscibility (partially miscible); case 3 (dotted line), microheterogeneous (partially miscible); case 4 (solid line), heterogeneous.

with polymer blends. Examples of these are numerous and include the polystyrene–polybutadiene simple mixtures and block copolymers. Several reviews have listed some of the examples of these systems [27–29], and a review of the endless number of these cases will not be attempted in this treatise.

Modulus–temperature data can be analyzed to determine the relative heterogeneity of the blend. For heterogeneous blends, the theories of Kerner [30], Takayanagi [31], and Hashin and Shtrikman [32] have been applied to polymer–polymer two-phase mixtures. Kerner's equation is designed to predict the modulus of a composite composed of a matrix in which spherical particles of a different modulus are incorporated. Kerner's equation is

$$E = E_c \frac{\{\phi_d E_d/[(7 - 5v_c)E_c + (8 - 10v_c)E_d]\} + [\phi_c/15(1 - v_c)]}{\{\phi_d E_c/[(7 - 5v_c)E_c + (8 - 10v_c)E_d]\} + [\phi_c/15(1 - v_c)]} \quad (6.7)$$

where E, E_c, and E_d are the respective moduli of the composite, the continuous phase, and the dispersed phase; ϕ_c and ϕ_d are the volume fractions of the continuous and dispersed phases; and v_c is Poisson's ratio for the continuous phase. A modification of Kerner's equation was used to determine the contribution of the respective components of a polysulfone–poly(dimethyl siloxane) $(AB)_n$ block copolymer to the continuous as well as dispersed phase structure [33]. By letting X_a and X_b represent the fractional contribution of components 1 and 2 to the continuous phase, with the provision that $X_a + X_b = 1$, Kerner's equation was modified to

$$E = X_a E_1 \frac{\{\phi_2 E_2/[(7 - 5v_1)E_1 + (8 - 10v_1)E_2]\} + [\phi_1/15(1 - v_1)]}{\{\phi_2 E_1/[(7 - 5v_1)E_1 + (8 - 10v_1)E_2]\} + [\phi_1/15(1 - v_1)]}$$

$$+ X_b E_2 \frac{\{\phi_1 E_1/[(7 - 5v_2)E_2 + (8 - 10v_2)E_1]\} + [\phi_2/15(1 - v_2)]}{\{\phi_1 E_2/[(7 - 5v_2)E_2 + (8 - 10v_2)E_1]\} + [\phi_2/15(1 - v_2)]} \quad (6.8)$$

where case A = component 1 continuous ($E_c = E_1$; $E_d = E_2$; $v_c = v_1$), case B = component 2 continuous ($E_c = E_2$; $E_d = E_1$; $v_c = v_2$), and ϕ_1 and ϕ_2 are the volume fractions of components 1 and 2.

From experimental data for the block copolymer, X_a, and likewise X_b, can be calculated. As an experimental check on this analysis, permeability data for the same block copolymers were analyzed using Maxwell's equation to determine X_a and X_b. Both Kerner's and Maxwell's modified equations predicted equal phase inversion points (i.e., $X_a = X_b$) at 0.53 volume fraction polysulfone for tetrahydrofuran cast films. The values of X_a and X_b were found to be dependent upon the method of sample preparation.

Hashin's analysis can also be used to predict the modulus of material composed of a matrix in which spherical inclusions of a different modulus are present [32]. Hashin's equation is

$$E = E_c\{1 - 15[(1 - v_c)/(7 - 5v_c)]\phi_d\} \quad (6.9)$$

Both Kerner's and Hashin's equations have been utilized to predict the modulus of impact polystyrene as a function of rubber phase volume [34]. Hashin's equation yielded better agreement with experimental results at higher rubber phase volumes.

The Takayanagi model was initially developed to predict the viscoelastic behavior of two-phase polymers [31]. Extension of the theory to predict results for isotropic as well as oriented crystalline polymers was then investigated [35].

In the initial form of the Takayanagi model to determine the temperature response of a phase-separated polymer–polymer blend, the parallel or series models were used. For the parallel model (Fig. 6.5) (results similar to those

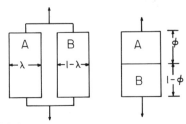

Fig. 6.5. Parallel (*left*) [Eq. (6.10)] and series (*right*) [Eq. (6.11)] mechanical models.

6.1. Thermal and Thermomechanical

in which the higher modulus phase is continuous), the complex modulus, E^*, is given by

$$E^* = \lambda E_a^* + (1 - \lambda)E_b^* \quad (\lambda = \text{volume fraction of a}) \quad (6.10)$$

which represents a Voigt modulus. The series mode (Fig. 6.5) yielding a Reuss modulus is

$$1/E^* = (\phi/E_a^*) + (1 - \phi)/E_b^* \quad (6.11)$$

These models, of course, fit the response when bonded films are attached either in parallel or in series. However, for films in which the components are well mixed, the predicted response from the above models poorly represents the experimental data. Takayanagi then proposed two alternative models, shown in Fig. 6.6. The complex modulus of model 1 is satisfied by Eq. (6.12) and model 2 by Eq. (6.13):

$$\frac{1}{E^*} = \frac{\phi}{\lambda E_a^* + (1 - \lambda)E_b^*} + \frac{1 - \phi}{E_b^*} \quad (6.12)$$

$$E^* = \lambda\left(\frac{\phi}{E_a^*} + \frac{1 - \phi}{E_b^*}\right)^{-1} + (1 - \lambda)E_b^* \quad (6.13)$$

For these models, λ and ϕ represent adjustable parameters used to fit experimental data. The Takayanagi model has been utilized to assess the behavior of two-phase blends from component property data. Application of the model to partially miscible or miscible systems, however, cannot be made. For partially miscible blends, Manabe et al. [24] proposed a model in which the blend was assumed to consist of many components with different values of free volume fraction (therefore different T_g values). A distribution function of free volume fraction, $F(f)$, was introduced to explain the viscoelastic behavior of the partially miscible blends.

In order to experimentally determine the value of $F(f)$, specific volume versus temperature data were required. For the individual cells, the prob-

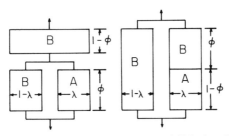

Fig. 6.6. Mechanical models proposed by Takayanagi [31], described by Eqs. (6.12) and (6.13).

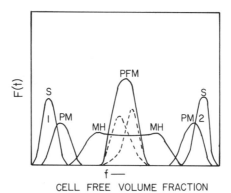

Fig. 6.7. Schematic generalization of the distribution function of free volume fraction $F(f)$ for various polymer blends. Curves 1 and 2 illustrate the behavior of the components. S, separated phases; PM, partially miscible; MH, microheterogeneous; PFM, perfectly miscible. [Reprinted by permission from S. Manabe, R. Murakami, and M. Takayanagi, *Mem. Fac. Eng., Kyushu Univ.* **28**, 295 (1969).]

ability that the value of f is between f and $f + df$ is given by $F(f)\,df$, where $F(f)$ is the probability density of f. The experimental free volume fraction f is given as

$$f(T) = \int_0^\infty f F(f)\,df \quad \text{where} \quad \int_0^\infty F(f)\,df = 1 \tag{6.14}$$

From specific volume versus temperature data, it was shown how to determine experimental values of $F(f)$.

Figure 6.7 illustrates the distribution function of free volume fraction $F(f)$ versus component free volume fraction as a function of the type of blend behavior.

Manabe et al. [24] differentiated between the separated phases, partial miscible, microheterogeneous, and perfectly miscible systems in a manner analogous to that mentioned previously for modulus–temperature behavior (cases 1–4, Fig. 6.4) where the mechanical loss spectra are analogous to the distribution function $F(f)$ shown in the above curve.

Experimental data for the butadiene/acrylonitrile rubber and poly(vinyl chloride) blend were obtained with the $\Delta\alpha F(T_g)$ data versus T_g, shown in Fig. 6.8. In terms of the cases presented in Fig. 6.7, this corresponds to the microheterogeneous case.

Manabe et al. [24] developed expressions for the complex modulus for both series and parallel models using $F(T_g)$ data to predict the temperature dependence of the dynamic viscoelasticity for polymer blends. For the detailed development and utility of this approach for analyzing the behavior of polymer blends, the interested reader should consult the original reference.

6.2. Mechanical

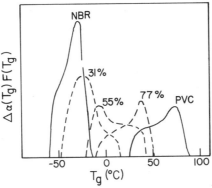

Fig. 6.8. $\Delta\alpha \cdot F(T_g)$ for the blend of poly(vinyl chloride) with butadiene/acrylonitrile copolymer. Percentages refer to weight percent poly(vinyl chloride). [Reprinted by permission from S. Manabe, R. Murakami, and M. Takayanagi, *Mem. Fac. Eng., Kyushu Univ.* **28**, 295 (1969).]

Miyata and Hata [36] studied the viscoelastic properties of poly(methyl methacrylate) and poly(vinyl acetate) as a function of the method of sample preparation. Distinct differences in the modulus–temperature behavior were noted as a function of preparation, going from heterogeneous when cast in solvent to fairly homogeneous when freeze-dried. On assuming the blend consisted of homogeneous microphases, the modulus data were analyzed to determine the distribution function of the microphases. The results illustrated a maximum distribution function value at 50/50 volume ratio for the freeze-dried blends, whereas solvent-cast samples yielded maximum distribution function values near the 0 or 100% V_f extremes.

Other references concerning the modulus behavior of two-phase polymeric blends include Nielsen [37, 38] and Hashin [39].

6.2 MECHANICAL

Dynamic mechanical analysis, used to characterize polymer miscibility (shear modulus, loss modulus, mechanical loss, tensile modulus, resilience), measures properties associated with nondestructive testing. The dynamic mechanical properties of miscible polymer blends have already been discussed in this treatise (Chapter 3 and earlier in this chapter). Ultimate mechanical properties such as tensile strength, impact strength, abrasion resistance, environmental stress crack or crazing resistance, and fatigue resistance are mechanical properties covered much less frequently in discussions of miscible polymer–polymer blends. These properties in application are obviously of great importance and therefore will be discussed here.

Ultimate properties have been most extensively covered for butadiene/acrylonitrile copolymer blends with poly(vinyl chloride). Emmett [40] studied the addition of plasticized poly(vinyl chloride) to nitrile rubber vulcanizates. The mixture exhibited advantages over unblended nitrile rubber in flex crack resistance, tear strength, and property retention in sunlight and ozone environments. Lower tensile strength and ultimate elongation were noted, as well as higher compression set. A detailed optimization study by Schwarz and Edwards [41] of the vulcanized nitrile rubber–poly(vinyl chloride) blend covered blend properties as a function of blend ratio, poly(vinyl chloride) molecular weight, acrylonitrile content, and nitrile rubber viscosity. A computer analysis of the factorial experiment showed all variables to yield a significant influence on the property balance, and optimized formulations for specific applications were discussed.

Most of the property studies of nitrile rubber–poly(vinyl chloride) blends have been related to the use of nitrile rubber as a permanent plasticizer for poly(vinyl chloride). The utility of these blends as compared to conventional plasticized poly(vinyl chloride) compositions lies in their resistance to plasticizer extraction and high temperature resistance to plasticizer volatilization. Pittenger and Cohan [42] illustrated this permanence in comparisons with conventional plasticized poly(vinyl chloride) by showing that high temperature aging of the nitrile rubber–poly(vinyl chloride) blends resulted in superior retention of original properties, such as tensile strength and ultimate elongation. Reed observed that lower brittle temperatures were obtained with nitrile rubber as the plasticizer for poly(vinyl chloride) than with typical low molecular weight plasticizers (compared at similar room temperature modulus) [43].

A miscible blend of poly(vinyl chloride) and an α-methylstyrene/methacrylonitrile copolymer investigated by Kenney [44] exhibited higher tensile and flexural strengths than either of the constituents, as shown by the data in Table 6.1. The α-methylstyrene/methacrylonitrile copolymer is a brittle polymer with failure before a yield point is attained, whereas poly(vinyl chloride) has a definite yield in the range of 5–7% elongation. Addition of poly(vinyl chloride) to the copolymer yields a material which approaches closer to the yield condition and exhibits a higher strength than the copolymer. The higher modulus of the blend relative to PVC is reflected in a higher strength than poly(vinyl chloride), as depicted in Fig. 6.9.

Blends of ethylene/vinyl acetate copolymer with poly(vinyl chloride) at 45% vinyl acetate in the copolymer were characterized as miscible blends by Feldman and Rusu [45] from mechanical loss and dielectric loss data. The composition dependence of the tensile strength and ultimate elongation

6.2. Mechanical

TABLE 6.1

Mechanical Properties of Poly(vinyl chloride)–α-Methylstyrene/Methacrylonitrile Copolymer Blend[a]

	PVC	50/50 Blend	α-MS/MAN
Notched Izod impact strength (ft-lb/in. of notch)	0.7	0.5	0.3
Ultimate elongation (%)	110	5	3
Tensile strength (psi)	8,000	10,180	6,800
Flexural strength (psi)	15,000	17,375	14,600
Flexural modulus (psi)	430,000	500,000	638,000

[a] Reprinted by permission from J. F. Kenney, *in* "Recent Advances in Polymer Blends, Grafts, and Blocks" (L. H. Sperling, ed.), p. 117. Copyright 1974 by Plenum Press.

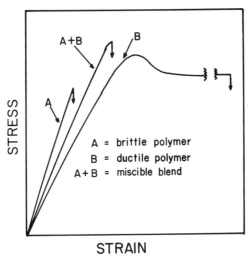

Fig. 6.9. Typical stress–strain behavior for a miscible mixture of a brittle polymer with a ductile polymer. Note that the miscible blend exhibits higher tensile strength than either component.

for these blends exhibited the characteristics of mechanical compatibility expected of partially miscible or even miscible polymer systems.

Partial miscibility of the poly(butylene terephthalate)–poly(tetrahydrofuran) $(AB)_n$ block copolymer with poly(vinyl chloride) has been previously cited (Chapter 5). Nishi and Kwei [46] investigated the impact strength of these blends as a function of annealing at 130°C. Annealing was found to increase the level of phase separation of this blend, with a resultant increase (sevenfold) in the notched impact strength.

Hammer [47] compared the mechanical properties of polymeric plasticized PVC relative to that containing conventional low molecular weight plasticizers. The primary mechanical property advantage noted for the high M_w plasticizers was low temperature toughness. Improvements in ultimate elongation, tensile strength, and tear strength were also cited relative to dioctyl phthalate-plasticized PVC.

The impact strength and resistance to thermal cycling in a thermoset block copolymer of a cycloaliphatic epoxy and poly(ε-caprolactone) were investigated by Noshay and Robeson [48–50] as a function of the poly-(ε-caprolactone) block molecular weight. The single-phase thermoset block copolymer exhibited improved ultimate elongations and notched impact strength over the two-phase systems containing poly(ε-caprolactone) blocks of higher molecular weight (at high levels of addition). However, this advantage was achieved at a considerable loss in heat distortion temperature. When comparisons were made at similar heat distortion temperatures, the two-phase blends exhibited superior performance.

Miscible polymer systems containing a crystalline component, in which crystallization is depressed or eliminated, can yield properties markedly different than their constituents. When mixed, poly(ethylene oxide) and poly(acrylic acid) (both water-soluble polymers) form a complex that resists water solubility. The properties of the constituents and their 50/50 blend are listed in Table 6.2. The low modulus of the complex is due to the elimination of poly(ethylene oxide) crystallinity, combined with a T_g of the blend below room temperature.

Yee investigated the failure properties of polystyrene–poly(2,6-dimethyl-1,4-phenyl oxide) blends as a function of composition and testing rate [51]. The plot of the maximum tensile stress versus composition as a function of strain rate (Fig. 6.10) indicates that the brittle–ductile failure envelope shifts

TABLE 6.2

Mechanical Properties of Poly(ethylene oxide)–Poly(acrylic acid) Complexes[a]

	Poly(acrylic acid)	50/50 Blend	Poly(ethylene oxide)
Tensile modulus (psi)	400,000	200	50,000
T_g (°C)	100	5	−55
Water extractables (%)	100	10	100
Tensile strength (psi)	9,000	1,500	2,100
Ultimate elongation (%)	4	600	1,000

[a] Reprinted by permission from K. L. Smith, A. E. Winslow, and D. E. Petersen, *Ind. Eng. Chem.* **51**, 136 (1959). Copyright by the American Chemical Society.

6.2. Mechanical

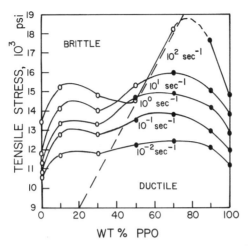

Fig. 6.10. Effect of PPO [poly(2,6-dimethyl-1,4-phenylene oxide)]–polystyrene composition on the tensile strength at various strain rates. Open circles denote brittle fracture and filled circles denote ductile yield. The dashed line separates the ductile and brittle regions. [Reprinted by permission from A. F. Yee, *Polym. Eng. Sci.* **17**, 213 (1977). Copyright by the Society of Plastics Engineers, Inc.]

to higher PPO contents as the strain rate is increased. Within the ductile region, a maximum in tensile strength is observed. Maximum values also occur in the brittle region at a composition of 15–20% PPO. The maximum tensile strength in the brittle region corresponded to the modulus of the 90/10 polystyrene–PPO blend, which was the highest modulus of all blends including 100% polystyrene.

In Fig. 6.11, the tensile yield stress versus strain rate for the higher PPO content blends shows anomalous behavior as a function of strain rate. This can be explained: The lower value at higher strain rates represents the change from ductile to brittle failure (see Fig. 6.10). Yee [51] observed that the addition of polystyrene to PPO also increased the sensitivity of yield stress to the strain rate. He rationalized this behavior as being the consequence of the elimination of the broad PPO β peak (90°C) with the addition of polystyrene, which also results in a negative excess volume of mixing. The Ree–Eyring equation for stress-activated flow, where v is the activated flow value,

$$d\sigma_y/d \ln \varepsilon = 2kT/v \tag{6.15}$$

was utilized to explain the experimental results. The negative excess volume of mixing was cited as being consistent with the predictions of the Ree–Eyring equation [52].

Polyelectrolyte complexes can exhibit a total spectrum of mechanical

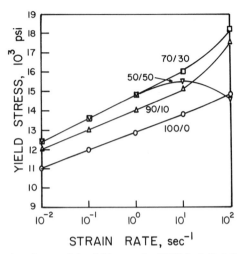

Fig. 6.11. Effect of strain rate of the yield strength of PPO [poly(2,6-dimethyl-1,4-phenylene oxide)]–polystyrene mixtures. [Reprinted by permission from A. F. Yee, *Polym. Eng. Sci.* **17**, 213 (1977). Copyright by the Society of Plastics Engineers, Inc.]

properties dependent upon water sorption [53–55]. When dry, the polyelectrolyte complexes are transparent, high modulus, and brittle. Upon limited water, alcohol, or glycol sorption, leathery or rubbery behavior is observed. The equimolar complex absorbs up to 30% (by weight) water when saturated. In aqueous electrolyte solutions, nonstoichiometric complexes can sorb several times their weights in water and thus exhibit hydrogel-type behavior.

6.3 ELECTRICAL

In Chapter 3, the dielectric loss measurement was described as a useful technique for the determination of polymer blend miscibility. This technique has been cited as evidence for miscibility in blends of poly(vinyl nitrate)–poly(vinyl acetate) [17], poly(vinyl nitrate) with ethylene/vinyl acetate copolymer (86% vinyl acetate) [17], polystyrene–poly(vinyl methyl ether) [56], polybutadiene with styrene/butadiene copolymers [57], and ethylene/vinyl acetate copolymer with poly(vinyl chloride) [45]. Generally, dielectric loss results are similar to other methods of miscibility determination. However, MacKnight *et al.* [58] observed differences between dielectric loss, dynamic mechanical, and calorimetric data with the poly(2,6-dimethyl-1,4-phenylene oxide)–polystyrene mixture. The composition dependence of the dynamic mechanical data indicated the presence of a polystyrene-rich

6.3. Electrical

phase and a PPO-rich phase. At all concentrations the calorimetric data yielded single T_g's, similar in value to those for the polystyrene-rich phase determined from dynamic mechanical data. The T_g determined from dielectric data yielded a sigmoidal curve as a function of concentration. Shultz and Beach [59], however, concluded that PPO and polystyrene were miscible at all compositions and attributed the observation of nonhomogeneity reported in other studies to inefficient mixing techniques or segregation of PPO from the amorphous phase of the miscible blend by crystallization.

The bulk of the electrical property data for miscible polymer blends has been limited to dielectric loss. A few papers dealing with other electrical property observations will be briefly discussed below.

The electrical conductivity of blends of polymers capable of charge-transfer complexation was reported by Sulzberg and Cotter [60]. When mixed, electron acceptor polyesters based on nitrophthalic acids and electron donor polymers based on aryliminodiethanol yielded yellow to orange transparent blends (films of unblended constituent polymers were colorless and transparent) with T_g's intermediate between the component T_g's. The charge-transfer polymer blends were cited to have electrical conductivities 10^3 (ohm-cm)$^{-1}$ times greater than that of the unblended donor polymer. A copolymer containing both the electron acceptor and electron donor groups, however, exhibited even higher conductivities than the electron donor polymer [10^5 (ohm-cm)$^{-1}$ times higher].

The electrical properties of the polyelectrolyte complex of poly(styrene sulfonate) and poly(vinyl benzyl trimethyl ammonium) were studied by Michaels and co-workers in detail [61-63]. Dilute solution conductivity measurements [62] were used to assess the stoichiometry of the complex. In the reaction of the complex of sodium poly(styrene sulfonate)–poly(vinyl benzyl trimethyl ammonium chloride), an increase in conductance was observed relative to the conductances of the unblended polymers (Fig. 6.12). This increase was due to the higher equivalent conductance of the released microions (i.e., NaCl) than the polyions removed by complexation. The maximum conductance was observed at equimolar amounts of the constituent polyions, identifying it as an equimolar stoichiometric complex. In the reaction of the polyanion, poly(styrene sulfonate), with the polycation, poly(vinyl benzyl trimethyl ammonium), a decrease in conductance is observed relative to the unblended constituent solutions, as expected from the neutralization occurring with complex formation (Fig. 6.13). An equimolar stoichiometric complex, evidenced by a minimum in solution conductance, was observed at equimolar amounts of the polyacid and polybase.

In a complementary study [63], the dielectric properties of the stoichiometric poly(vinyl benzyl trimethyl ammonium)–poly(styrene sulfonate) complex were studied as a function of water and NaBr content. Without

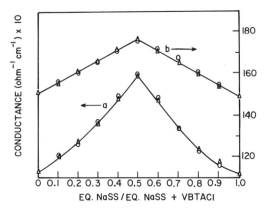

Fig. 6.12. Conductances of reaction mixtures (○) and control solutions (△). Curve a, 10^{-3} N NaSS [sodium poly(styrene sulfonate)]–10^{-3} N VBTACl [poly(vinyl benzyl trimethyl ammonium chloride)] salt-free solutions; curve b, 10^{-2} N NaSS–10^{-2} N VBTACl in 10^{-2} N NaCl. [Reprinted by permission from A. S. Michaels, L. Mir, and N. S. Schneider, *J. Phys. Chem.* **69**, 1447 (1965). Copyright by the American Chemical Society.]

the presence of NaBr, the equimolar complex exhibits dielectric constants of 5 (dry) and 50 (equilibrium water content), with a resistivity of $\sim 10^{10}$ ohm-cm. With addition of up to 0.50 equivalent of NaBr per equivalent of complex, the dielectric constants at low frequency was in excess of 10^5 with a resultant higher dielectric loss factor. The dielectric results indicate that microions are isolated in the polyelectrolyte complex within noninterconnecting domains.

The conductance of the equimolar complex of poly(vinyl benzyl trimethyl ammonium)–poly(styrene sulfonate) was observed to be zero, thus indicating complete interaction of ionizable groups [62]. Studies with poly(ethyleneimine) and carboxymethyl cellulose, however, yielded a measurable conductance at the minimum point, indicating the presence of ionizable groups on poly(ethyleneimine) [64]. This was attributed to the rigid nature of the carboxymethyl cellulose chain, the high flexibility of poly(ethyleneimine), and/or the branching characteristics of poly(ethyleneimine). The distance between ionizable groups of poly(ethyleneimine) was shown to be less than that of carboxymethyl cellulose; thus, the geometrical hinderance of the rigid carboxymethyl cellulose molecules prevented complete neutralization of the ionizable groups on poly(ethyleneimine).

While dielectric loss data have been studied and reported in detail, electrical conductivity, dielectric strength, arc track, and treeing resistance data for miscible polymer blends are generally unreported and leave an important property gap wide open for future investigation. Pyro- and piezoelectricity

6.4. Rheological and Viscoelastic

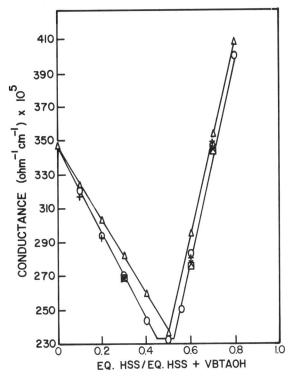

Fig. 6.13. Conductances of reaction mixtures and control solutions: 10^{-2} N HSS [poly-(styrene sulfonic acid)]–10^{-2} N VBTAOH [poly(vinyl benzyl trimethyl ammonium hydroxide)] in 2×10^{-2} N $CaCl_2$. △, control solutions. The reaction mixtures are ○, VBTAOH into HSS; □, HSS into VBTAOH; +, VBTAOH into HSS, centrifuged supernatant; ×, HSS into VBTAOH, centrifuged supernatant. [Reprinted by permission from A. S. Michaels, L. Mir, and N. S. Schneider, *J. Phys. Chem.* **69**, 1447 (1965). Copyright by the American Chemical Society.]

data for poly(vinylidene fluoride)–poly(methyl methacrylate) blends have recently been reported by Lee *et al.* [65]. PMMA, as well as poly(vinyl fluoride), was added to PVF_2 in order to vary the crystallinity and the crystalline form.

6.4 RHEOLOGICAL AND VISCOELASTIC

A major characteristic of a single-phase polymeric system is that it exhibits only one primary glass transition. The transition is from the glassy state, where only limited main-chain bending or side-chain rotation occurs, to the rubbery state, where unencumbered motion of many segments takes place, but where entanglements still exist. The transition from the rubbery state to the liquid state is also rather distinct for narrow-distribution, amorphous, uncrosslinked polymers. In this transition the last encumberance, entanglements, is thrown off. Time and temperature are the variables of importance in moving through these kinetic transitions.

Little enough is known about the transitions of pure polymers, or pure polymers plus low molecular weight plasticizers, to say a great deal about the properties expected for miscible polymer systems near or above the glass transition temperature. But, as with low molecular weight plasticizers, it is reasonable that miscible systems should show one intermediate glass transition. Furthermore, it is reasonable that the rubbery plateau moduli should average and that the entanglement relaxation times should average. It is not clear what the exact averaging rules should be (regrettably the same situation exists for mixtures of polymers with the *same* structure), but the maximum relaxation time, τ_m, and the viscosity at vanishing shear rate, η_0, for the mixture should at least fall between those of the components unless they are strong, positive interactions which would act to increase the resistance to flow.

6.4.1 Steady Viscometric Functions

A class of flows known as viscometric flows is widely used to study the rheological properties of fluids. In these flows the streamlines are parallel and steady. Furthermore, it is assumed that the flow has been continuing long enough so that each material element has developed a steady stress—an important consideration for time-dependent fluids. A flow of this type will occur in the commonly used bob and cup viscometer (Fig. 6.14).

Under circumstances of steady viscometric flow, it can be shown that three material functions are necessary and sufficient to describe most fluids [66]. These functions, η, Ψ_1, and Ψ_2, depend on the velocity gradient, the only kinematic variable; temperature; pressure; and concentration. They are, effectively, the proportionality constants between the components of stress and the velocity gradient (or shear rate). The viscosity function is defined as the ratio of shear stress to shear rate $\dot{\gamma}$. The second, Ψ_1, is defined as the ratio of the first normal stress difference to $\dot{\gamma}^2$, while Ψ_2 relates the second normal stress difference to $\dot{\gamma}^2$.

No complete study of the viscometric functions for a miscible system has been reported. The viscosity function $\eta(\dot{\gamma},\phi_2)$ has been measured for a few, however, with Schmidt [67] recently describing both the viscosity and first-normal stress functions for PPO–HIPS (high impact polystyrene), finding both functions to be completely monotonic with composition. Slonimskii

6.4. Rheological and Viscoelastic

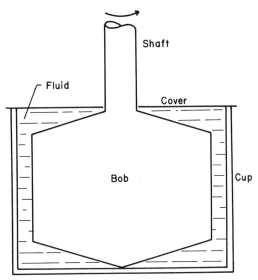

Fig. 6.14. Schematic of a bob and cup viscometer suitable for determining the viscosity of a fluid at various rates of shear and temperatures.

and Komskaya [68] measured a viscosity at low shear rate and found that $\eta(\phi_2)$ was a monotonic function of ϕ_2 for the systems polybutadiene–natural rubber and SBR (10% styrene)–SBR (30% styrene), whereas a maximum was observed when SBR (30% styrene) was mixed with either NR or BR. This behavior correlated with the ternary phase behavior of 5% solutions.

The viscosity functions for the insoluble mixture polyethylene (PE)–polypropylene (PP) were reported in the range 20–10,000 sec^{-1} by Noel and Carley [69]. They found *minima* in the function $\eta(\phi_2)$ at various, constant shear rates. Similar behavior was found by Plochocki [70] for PE–PP and by Natov *et al.* for polyethylene–natural rubber (NR). Kongarov and Bartenev [71] studied two systems—NR–nitrile rubber and polyisobutylene–polyethylene—and found minima in $\eta(\phi_2)$ as well. Extremely strong viscosity minima have been noted in very immiscible systems, such as Hypalon–EPR [72]. There has been no satisfactory explanation of this phenomenon.

In the case of systems with strong interactions, such as PVC–ester urethanes, a soluble system can easily show a *maximum* in $\eta(\phi_2)$ at low, constant shear rate. The explanation is the presence of a rather strong, hydrogen-bonded network. Galimov *et al.* [73] found that a polyurethane based on ethylene glycol adipate, when mixed with PVC, led to concomitant maxima in the viscosity and density versus composition relationships. With poly-(tetrahydrofuran)-based urethanes the composition dependence of the viscosity was more complex [74], although an initial drop in viscosity and T_g with the first 1% of the urethane indicated limited miscibility.

6.4.2 Linear Viscoelastic Functions

An area of rheology distinct from the determination and explanation of the viscometric functions (η, Ψ_1, Ψ_2) is linear viscoelasticity. The scientist working in this area attempts to determine linear viscoelastic functions (G, G', η', etc.) and explain them in terms of the structure of the material. The two areas of rheology meet in the regime of low shear rates; e.g., the zero shear viscosity η_0 is a viscometric and a linear viscoelastic parameter.

The fundamental linear viscoelastic function is the relaxation modulus, $G(t)$, which can be used to derive all other linear functions. The relaxation modulus can be measured, in principle, by quickly shearing the material to a fixed, small strain and observing the decay of the stress with time. Then

$$G(t) = \sigma(t)/\gamma \tag{6.16}$$

where $\sigma(t)$ is the stress and γ is the strain. To cover a broad time scale, it is often necessary to run the experiment at different temperatures, superimposing the results into one smooth curve. This is known as time–temperature superposition, with the amount of shift along the log time scale being called the shift factor, a_T.

The dynamic functions, $G'(\omega)$ and $G''(\omega)$, are often more convenient to measure than $G(t)$. $G'(\omega)$ and $G''(\omega)$ are the storage and loss moduli, respectively, and are functions of the frequency, ω. The relationships between the dynamic functions and the relaxation modulus are

$$G'(\omega) = \omega \int_0^\infty \sin \omega s \, G(s) \, ds \tag{6.17}$$

$$G''(\omega) = \omega \int_0^\infty \cos \omega s \, G(s) \, ds \tag{6.18}$$

In these equations, s is a dummy integration variable.

The great appeal of linear viscoelasticity is that several molecular theories for the behavior of pure materials in solution have been formulated [75]. Thus, in principle, changes in the fundamental motions of macromolecules will be revealed by the corresponding changes in the linear viscoelastic functions. It is natural that this discipline should be applied to polymer–polymer systems.

A good example of the possibilities in this area is the work of Prest and Porter [76] on the system polystyrene–PPO. They found that the dynamic loss and storage functions for mixtures of PS and PPO were linear (strain

6.4. Rheological and Viscoelastic

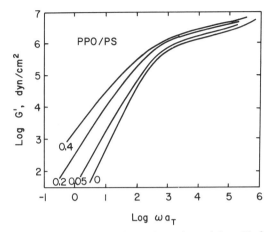

Fig. 6.15. Composition dependence of the dynamic modulus, G', for PPO–PS blends reduced to the iso-free-volume state, $f = 0.115$. The numbers on the curves denote the weight fraction PPO. [Reprinted by permission from W. M. Prest, Jr. and R. S. Porter, *J. Polym. Sci., Part A-2* **10**, 1639 (1972). Copyright by John Wiley and Sons, Inc.]

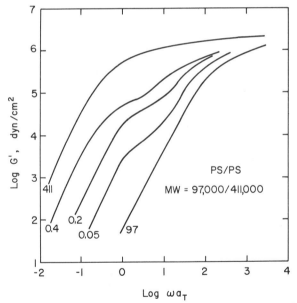

Fig. 6.16. Composition dependence of the dynamic modulus, G', for blends of two polystyrene resins, weight average molecular weights equal to 411,000 and 97,000. The numbers on the middle three curves represent the weight fraction of the higher molecular weight polystyrene in the blend. [Reprinted by permission from W. M. Prest, Jr. and R. S. Porter, *Polym. J.* **4**, 154 (1973).]

independent) and reducible by time–temperature superposition. Two-phase blends are known to fall short in both these tests. Furthermore, it was found that the variation of shift parameter with temperature and composition followed a WLF-like relationship [77], assuming a composition-independent free volume at T_g. Comparison of the viscoelastic functions at equal free volume effectively put all blends on the same time scale. The results of the comparison were somewhat different than found with blends of polystyrene of low molecular weight with polystyrene of high molecular weight [78], as shown in Figs. 6.15 and 6.16. In these figures, $G'(\omega)$ is plotted against reduced frequency.

In the system PS–PS, a slight decrease of slope in the rubbery plateau region was found on mixing in a higher molecular weight component, while the system PS–PPO showed a slight increase. The falloff of modulus with decreasing frequency due to the PS relaxation processes was barely preserved in the PPO–PS system, but was quite evident in the PS–PS system. At longer times the two behaved the same. The PPO component was actually *lower* in molecular weight than the PS, the longer relaxation times being explainable by the *lower* entanglement molecular weight of PPO. The compositional dependence of the zero-shear viscosity at equal free volume was in agreement with these results.

The conclusion of this study was that PPO influences the motions of PS on a level of fineness down to and including the scale of entanglements.

Other studies of the viscoelastic properties of soluble systems in the rubbery plateau and rubber to liquid transition regions are rare, as most workers have not included the modulus region below 10^8 dyn/cm^2 in their studies. The behavior of miscible systems in the high modulus region is covered in Sections 3.2.1 and 6.1.

6.5 TRANSPORT

The transport properties of miscible polymers will be discussed with primary emphasis on gas or vapor permeability. The typical permeability characteristics of miscible polymer mixtures versus the two-phase polymer mixture counterparts have analogies in other physical property counterparts (e.g., thermal conductivity and electrical conductivity). For review purposes, the theory of gas or vapor permeability, solubility, and diffusivity in polymers can be found in the listed references [79–81].

While brief citations are made in the above references concerning two-phase polymer mixtures, miscible polymer blends have not been covered. Low molecular weight additives in polymers [e.g., plasticized poly(vinyl chloride)] have been mentioned in many references concerning permeability

6.5. Transport

[82, 83]. With plasticized compositions, an increase in the gas permeability is generally observed, as with oil-extended rubber and plasticized poly(vinyl chloride). In rare cases, permeability can markedly decrease, as is the case with low molecular weight compounds which act as antiplasticizers—substances causing an increased modulus [84]. This has been hypothesized to be the result of free volume reduction, as shown by density increase and the elimination of prominent low temperature relaxations. In this section, we will be concerned with miscible blends where the components are of high molecular weight.

The principles of permeability have been adequately treated in the above-cited references and will only be alluded to in this section where pertinent to the discussion. The permeability constant, P, is related to the solubility constant, S, and the diffusion coefficient, D, by the simple equation

$$P = DS \tag{6.19}$$

where the typical units are P, cc(STP)/cm-sec-cm Hg; D, cm^2/sec; and S, cc(STP)/cc film-cm Hg. Diffusion is an activated energy-type process and therefore can be related to an activation energy of diffusion, E_d, where

$$D = D_0 \exp(-E_d/RT) \tag{6.20}$$

The activation energy of diffusion changes at the glass transition temperature, which means that the data of log D versus $1/T$ can be used to determine the T_g. Therefore, accurate diffusion data can be used to determine whether a polymer–polymer blend exhibits miscibility, assuming the T_g's of the constituents are sufficiently different that resolution of the transitions is possible. The solubility constant can likewise be represented by

$$S = S_0 \exp(-\Delta H/RT) \tag{6.21}$$

where ΔH is the heat of mixing.

The composition dependence of permeability in two-phase systems compared to single-phase systems can be represented by the generalized curves shown in Fig. 6.17. Note that the single-phase systems follow the behavior expected of random copolymers.

The permeability of two-phase polymer–polymer mixtures or phase-separated block and graft copolymers can be determined by a modification [85] of Maxwell's equation applied to permeability data:

$$P = P_c \frac{P_d + 2P_c - 2\phi_d(P_c - P_d)}{P_d + 2P_c + \phi_d(P_c - P_d)} \tag{6.22}$$

where P_c, permeability of continuous phase; P_d, permeability of discontinuous phase; ϕ_c, volume fraction of continuous phase; ϕ_d, volume frac-

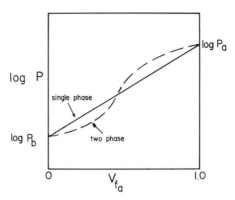

Fig. 6.17. Generalized behavior for gaseous permeability in polymer blends; miscible versus phase separated.

tion of discontinuous phase. Maxwell's equation is concerned with spherical inclusions in a continuous matrix. To extend the applicability of this equation over the total concentration range, the modified Maxwell equation to be used is [33]

$$P = X_a P_1 \frac{P_2 + 2P_1 - 2\phi_2(P_1 - P_2)}{P_2 - 2P_1 + \phi_2(P_1 - P_2)} + X_b P_2 \frac{P_1 + 2P_2 - 2\phi_1(P_2 - P_1)}{P_1 + 2P_2 + \phi_1(P_2 - P_1)}$$

(6.23)

where $X_a + X_b = 1$; case A = component 1 continuous ($P_c = P_1$; $P_d = P_2$); case B = component 2 continuous ($P_c = P_2$; $P_d = P_1$); ϕ_1 = volume fraction of component 1; ϕ_2 = volume fraction of component 2.

This modification of Maxwell's equation consists of an addition of the permeabilities of both phases weighted by their fractional contribution (X_a or X_b) to the continuous phase. Phase inversion is considered the position where the fractional contribution to the continuous phase structure is equal ($X_a = X_b = 0.5$). For phase-separated polymer blends, the phase inversion is not a sharp transition, but rather occurs over a broad volume fraction range. The use of the modified Maxwell equation for permeability data for the phase-separated polysulfone–poly(dimethyl siloxane) block copolymers agreed quite well with modulus data using a similar modification of Kerner's equation [33]. The phase inversion points using the analyses of the modulus and permeability data were found to be similar.

Only recently has the gas permeability of polymer–polymer miscible blends been experimentally investigated. Gas molecules are basically a probe by which the phase behavior can be determined by its influence on the transport properties. Gas permeability and diffusion constitute a rather unique

6.5. Transport

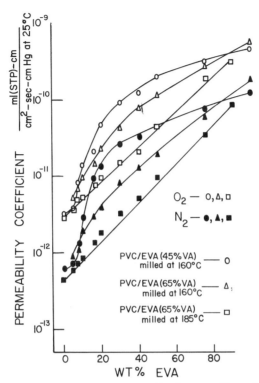

Fig. 6.18. Permeability coefficients versus blend composition of blends of poly(vinyl chloride) with ethylene/vinyl acetate copolymer. [Reprinted by permission from B. G. Ranby, *J. Polym. Sci., Polym. Symp.* **51**, 89 (1975). Copyright by John Wiley and Sons, Inc.]

method in that the "probe" is smaller than any conceivable level of phase separation. Shur and Ranby [86–89] have used gas permeation data to elucidate the structure of various poly(vinyl chloride) blends. Good agreement of this technique for characterizing blend phase behavior was noted by comparison with other accepted techniques (e.g., dynamic mechanical). The miscible blends of poly(vinyl chloride) and ethylene/vinyl acetate (35/65 by weight) copolymer were investigated over a range of compositions and compared to a two-phase blend of poly(vinyl chloride) and ethylene/vinyl acetate (55/45) copolymer [86]. At high processing temperatures, the blend containing the 65% vinyl acetate copolymer displayed a linear relationship of permeability versus composition, as expected for a miscible blend. With the 45% vinyl acetate copolymer blend, the observed sigmoidal relationship was shifted toward the low ethylene/vinyl acetate copolymer compositions (phase inversion at $\sim 7.5\%$ ethylene/vinyl acetate copolymer). The O_2 and N_2 permeability results for these blends are illustrated in Fig.

6.18. The position of phase inversion at ~7.5% ethylene/vinyl acetate copolymer content in the two-phase blends with poly(vinyl chloride) has been observed by other investigators. This low value is believed to result from incomplete fusion of the poly(vinyl chloride) particles due to the lubricating behavior of the ethylene/vinyl acetate copolymer, yielding a continuous EVA phase above ~7.5 wt% [90].

For all these blends, the activation energy of diffusion goes through a maximum at 30 wt% EVA, with the more miscible systems exhibiting higher activation energies. The increase in E_d as miscibility is increased was interpreted as the result of a decrease in the mobility of the polymer chain segments due to the interaction of poly(vinyl chloride) with ethylene/vinyl acetate copolymer. This was supported by experimental evidence; the densities for the miscible blends were higher than calculated values based on volume additivity of the components.

Ranby also investigated the gas permeability of blends of butadiene/acrylonitrile copolymers with poly(vinyl chloride) [91]. Acrylonitrile contents of the copolymers studied were 21.7, 29.6, and 41.6 wt%. These values are in the range of values (21–45 wt%) cited by Zakrzewski [8] as yielding miscible blends with poly(vinyl chloride). The blends of these copolymers with poly(vinyl chloride) at the 50/50 weight ratio gave measured densities of 0.033, 0.050, and 0.053 g/cc, respectively, higher than calculated values assuming volume additivity of pure component values. Permeability measurements with oxygen and nitrogen for these blends showed no indication of phase inversion. For helium and carbon dioxide, some indication of phase inversion for the lower acrylonitrile content copolymers was present. The activation energies of diffusion for these blends clearly increased (Fig. 6.19) as the acrylonitrile content was increased in the copolymer. As with the ethylene/vinyl acetate copolymer blends with poly(vinyl chloride), this was hypothesized to be the result of increased interaction of the components of the blend as demonstrated by density data.

Vapor sorption studies for the blend of polystyrene and poly(2,6-dimethyl-1,4-phenylene oxide) have been discussed in several papers with n-hexane as the penetrant [92–94]. Case II transport with relaxation-controlled sorption was observed. (Case II transport kinetics are the result of a diffusion rate controlled by the swelling of the outer surface of the polymer or blend, with a sharp boundary between the swollen outer surface and an unpenetrated central portion.) Sorption rates increased monotonically with increasing PPO content at n-hexane vapor activity of 0.775. The hexane sorption rate of PPO was observed to be 4 orders of magnitude greater than that of polystyrene. The monotonic increase is indicative of a single-phase blend; a sigmoidal-type curve would be expected with a two-phase mixture. At higher n-hexane vapor activities (>0.925), a definite minimum in the sorp-

6.5. Transport

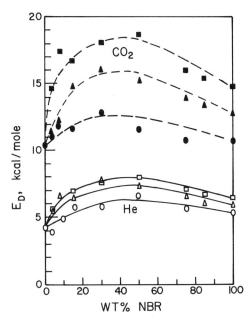

Fig. 6.19. Activation energy for diffusion (E_D) for helium (open symbols) and carbon dioxide (filled symbols) as a function of polymer blend composition. ○,●, poly(vinyl chloride) (PVC)–NBR-1 [butadiene/acrylonitrile copolymer, 21.7% acrylonitrile–(AN)]; △,▲, PVC–NBR-2 (29.6% AN); □,■, PVC–NBR-3 (41.6% AN). [Reprinted by permission from B. G. Ranby, *J. Polym. Sci.; Polym. Symp.* **51,** 89 (1975). Copyright by John Wiley and Sons, Inc.]

tion rate occurred with blend composition. This was explained to be due to coupling between sorption equilibria and sorption rate. Specifically, equilibrated polystyrene-rich films were above the T_g, whereas equilibrated PPO films were below the T_g. As an increase in activation energy of diffusion occurs when going from the glassy state to the rubbery state, these results could, therefore, be rationalized.

Stallings *et al.* [95, 96] studied the permeability of the noble gases in polystyrene–poly(2,6-dimethyl-1,4-phenylene oxide) blends. The permeability of argon and krypton varied monotonically with blend composition; however, helium and neon permeability reached a minimum at a 50/50 blend composition. The neon and helium permeability at high PPO levels satisfied heterogeneous systems analysis, assuming polystyrene is the dispersed phase. The characteristic features of the permeability–composition curves for the noble gas penetrants were similar in that they were concave at low PPO levels and convex at high PPO content. This sigmoidal-type behavior suggests phase structure in these blends. Additional data indicated miscibility as single T_g values were determined from calorimetric measurements and a minimum in specific volume is reached at an intermediate blend composition.

In studies of nylon 6 grafted onto ethylene copolymers, significant differences in O_2 permeability were observed with different ethylene copolymers [97]. With nylon 6 grafted onto ethylene/ethyl acrylate copolymers, a two-phase system was observed, judging from modulus–temperature data. The permeability data confirmed this as the O_2 permeability was much lower at high nylon-6 content than would be predicted for a single-phase system. Conversely, nylon 6 grafted onto an ethylene/acrylic acid copolymer yielded O_2 permeability results similar to those expected for one-phase systems. Dynamic mechanical data yielded a single T_g intermediate between the constituent glass transition temperatures for the graft of nylon 6 with ethylene/acrylic acid copolymer. The one-phase behavior for this graft copolymer is indicative of an acid–base interaction.

6.6 CRYSTALLIZATION

In several of the earlier reviews, the potential of achieving miscible polymer blends in which one or both of the components were crystalline was believed to be quite low due to the heat of fusion which would have to be overcome to achieve the necessary thermodynamic criteria for mixing. This generalization appears to be erroneous; the comprehensive survey of known miscible polymer blends (Chapter 5) listed many systems containing crystallizable components. In these blends, the crystalline component generally retained the ability to crystallize; however, miscibility was judged on the nature of the residual amorphous phase. The crystallization behavior of these blends is quite important for a review of the properties of miscible polymer blends and will be covered in this section.

For two-phase systems, the crystallization behavior, including crystallization kinetics, is expected to be equivalent to that of the unblended state. The exception to this is the heterogeneous nucleation which has been observed in several two-phase polymer mixtures [98, 99]. In the cases where crystallization has been studied in miscible blends, the crystallizable constituent displays certain characteristics similar to its unblended crystalline state (i.e., crystal lattice); however, several differences do occur. The primary change is observed with the crystallization kinetics and a secondary change is observed with the lowering of the crystalline melting point as with polymer and low molecular weight diluent mixtures. This discussion will not provide a background on crystallization in polymers. The interested reader can consult the listed references [100, 101].

Crystallization kinetics can be studied by utilizing the spherulitic growth

6.6. Crystallization

rate equation [102] to predict the effect of a miscible polymer diluent on the crystallization rate of another component. The spherulitic growth rate equation is

$$G = G_0 \exp(-\Delta F^*/RT) \exp(-4b_0\sigma\sigma_e T_m°/\Delta H_f(\Delta T)kT) \quad (6.24)$$

where ΔF^* is representative of the barrier restricting polymer diffusion to the crystallizing surface and has been suggested by Hoffman and Weeks [103] to be satisfied by the Williams, Landel, and Ferry equation:

$$\Delta F^* = 4120T/(51.6 + T - T_g) \quad (6.25)$$

The other variables are b_0 (monolayer thickness), σ (lateral interfacial free energy). σ_e (interfacial free energy of the chain-folded surface), $T_m°$ (equilibrium melting temperature), ΔH_f (heat of fusion), and $\Delta T(T - T_m)$. G is equal to the radial growth of the spherulite and, thus, is dr/dt.

The values of the second exponential of the spherulitic growth rate equation are rather insensitive to the composition of the blend, as has been shown by limited experimental data. The major change occurs with the first exponential term for those cases where a strong dependence of T_g on the spherulitic growth rate is observed. Analysis of this equation predicts that the addition of a miscible polymer constituent to a crystallizable polymer will lower the crystallization rate if the blend T_g is higher than that of the crystallizable polymer. Conversely, a higher crystallization rate will be expected if the resultant blend T_g is lower. One further modification of the spherulitic growth rate equation must be made to correct for concentration changes: Equation (6.24) must be multiplied by $(1 - W_d)$, where W_d is the concentration of the diluent in the weight fraction. The linear dependence of the spherulitic growth rate on concentration (other variables constant) is experimentally verified by studies involving atactic/isotactic polystyrene blends [104].

The melting point will be expected to be slightly lowered in a miscible polymer blend, similar to the lowering in blends of polymers with low molecular weight diluents, as exemplified by the following equation:

$$(1/T_m) - (1/T_m°) = (RV_2/\Delta H_f°V_1)\{(1 - \phi_2) - \chi_{12}(1 - \phi_2)^2\} \quad (6.26)$$

For a polymer diluent added to a crystallizable polymer in which it is miscible, the equation is slightly different, as has been shown by Nishi and Wang [105]:

$$(1/T_m) - (1/T_m°) = (-RV_2/\Delta H_f°V_1)\{\chi_{12}(1 - \phi_2)^2\} \quad (6.27)$$

Several studies of the crystallization of isotactic polystyrene from blends containing atactic polystyrene have been reported [104, 106]. The spherulitic growth rate was found to linearly decrease with increasing concentration of atactic polystyrene, thus experimentally verifying the $(1 - W_d)$ term required for the spherulitic growth rate equation previously mentioned. As this experiment involved a polymer–polymer miscible blend in which the T_g is invariant with composition, verification of the T_g dependence was not possible.

The most comprehensive study of the crystallization characteristics of a crystallizable constituent of a miscible polymer blend has involved the blend of poly(ε-caprolactone) (PCL)–poly(vinyl chloride) (PVC). In this system, PCL has a much lower T_g ($-72°C$, amorphous) than PVC (80°C); thus, the addition of PVC to PCL results in a dramatic lowering of the crystallization rate as the T_g is increased. Robeson [107] investigated a series of blends containing from 30 to 60% PCL using modulus data as an experimental means of ascertaining the rate of crystallization. As this study represents a clear case for the utility of the spherulitic growth rate equation, the results will be described in some detail. Note that in pure-component crystallization rate studies the crystallization temperature is the variable as the T_g is obviously invariant. With the PCL–PVC system discussed here, the crystallization temperature was held constant while composition changes yielded variant T_g's.

Modulus versus time data were obtained on samples initially raised above the melting point of PCL. An interesting observation in this study deals with the question of adequate molecular mixing of the PCL spherulites with the amorphous phase of PVC–PCL once melting has occurred. Even though PCL spherulites result in dimensions adequate to yield film opacity, a time

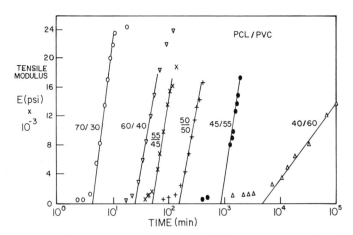

Fig. 6.20. Crystallization of poly(ε-caprolactone) (PCL)–poly(vinyl chloride) (PVC) blends as determined by modulus–time measurements at 25°C. [Reprinted by permission from L. M. Robeson, *J. Appl. Polym. Sci.* **17**, 3607 (1973). Copyright by John Wiley and Sons, Inc.]

6.6. Crystallization

period of only 2 min above the melting point (or possibly an even shorter time interval) was needed to yield a homogeneous system with crystallization kinetics equal to those predicted for a homogeneous, single-phase system [107]. This indicates that the phase-separated system (by crystallization of one of the components) mixes back to an equilibrium single-phase mixture in a relatively short time period. The modulus versus log time data for these blends are shown in Fig. 6.20 and illustrate the dramatic decrease in crystallization with increasing PVC concentration in the PVC–PCL blend.

To analyze the results with the spherulitic growth rate equation, the following analysis was required. The modulus was assumed proportional to the volume fraction of spherulites; that is,

$$E = k_2 \tfrac{4}{3}\pi r^3, \qquad E^{1/3} = (k_2 \tfrac{4}{3}\pi)^{1/3} r \qquad (6.28)$$

$$\frac{dE^{1/3}}{dt} = (k_2 \tfrac{4}{3}\pi)^{1/3}\frac{dr}{dt} \qquad \text{where} \quad \frac{dr}{dt} = G \qquad (6.29)$$

$$\frac{dE^{1/3}}{dt} = k'(1 - W_d)\exp\!\left(-\frac{4120}{(51.6 + T - T_g)R}\right) \qquad (6.30)$$

where

$$k' = (k_2 \tfrac{4}{3}\pi)^{1/3} G_0 \exp\!\left(-\frac{4b_0 \sigma \sigma_e T_m^\circ}{\Delta h_f (\Delta T) kT}\right)$$

From $dE^{1/3}/dt$ data for the respective blends, the experimental data are shown in Fig. 6.21; the calculated curve was determined by establishing k' from the center position of the experimental curve.

Because good agreement between the experimental and calculated results of the spherulitic growth rate equation was observed, the form of this equation appears to be adequate for projecting the crystallization kinetics for miscible polymer systems. An assumption which would require corrections in a more rigorous treatment is that the T_g remains constant (WLF equation). As PCL crystallizes out of the PVC–PCL blend, the ratio of PCL to PVC changes, thus resulting in a T_g change of the amorphous phase. An analysis of the T_g change in the concentration range in which the values of $dE^{1/3}/dt$ were obtained indicated the effect on the predictions of equation to be small but not totally negligible. The melting point depression which was observed ($\sim 5°C$ over the range of compositions investigated) was also not taken into account. However, the effect of melting point depression on the spherulitic growth rate was much lower than that for the T_g change as a function of composition.

With the T_g change that occurs with crystallization of a single component from a miscible polymer blend (assuming the T_g's of the constituents are

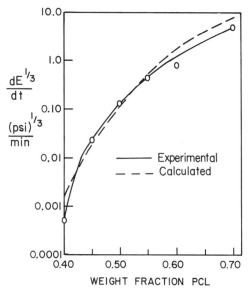

Fig. 6.21. Crystallization rate as a function of poly(ε-caprolactone) concentration in poly(ε-caprolactone)–poly(vinyl chloride) blends. [Reprinted by permission from L. M. Robeson, *J. Appl. Polym. Sci.* **17**, 3607 (1973). Copyright by John Wiley and Sons, Inc.]

measurably different), the amount of the crystalline phase may be capable of being determined. Glass transition data on amorphous versus crystalline samples for the polymer blend can yield directly the degree of crystallinity for the blend via a simple material balance. Once this is obtained and heat of fusion of the blend is experimentally determined, ΔH_f^0 (heat of fusion of 100% crystallinity for the crystallizable component) can be calculated. This is a unique means for determining the degree of crystallinity and has shown good agreement with the classical methods when comparisons with poly(ε-caprolactone) were made, as will now be discussed.

The degree of crystallinity of PCL in PCL–PVC blends, as well as in unblended PCL, was determined by Robeson and Joesten [108] using the approach described above. Blends of PCL and PVC were prepared, and the T_g's of both quenched (amorphous) and annealed (crystalline) samples were determined with a torsion pendulum using the peak of the loss modulus, G'', as the criterion for T_g determination.

The following material balance was used:

$$W_{PCL} = W_{aPCL} X_a + X_c \qquad (6.31)$$

where W_{PCL}, weight fraction of PCL in blend; W_{aPCL}, weight fraction of PCL in the amorphous phase; X_a, weight fraction of amorphous phase;

6.6. Crystallization

X_c, weight fraction of crystalline phase. Since $X_a + X_c = 1.0$, the equation reduces to

$$X_c = W_{PCL} - W_{aPCL}(1 - X_c)$$

W_{aPCL} can be determined from T_g data but only after an appropriate correction has been made. This correction is due to the T_g shift resulting from the stiffening of the amorphous phase by the crystalline phase, as is observed with unblended PCL. The experimental T_g change is represented by

$$\Delta T_{g(\exp)} = \Delta T_{gc} + \Delta T_{gs} \tag{6.32}$$

where $\Delta T_{g(\exp)}$, experimental T_g change from amorphous to crystalline state; ΔT_{gc}, T_g change due to composition change in amorphous phase with poly-(ε-caprolactone) crystallization; ΔT_{gs}, T_g change due to amorphous phase stiffening by crystallization. ΔT_{gs} was calculated from heat of fusion data on the crystalline blend compared to that of unblended PCL. The ratio of those values times the T_g change of amorphous versus crystalline unblended PCL (16°C) was used to determine ΔT_{gs}. ΔT_{gc} is then added to the T_g of the initially amorphous blend to determine, from T_g versus composition data (amorphous blend data), the composition of the amorphous phase in the crystalline blend.

The results for X_c and ΔH_f^0 are listed in Table 6.3 for melt crystallized PCL, annealed PCL, and PCL in PVC blends. The results for unblended PCL have been compared with results of other investigators [109, 110] using classical methods for the determination of the degree of crystallinity. The results agree quite well when the effect of annealing is taken into consideration (PCL slowly crystallizes with time). Thus, not only does this technique allow one to determine the degree of crystallinity of a crystallizable con-

TABLE 6.3

Degree of Crystallinity Results from T_g Shift Determination Method

PCL–PVC blend	T_{gs}^a (°C)	T_{gc}^a (°C)	$W_{a(PCL)}$	X_c^b PCL–PVC blend	ΔH°_f (PCL = 100% crystalline) (cal/g)	X_c (PCL annealed 4 weeks)	X_c (PCL melt-crystallized)
60/40	8.4	25.1	0.420	0.52	34.8	0.59	0.53
55/45	7.5	24.5	0.375	0.50	35.0	0.59	0.53
50/50	6.2	21.8	0.355	0.45	36.0	0.57	0.52
45/55	5.0	22.0	0.315	0.43	32.5	0.63	0.57
Average					34.6	0.59	0.53

[a] Average of three specimens.
[b] Based on amount of PCL (not on total blend).

stituent of a polymer blend, but it can be used to determine the degree of crystallinity of the unblended crystalline polymer.

The crystallization of poly(ε-caprolactone) form poly(ε-caprolactone)–poly(vinyl chloride) blends was also extensively studied by Ong in his Ph.D. dissertation [111]. Qualitative agreement with the spherulitic growth rate equation was observed. Optical and electron microscopy on the crystallized blends revealed that the spherulites became increasingly coarse and open with increasing PVC content. The X-ray spectrometer results showed a uniform distribution of the blend. This implied that the PVC molecules were trapped within the growing spherulite, but, with further crystallization, rejection into the interlamellar regions occurred. Dynamic mechanical data indicated that the amorphous regions (i.e., interlamellar regions) were one phase, consisting of PVC and amorphous PCL. A melting point depression was observed with the addition of PVC to PCL, implying a negative interaction coefficient. Ong analyzed the melting point depression using the classic diluent–polymer melting point depression equation. This analysis indicated that 8 to 9 PVC monomer units acted as a single diluent unit to depress the observed melting point.

Small-angle X-ray and light scattering studies of the morphology of poly(ε-caprolactone)–poly(vinyl chloride) blends were the subject of an extensive effort of Khambatta et al. [112]. From small-angle X-ray studies, the amorphous layer thickness of the lamellar structure was found to increase with PVC content, thus accommodating PVC in the amorphous layer of the spherulite. The level of PCL crystallinity was virtually unchanged down to 50% PCL. A slight increase in the spherulite radius of PCL was observed with the initial addition of PVC. This was believed to be a result of crystallization occurring at lower supercooling, thus a greater critical nucleus size due to the melting point depression. The X-ray scattering results were interpreted in terms of the Debye–Bueche theory, providing values for a correlation distance and the mean-square electron density fluctuation. The fluctuation was found to be larger than one would expect for a two-phase model with phases composed of pure components. The data presented indicated a transition zone of 30 Å between the phases. At low concentrations of PCL, the correlation distance is in the range of end-to-end unperturbed chain dimensions, but the sizes of the PCL-rich regions were shown to increase with concentration. It was pointed out that PVC exhibits some structure (correlation distance of 22 Å) but less than the amorphous blend. While these results may seem inconsistent with the previous investigations which concluded miscibility, Khambatta et al. [112] concluded partial miscibility for this blend with some level of heterogeneity observed in the range of molecular dimensions.

As with atactic polystyrene, isotactic polystyrene is also miscible with

6.6. Crystallization

poly(2,6-dimethyl-1,4-phenylene oxide) (PPO). The crystallization rate of isotactic polystyrene is noticeably lower than most commercial crystalline polymers and is even further depressed by the addition of PPO. PPO can also be crystallized via solvent addition; thus, the isotactic polystyrene–PPO blends have the potential of forming a triphase blend consisting of crystalline regions of the individual components and an amorphous region of the miscible components. In a study by Wenig *et al.* [113], the thickness of the isotactic polystyrene crystal lamellae decreased with increasing PPO content, which is the probable reason for the melting point depression observed by the addition of PPO to isotactic polystyrene. Light scattering results indicated that the radius of the isotactic polystyrene spherulites increased with the addition of small amounts of PPO. The spherulitic structure above 30% PPO content was found to be discontinuous in the crystallized samples.

The subject of the Ph.D. dissertation of Lemos [114] was the isotactic polystyrene–PPO blend. In solvent-crystallized blends, crystallinity of both isotactic polystyrene and PPO was observed with a depression in melting points with increasing content of the counterpart diluent polymers. For thermally crystallized blends, compositions containing less than 50% isotactic polystyrene yielded no detectable crystallinity. A melting point depression was observed for isotactic polystyrene with PPO addition for the thermally crystallized samples. For the solvent-crystallized blends, methyl ethyl ketone and/or acetone vapor led to isotactic polystyrene crystallization, whereas PPO crystallization required the combination of solvent plus thermal treatment. The melting depression of PPO was larger with the addition of either atactic or isotactic polystyrene than that observed for isotactic polystyrene with the addition of PPO.

Poly(vinylidene fluoride), a crystalline polymer with a melting point of 168°C, exhibits miscibility with poly(methyl methacrylate) (PMMA) and poly(ethyl methacrylate) (PEMA) [115–118]. PVF_2 crystallization was observed in blends containing up to 50 wt% PVF_2 [105]. Paul and Altamirano [116] observed similar results and noted a melting point depression of PVF_2 linear with increasing PMMA content. Thermal analysis of solution cast films of PVF_2–PMMA blends, studied by Nishi and Wang [105], also yielded a melting point depression. They developed an equation [Eq. (6.27)] to determine the interaction coefficient which yielded a value of -0.295 at 160°C for this blend. The crystallization rate of PVF_2 was shown to be significantly decreased with the addition of PMMA. This is presumably due to the noticeable increase in the T_g of PVF_2 with the addition of PMMA, thus lowering the diffusion rate of PVF_2 to the crystallizing interface. Crystallization of blends containing less than 50% PVF_2 was observed in annealing studies at 85°C. Kwei *et al.* [118] calculated an interaction coefficient of -0.34 for

PEMA–PVF$_2$ blends. Mixtures of less than 20% PVF$_2$ were amorphous. At low PEMA content, the PVF$_2$ crystalline region was stated to exist, with two conjugate amorphous regions containing 45 and 100% PVF$_2$. Imken *et al.* [119], however, concluded from their studies that PEMA–PVF$_2$ blends were miscible over the entire concentration range.

Poly(butylene terephthalate)–poly(tetrahydrofuran) (AB)$_n$ block copolymers are recent commercial thermoplastic elastomers with properties similar to those of the thermoplastic polyurethanes. The amorphous phase of this block copolymer has been shown to be single phase [120]. This material has been touted to be useful as a permanent plasticizer for PVC, thus implying a degree of miscibility [121]. Studies of PBT–PTHF/PVC blends by Nishi *et al.* [122] indicated that the soft PTHF segments mix extensively with PVC. Recent work by Robeson [123] illustrated the miscibility of PBT and PVC, thus questioning the conclusions reached by Nishi *et al.* The hard PBT block segments crystallize out of the miscible blend when cooled under 150°C. The glass transition peaks for the blends are considerably broader than the pure-component peaks, indicating some level of microheterogeneity.

Hubbell and Cooper studied the segmental orientation of the components of blends of poly(ε-caprolactone)–poly(vinyl chloride) and poly(ε-caprolactone)–nitrocellulose using dynamic, differential infrared dichroism [124]. In blends containing crystalline regions of poly(ε-caprolactone) (PCL), the PCL segments exhibited orientations significantly different than those of either nitrocellulose or PVC, unlike amorphous blends where the orientations of all components were similar. Further studies by Hubbell and Cooper [125] discussed the crystallization behavior of PCL from the above-mentioned blends as well as PCL–cellulose acetate butyrate blends.

6.7 DEGRADATION

Miscible polymer systems present the important possibility of permanently introducing substantial concentrations of chemically active substances into a polymer matrix without sacrificing physical or thermomechanical properties. The active substances could include agents designed to protect or encourage degradation, depending on the application. The thermally unstable commodity resin, PVC, has understandably received the most attention in this area. PVC is traditionally protected by very efficient, but low molecular weight, stabilizers, which may leach out under service.

The system atactic PMMA and PVC has limited miscibility, perhaps a few percent at most [126]. From the thermal volatilization work of McNeill and Neil [127], it appears that this is enough to delay the release of HCl from PVC. However, on the other end of the composition spectrum, a small

6.7. Degradation

amount of PVC in PMMA facilitated unzippering of the PMMA to monomer. (Small amounts of PVC added to PBD increase the oxidation rate of the latter [128] in spite of the two-phase nature of the system. Thus, cooperative effects are not limited to soluble systems.)

The influence of nitrile rubber on the thermal stability of PVC has not received extensive study in spite of the commercial importance of the system. Woods and Frazer [129] state that the thermal stability of PVC and nitrile rubber is "strongly influenced by the type of rubber used." No explanation was given or quantitative comparisons made. At the other end of the scale, it has been found that the ozone resistance of nitrile rubber is increased by PVC addition [130].

Clear examples of exchange reactions in miscible polymer systems have been noted [9]. Phenoxy–poly(butylene terephthalate) is one system which cross-links easily under normal processing conditions, forming structures of the type

Phenoxy–poly(ethylene terephthalate) behaves in a similar manner [19], whereas phenoxy–PCL would be expected to graft only [20]. Copolymer formation in miscible systems by exchange reaction is also possible, but not certain, in systems such as poly(ethylene terephthalate)–polycarbonate [131]. Exchange and grafting reactions such as described can be regarded as thermal degradation processes, but may actually improve the properties of the system. The cross-linking reactions would, of course, eliminate the possibility of reprocessing the material.

The biodegradability of soluble polymer systems offers possibilities in application areas such as packaging, where fast degradation by the environment is desired. Poly(ε-caprolactone) is biodegradable and forms many soluble systems (Section 5.2.6), but its influence on the degradability of these systems does not appear to be significant, judging from the lack of patent literature on the subject. *Protection* of PVC against ultraviolet degradation by semisoluble acrylic additives is a well-known art [132]. The influence of polystyrene on the photooxidative stability of PPO has recently been investigated [133]. The addition of PS to PPO had a beneficial effect, as judged from changes in solution viscosity and ultimate mechanical properties.

REFERENCES

1. M. L. Williams, R. F. Landel, and J. D. Ferry, *J. Am. Soc.* **77**, 3701 (1955).
2. V. A. Kargin, *J. Polym. Sci., Part C* **4**, 1601 (1963).
3. G. M. Bartenev and G. S. Kongarov, *Rubber Chem. Technol.* **36**, 668 (1963).
4. T. G. Fox, *Bull. Am. Phys. Soc.* [2] **2**, 123 (1956).
5. M. Gordon and J. S. Taylor, *J. Appl. Chem.* **2**, 493 (1952).
6. F. N. Kelley and F. Bueche, *J. Polym. Sci.* **50**, 549 (1961).
7. A. V. Tobolsky, "Structure and Properties of Polymers," p. 85. Wiley, New York, 1960.
8. G. A. Zakrzewski, *Polymer* **14**, 348 (1973).
9. L. M. Robeson and A. B. Furtek, *Am. Chem. Soc., Div. Org. Coat. Plast. Chem., Pap.* **37** (1), 136 (1977).
10. J. J. Hickman and R. M. Ikeda, *J. Polym. Sci., Polym Phys. Ed.* **11**, 1713 (1973).
11. S. Ichihara, A. Komatsu, and T. Hata, *Polym. J.* **2**, 640 (1971).
12. L. Y. Zlatkevich and V. G. Nikolskii, *Rubber Chem. Technol.* **46**, 1210 (1973).
13. J. V. Koleske and R. D. Lundberg, *J. Polym. Sci., Part A-2* **7**, 795 (1969).
14. T. Nishi and T. T. Wang, *Macromolecules* **8**, 909 (1975).
15. J. S. Noland, N. N. C. Hsu, R. Saxon, and J. M. Schmitt, *Adv. Chem. Ser.* **99**, 15 (1971).
16. R. L. Imken, D. R. Paul, and J. R. Barlow, *Polym. Eng. Sci.* **16**, 593 (1976).
17. S. Akiyama, J. Komatsu, and R. Kaneko, *Polym. J.* **7**, 172 (1975).
18. K. L. Smith, A. E. Winslow, and D. E. Peterson, *Ind. Eng. Chem.* **51**, 1361 (1959).
19. L. M. Robeson and A. B. Furtek, *J. Appl. Polym. Sci.* **23**, 645 (1979).
20. G. L. Brode and J. V. Koleske, *J. Macromol. Sci., Chem.* **6** (6), 1109 (1972).
21. K. Marcincin, A. Romanov, and V. Pollak, *J. Appl. Polym. Sci.* **16**, 2239 (1972).
22. L. H. Sperling and D. W. Friedman, *J. Polym. Sci., Part A-2* **7**, 425 (1969).
23. L. Bohn, *Rubber Chem. Technol.* **41**, 495 (1968).
24. S. Manabe, R. Murakami, and M. Takayanagi, *Mem. Fac. Eng., Kyushu Univ.* **28** (4), 295 (1969).
25. V. Huelck, D. A. Thomas, and L. H. Sperling, *Macromolecules* **5**, 348 (1972).
26. D. I. Livingston and J. E. Brown, Jr., *Proc. Int. Congr. Rheol. 5th, 1968* Vol. 4, p. 25 (1970).
27. D. J. Buckley, *Trans. N. Y. Acad. Sci.* [2] **29**, 735 (1967).
28. S. Krause, *J. Macromol. Sci., Rev. Macromol. Chem.* **7** (2), 25 (1972).
29. J. A. Manson and L. H. Sperling, "Polymer Blends and Composites." Plenum, New York, 1976.
30. E. H. Kerner, *Proc. Phys. Soc., London, Ser. B* **69**, 808 (1956).
31. M. Takayanagi, M. Harima, and Y. Iwatu, *Mem. Fac. Eng., Kyushu Univ.* **23**, 1 (1963).

References

32. Z. Hashin and S. Shtrikman, *J. Mech. Phys. Solids* **11**, 127 (1963).
33. L. M. Robeson, A. Noshay, M. Matzner, and C. N. Merriam, *Angew. Makromol. Chemie* **29/30**, 47 (1973).
34. E. Wagner and L. M. Robeson, *Rubber Chem. Technol.* **43**, 1129 (1970).
35. M. Takayanagi, K. Imada, and T. Kajiyama, *J. Polym. Sci., Part C* **15**, 263 (1966).
36. S. Miyata and T. Hata, *Proc. Int. Congr. Rheol., 5th, 1968* Vol. 3, p. 71 (1970).
37. L. E. Nielsen, *J. Compos. Mater.* **1**, 100 (1967).
38. L. E. Nielsen, "Mechanical Properties of Polymers." Dekker, New York, 1974.
39. Z. Hashin, *Bull. Res. Counc. Is., Sect. C* **5**, 46 (1955).
40. R. A. Emmett, *Ind. Eng. Chem.* **36**, 730 (1944).
41. H. F. Schwarz and W. S. Edwards, *Appl. Polym. Symp.* **25**, 243 (1974).
42. J. E. Pittenger and C. F. Cohan, *Mod. Plast.*, p. 81 (1947).
43. M. C. Reed, *Mod. Plast.* **27**, 117 (1949).
44. J. F. Kenney, *in* "Recent Advances in Polymer Blends, Grafts, and Blocks" (L. H. Sperling, ed.), p. 117. Plenum, New York, 1974.
45. D. Feldman and M. Rusu, *Eur. Polym. J.* **10**, 41 (1974).
46. T. Nishi and T. K. Kwei, *J. Appl. Polym. Sci.* **20**, 1331 (1976).
47. C. F. Hammer, *Am. Chem. Soc., Div. Org. Coat. Plast. Chem., Pap.* **37** (1), 234 (1977).
48. A. Noshay and L. M. Robeson, *J. Polym. Sci., Polym. Chem. Ed.* **12**, 689 (1974).
49. A. Noshay and L. M. Robeson, *Polym. Prepr., Am. Chem. Soc., Div. Polym. Chem.* **15** (1), 613 (1974).
50. A. Noshay, M. Matzner, and L. M. Robeson, *8th Reg. Am. Chem. Soc. Meet. 1976*.
51. A. F. Yee, *Polym. Prepr., Am. Chem. Soc., Div. Polym. Chem.* **17** (1), 145 (1976).
52. T. Ree and H. Eyring, *J. Appl. Phys.* **26**, 793 (1955).
53. A. S. Michaels, *Ind. Eng. Chem.* **57** (10), 32 (1965).
54. A. S. Hoffman, R. W. Lewis, and A. S. Michaels, *Am. Chem. Soc., Div. Org. Coat. Plast. Chem. Pap.* **29** (2), 236 (1969)
55. M. J. Lysaght, *in* "Ionic Polymers" (L. Holliday, ed.), p. 281. Wiley, New York, 1975.
56. M. Bank, J. Leffingwell, and C. Thies, *Macromolecules* **4**, 43 (1971).
57. K. Fujimoto and N. Yoshimiya, *Rubber Chem. Technol.* **41**, 669 (1968).
58. W. J. MacKnight, J. Stoelting, and F. E. Karasz, *Adv. Chem. Ser.* **99**, 29 (1971).
59. A. R. Shultz and B. M. Beach, *Macromolecules* **7**, 902 (1974).
60. T. Sulzberg and R. J. Cotter, *J. Polym. Sci., Part A-1* **8**, 2747 (1970).
61. A. S. Michaels and R. G. Miekka, *J. Phys. Chem.* **65**, 1765 (1961).
62. A. S. Michaels, L. Mir, and N. S. Schneider, *J. Phys. Chem.* **69**, 1447 (1965).
63. A. S. Michaels, G. L. Falkenstein, and N. S. Schneider, *J. Phys. Chem.* **69**, 1456 (1965).
64. H. Sato and A. Nakajima, *Polym. J.* **7**, 241 (1975).
65. H. Lee, R. E. Salomon, and M. M. Labes, *Macromolecules* **11**, 171 (1978).
66. S. Middleman, "The Flow of High Polymers," p. 91. Wiley (Interscience), New York, 1968.
67. L. R. Schmidt, *J. Appl. Polym. Sci.* **23**, 2463 (1979).
68. G. L. Slonimskii and N. F. Komskaya, *Rubber Chem. Technol.* **31**, 244 (1958).
69. O. F. Noel, III and J. F. Carley, *Polym. Eng. Sci.* **15**, 117 (1975).
70. A. Plochocki, *J. Appl. Polym. Sci.* **16**, 987 (1972).
71. G. S. Kongarov and G. M. Bertenev, *Rubber Chem. Technol.* **47**, 1188 (1974).
72. C. K. Shih, *Polym. Eng. Sci.* **16**, 742 (1976).
73. E. R. Galimov, G. G. Ushakova, R. G. Timergaleev, and V. A. Voskresenskii, *Int. Polym. Sci. Technol.* **2**, T/7 (1975).
74. M. Kh. Giniyatullin, R. G. Timergaleev, M. Kh. Khasanov, and V. A. Voskresenskii, *Sov. Plast.* (*Engl. Transl.*) No. 6, p. 8 (1973).

75. M. C. Williams, *AIChEJ.* **21**, 1 (1975).
76. W. M. Prest, Jr. and R. S. Porter, *J. Polym. Sci., Part A-2* **10**, 1639 (1972).
77. J. D. Ferry, "Viscoelastic Properties of Polymers," p. 292. Wiley, New York, 1970.
78. W. M. Prest, Jr. and R. S. Porter, *Polym J.* **4**, 154 (1973).
79. J. Crank and G. S. Park, eds., "Diffusion in Polymers." Academic Press, New York, 1968.
80. J. Crank, "Mathematics of Diffusion." Oxford Univ. Press, London and New York, 1956.
81. P. Meares, "Polymers: Structure and Bulk Properties." Van Nostrand-Reinhold, Princeton, New Jersey, 1965.
82. C. A. Kumins, C. J. Rolle, and J. Roteman, *J. Phys. Chem.* **61**, 1290 (1957).
83. Y. Ito, *Chem. High Polym.* **18**, 158 (1961).
84. L. M. Robeson, *Polym. Eng. Sci.* **9**, 277 (1969).
85. H. Herman, ed., "Advances in Materials Research," Vol. 3, p. 305. Wiley (Interscience), New York.
86. Y. J. Shur and B. G. Ranby, *J. Appl. Polym. Sci.* **19**, 1337 (1975).
87. Y. J. Shur and B. G. Ranby, *J. Appl. Polym. Sci.* **19**, 2143 (1975).
88. Y. J. Shur and B. G. Ranby, *J. Appl. Polym. Sci.* **20**, 3105 (1976).
89. Y. J. Shur and B. G. Ranby, *J. Appl. Polym. Sci.* **20**, 3121 (1976).
90. H. Storstrom and B. G. Ranby, *Adv. Chem. Ser.* **99**, 107 (1971).
91. B. G. Ranby, *J. Polym. Sci., Polym. Symp.* **51**, 89 (1975).
92. C. H. Jacques, H. B. Hopfenberg, and V. Stannett, *Polym. Eng. Sci.* **13**, 81 (1973).
92. C. H. Jacques and H. B. Hopfenberg, *Polym. Eng. Sci.* **14**, 441 (1974).
94. H. B. Hopfenberg, V. T. Stannett, and G. M. Folk, *Polym. Eng. Sci.* **15**, 261 (1975).
95. R. L. Stallings, Ph.D. Thesis, North Carolina State University, Raleigh (1975).
96. R. L. Stallings, H. B. Hopfenberg, and V. T. Stannett, *J. Polym. Sci., Polym. Symp.* **41**, 23 (1973).
97. M. Matzner, D. L. Schober, R. N. Johnson, L. M. Robeson, and J. E. McGrath, *in* "Permeability of Plastic Films and Coatings" (H. B. Hopfenberg, ed.), p. 125. Plenum, New York, 1975.
98. A. G. M. Last, *J. Polym. Sci.* **39**, 543 (1959).
99. M. Inoue, *J. Polym. Sci., Part A-1* **1**, 2013 (1963).
100. B. Wunderlich, "Macromolecular Physics," Vol. 2. Academic Press, New York, 1976.
101. L. Mandelkern, "Crystallization of Polymers." McGraw-Hill, New York, 1964.
102. F. Gornich and J. D. Hoffman, *in* "Nucleation Phenomena" (A. S. Michaels, ed.), p. 53. *Am. Chem. Soc.*, Washington, D.C., 1966.
103. J. D. Hoffman and J. J. Weeks, *J. Chem. Phys.* **37**, 1723 (1962).
104. G. S. Y. Yeh and S. L. Lambert, *J. Polym. Sci., Part A-2* **10**, 1183 (1972).
105. T. Nishi and T. T. Wang, *Macromolecules* **8**, 909 (1975).
106. H. D. Keith and F. J. Padden, Jr., *J. Appl. Phys.* **35**, 1286 (1964).
107. L. M. Robeson, *J. Appl. Polym. Sci.* **17**, 3607 (1973).
108. L. M. Robeson and B. L. Joesten, *Pap., N. Y. Acad. Sci.*, 1975.
109. V. Crescenzi, G. Manzini, G. Calzolan, and C. Borri, *Eur. Polym. J.* **8**, 449 (1972).
110. J. V. Koleske and R. D. Lundberg, *J. Polym. Sci., Part A-2* **7**, 795 (1969).
111. C. Ong, Ph.D. Thesis, University of Massachusetts, Amherst (1974).
112. F. B. Khambatta, F. Warner, T. Russell, and R. S. Stein, *J. Polym. Sci., Polym. Phys. Ed.* **14**, 1391 (1976).
113. W. Wenig, F. E. Karasz, and W. J. MacKnight, *J. Appl. Phys.* **46**, 4194 (1975).
114. R. N. Lemos, Ph.D. Thesis, University of Massachusetts, Amherst (1975).
115. D. J. Hourston and I. D. Hughes, *Polymer* **18**, 1175 (1977).
116. D. R. Paul and J. O. Altamirano, *Polym. Prepr., Am. Chem. Soc. Div. Polym. Chem.* **15** (1), 409 (1974).

References

117. E. Roerdink and G. Challa, *Polymer* **19**, 173 (1978).
118. T. K. Kwei, G. D. Patterson, and T. T. Wang, *Macromolecules* **9**, 780 (1976).
119. R. L. Imken, D. R. Paul, and J. W. Barlow, *Polym. Eng. Sci.* **16**, 593 (1976).
120. A. Lilaonitkul, J. C. West, and S. L. Cooper, *J. Macromol. Sci., Phys.* **12** (4), 563 (1976).
121. R. W. Crawford and W. K. Witsiepe, U. S. Patent 3,718,715, (1973) (assigned to E. I. duPont de Nemours and Co.).
122. T. Nishi, T. K. Kwei, and T. T. Wang, *J. Appl. Phys.* **46**, 4157 (1975).
123. L. M. Robeson, *J. Polym. Sci., Polym. Lett. Ed.* **16**, 261 (1978).
124. D. S. Hubbell and S. L. Cooper, *J. Polym. Sci., Polym. Phys. Ed.* **15**, 1143 (1977).
125. D. S. Hubbell and S. L. Cooper, *J. Appl. Polym. Sci.* **21**, 3035 (1977).
126. R. J. Kern, *J. Polym. Sci.* **33**, 524 (1958).
127. I. C. McNeill and D. Neil, *Eur. Polym. J.* **6**, 143 (1970).
128. T. Takahasi, T. Yasukawa, and K. Murakami, *Angew. Makromol. Chem.* **9**, 182 (1969).
129. M. E. Woods and D. G. Frazer, *Soc. Plast. Eng., Tech. Pap.* **32**, 426 (1974).
130. R. D. DeMarco, M. E. Woods, and L. F. Arnold, *Rubber Chem. Technol.* **45**, 1111 (1972).
131. D. R. Paul, J. W. Barlow, C. A. Cruz, R. N. Mohn, T. R. Nassar, and D. C. Wahrmund, *Am. Chem. Soc., Div. Org. Coat. Plast. Chem.* **37** (1), 130 (1977).
132. R. G. Bauer and M. S. Guilloid, *Adv. Chem. Ser.* **142**, 231 (1975).
133. Z. Slama and J. Majer, *Plaste Kautsch.* **24**, 423 (1977).

Chapter 7

Utilization of Miscible Polymers

7.1 INDUSTRIAL EXAMPLES

In this section, discussion of the commercial interest in polymer miscibility will be highlighted with examples of present utility where the unique characteristics of miscible polymer blends lead to a compromise of price/performance variables such that viable commercial products have been developed. In the field of material science, a vast amount of research and development effort with metals has provided an extremely large number of metal alloys with improved price/performance behavior resulting in innumerable commercial products. Blends from polymer mixtures or alloys have recently achieved an important position in polymer science and technology. The inherent reasons for this may be understood by summarizing the development of technology in polymer science as it relates to industrial application.

In the earlier days of polymer science, much of the effort was directed toward obtaining information on the synthesis and properties of a myriad of polymer structures. With the tremendous background and wealth of information derived from these studies, many commercial polymers were developed which form the basis of the plastics and elastomers industry. Recent polymer research and development have been less centered on new polymer compositions, but rather on the modification of existing polymer structures. These modifications include composites, block copolymers, interpenetrating networks, and polymer blends. Under polymer blends, two separate areas can be defined: miscible and immiscible polymer blends with, of course, the obvious intermediate cases.

In terms of property enhancement of specific polymer blends, miscibility, per se, is not a criterion for utility. The impact modification of polystyrene, styrene/acrylonitrile copolymers (ABS), poly(vinyl chloride), poly(methyl methacrylate), and nitrile-based barrier polymers all include examples where elastomeric modification yields desired toughness by the incorporation of an immiscible polymer phase. Other examples of useful (and commercial) polymer blends where miscibility is not achieved (although mechanical compatibility is observed) include bisphenol A polycarbonate–ABS (Cycoloy, Bayblend) [1], polypropylene–ethylene/propylene rubber (TPR, Somel, Telcar) [2, 3], chlorinated polyethylene–PVC [4], and poly(methyl methacrylate)–PVC (Kydex, DKE) [5]. Various elastomer blends of commercial utility have been cited in Dunn [6] and Corish and Powell [7], and polyolefin blends also comprise important commercial blends, as will be discussed in a later section of this chapter (Section 7.2.1).

Covalent bonding of different structures is used in block copolymers to achieve mechanical compatibility where phase separation is observed, and it is required to yield elastomeric properties. Examples include the following commercial polymers: styrene–diene–styrene ABA block copolymers [8], bisphenol A polycarbonate–silicone rubber $(AB)_n$ block copolymer [9], butadiene/acrylonitrile copolymer modification of epoxies [10], styrene–diene AB block copolymers [11], and styrene–ethylene/butylene–styrene ABA block copolymers [12].

Miscible polymers yield property behavior distinctly different from two-phase polymer mixtures in that the miscible blend exhibits a single T_g. As the glass transition temperature is related to physical properties, including modulus–temperature behavior, creep, viscosity, crystallization rate, and impact strength, the difference between a blend exhibiting a single T_g and a phase-separated blend with two T_g's can be quite pronounced. These differences have been alluded to in previous chapters and will be summarized later in this chapter.

Commercial polymer–polymer miscible blends include several high molecular weight plasticizers for PVC, polystyrene–PPO blends (Noryl), elastomer blends, and polyelectrolyte complexes. The patent literature contains many examples of the utility of polymer combinations which are listed as miscible in the comprehensive survey in this book (Chapter 5). Several of these cases will be discussed as they represent either present or potential commercial use.

7.1.1 PVC–Nitrile Rubber Blends

The first example of commercialization of polymer–polymer miscible blends occurred in the early 1940s with blends of poly(vinyl chloride) and

7.1. Industrial Examples

butadiene/acrylonitrile rubber [13]. The addition of plasticized PVC to nitrile rubber was discussed by Emmett [14] with a series of property comparisons on sulfur-vulcanized mixtures. The addition of PVC to nitrile rubber yielded advantages in ultraviolet and ozone resistance, tear and flex cracking resistance, thermal aging, and solvent resistance. Poorer abrasion resistance, higher compression set, and lower tensile strength were cited as disadvantages. The optimum range of addition of plasticized PVC to nitrile rubber was cited to be 20–40 wt%. Pittinger and Cohan [15] discussed the advantages of PVC addition to nitrile rubber and cited several applications for the blend, including wire and cable jacketing, printing roll covers, gaskets, valve disks, and football covers.

At higher PVC concentrations (>50 wt%), the nitrile rubber–PVC blend resembles conventionally plasticized PVC without the plasticizer migration problems associated with conventional low molecular weight plasticizers. These materials are not vulcanized and are processed using typical plasticized PVC processing techniques. Plasticizers can be added to yield lower modulus, easier processing materials. Pittinger and Cohan [15] listed various applications for this permanently plasticized vinyl, including shoe uppers, shoe welt, book bindings, vinyl adhesives, tubing, and hose. Reed [16] later discussed applications which required the permanent characteristics of this unique blend. These included oleomargarine bags as well as general purpose food contact film. Recent interest in the potential toxicity which could be caused by migration of lower molecular weight plasticizers into food products has catalyzed interest in permanent high molecular weight plasticizers for PVC [17].

The references cited thus far were prior to 1950. This blend is still of definite commercial utility in many of the above-cited applications as well as new applications (e.g., pit and pond liners for water and oil containment). Optimization of parameters available to the polymer processors of this blend was an extensive study by Schwarz and Edwards [18]. Structure/property studies involved nitrile rubber–PVC ratio, PVC molecular weight, nitrile rubber viscosity, and acrylonitrile level of the nitrile rubber. These results were discussed in reference to specific applications including fuel hose covers, wire and cable jacketing, and conveyor belt covers. Landi [19] investigated the properties of liquid nitrile rubber compared with solid nitrile rubber in PVC blends. Deanin *et al.* [20] compared the effect of various additives to plasticized PVC compositions and observed that only nitrile rubber addition did not impair the abrasion resistance. Jordan *et al.* [21–23] studied the mechanical and viscoelastic properties of nitrile rubber blends with PVC and vinyl chloride/vinyl stearate copolymers. The high volatility of dioctyl phthalate (DOP) and the extraction rate were noted relative to the negligible plasticizer loss with nitrile rubber as the plasticizer for PVC.

TABLE 7.1

Powdered Nitrile Rubber–PVC Blends; Physical Properties of High Hardness Blends[a]

	Compound number[b]						
	11	12	13	14	15	16	17
Plasticizers							
DOP	40	27	13	0	27	13	0
Hycar 1422 (nitrile rubber)	—	39	81	120	—	—	—
Hycar 1452P-50 (powdered nitrile rubber)	—	—	—	—	26	54	80
Total recipe parts	157	183	211	237	170	184	197
Physical properties							
Hardness, A	93	94	88	89	94	93	95
100% Modulus (psi)	1750	1350	1150	1150	1525	1400	1300
Tensile strength (psi)	2800	2825	2300	1950	2425	2000	1625
% Elongation	345	385	360	285	375	355	300
Graves tear (lb/in., with grain)	450	370	230	190	480	480	380
Brittle point (°C)	−27	−29	−35	−36	−33	−32	−38

[a] Reprinted by permission from M. E. Woods and D. G. Frazer, *Soc. Plast. Eng., Tech. Pap.* **32,** 426 (1974). Copyright by the Society of Plastics Engineers, Inc.
[b] 100 parts PVC, 117 parts of base resin.

Processing of nitrile rubber bales with PVC does not allow the use of several processing techniques typical for plasticized PVC compositions (e.g., ribbon blenders or Henschel blenders). Even granulated nitrile rubber particles do not perform as well as typical plasticized compositions using these processing techniques, due to migration and settling of one component relative to the other. To alleviate these problems, powdered nitrile rubber has been commercially introduced [24, 25]. Free-flowing dry blends have been demonstrated and touted to overcome problems previously encountered with nitrile rubber as well as with other "polymeric" plasticizers. Woods and Frazer [25] discussed the properties of the powdered nitrile rubber–PVC blends, and Table 7.1 lists the physical properties of high-hardness compounds from these blends.

7.1.2 PVC–Other High Molecular Weight Plasticizers

With the commercial demand for plasticized PVC compositions of greater permanency for the more demanding applications as well as for freedom from the ecological implications of plasticizer migration into the environment [26], recent emphasis has been placed on readily available intermediate

7.1. Industrial Examples

molecular weight plasticizers (2000–4000 M_n). These oligomers are generally polyester types (e.g., adipate of adipic acid and ethylene glycol). Examples of these oligomers are cited in references [27–34] and will not be discussed in detail because the molecular weights are generally lower than the systems covered in this treatise.

Of the high molecular weight plasticizers, ethylene/vinyl acetate copolymers (vinyl acetate content of 65–70 wt%) are of industrial interest primarily due to the low cost of this copolymer [35]. Potential applications include those cited from the section on nitrile rubber–PVC, as well as outdoor applications or in ozone and oxiding environments because of the fully saturated structure of EVA.

One of the basic problems of this lower cost high molecular weight plasticizer is its form: a gumstock-type material due to the absence of polyethylene crystallinity and a T_g below ambient temperatures at the ethylene–vinyl acetate composition required for miscibility with PVC. This results in significantly reducing the processing possibilities with typical plasticized PVC which is adapted to liquids or small-particle-size solids. Powder blends of PVC, plasticizer, stabilizers, and flow additives, to be delivered to processing equipment such as a Banbury mixer, are not possible with ethylene/vinyl acetate copolymers as the plasticizer. Liquids impregnate the porous PVC particles yielding a flowable powder blend; solids (with a small particle size) can be evenly distributed in a powder blend. Gumstock or highly viscous materials, however, are precluded from powder blends commonly prepared in a ribbon blender or Henschel mixer. Processing requires nonstandard approaches, which include addition of the highly viscous material to a prefluxed PVC powder blend on a two-roll mill or in a Banbury mixer.

To achieve a solid-particle plasticizer for PVC, modifications of the ethylene/vinyl acetate copolymer composition have been cited as forming miscible blends with PVC while retaining a low level of polyethylene crystallinity [36–38]. These structural modifications include incorporation of sulfur dioxide or carbon monoxide units into the EVA backbone. This allows a much higher concentration of ethylene because the sulfur dioxide or carbon monoxide units yield sites for specific interactions with PVC. A polymeric plasticizer commercially available from duPont (PB-3041) offers a highly permanent plasticized PVC composition along with excellent low-temperature toughness relative to conventional plasticized PVC formulations [39, 40]. Although duPont has not disclosed the composition of this polymeric plasticizer, it is believed to be a terpolymer containing ethylene, vinyl acetate, and carbon monoxide [41].

Mixtures of PVC and poly(butylene terephthalate)–poly(tetrahydrofuran) $(AB)_n$ block copolymers have been reported to exhibit at least partial miscibility [42–44]. These block copolymers, with the trade name Hytrel (duPont), have been found [42, 45] to be excellent permanent plasticizers for use in the more demanding plasticized PVC applications. These include the examples previously cited for plasticized PVC with particular emphasis on shoe components, where abrasion resistance is important. Hytrel–PVC blends have been used in agricultural insecticide hose jackets to obtain better low-temperature flexibility and abrasion resistance than typical plasticized PVC [46]. The patent describing Hytrel–PVC blends lists upholstery, interior automotive applications, food packaging film, wall covering, flooring, and wire and cable coatings as potential applications [45]. This patent claims excellent permanence, abrasion resistance, low-temperature flexibility, and impact resistance, along with improved scuff resistance. Advantages cited for the use of Hytrel over polyester urethanes also employed as permanent plasticizers for PVC were better thermal stability, better release from highly polished chrome-plated processing rolls, and the ability to use higher shear (more efficient) mixing screws in extrusion processing. Typical properties of Hytrel–PVC blends are listed in Table 7.2.

Polyester-based thermoplastic polyurethanes have also been cited as permanent plasticizers for PVC [47–49]. A specific example is the class of poly(ε-caprolactone)-based thermoplastic polyurethanes [50]. These prod-

TABLE 7.2

Modification of PVC with Hytrel (Injection Molded Slabs)[a]

PVC flexible compound	100	80	70	60
40D Hytrel	—	20	30	40
Hardness, durometer A	86	83	85	86
Tensile strength, MPa (psi)	16.9 (2450)	17.6 (2550)	18.6 (2700)	19.6 (2850)
Elongation of break (%)	300	450	520	560
100% Modulus, MPa (psi)	10 (1450)	7.6 (1100)	6.9 (100)	6.5 (950)
Torsional modulus, MPa (psi × 10^3)				
At 23°C	6.9 (1)	6.9 (1)	10.3 (1.5)	11.7 (1.7)
At −18°C	96.5 (14)	75.8 (11)	55 (8)	48.2 (7)
At −29°C	206 (30)	138 (20)	110 (16)	96 (14)
At −40°C	386 (56)	248 (36)	179 (26)	158 (23)
Ross flex				
Flexes × 10^6 at 23°C	1	1	1	1
Cut growth	0	0	0	0
Flexes × 10^3 at −18°C	1.5	51	51	51
Cut growth	5×	5×	2×	0
Taber abrasion (CS-17 wheel)				
mg/100 revolutions	33	19	10	13

[a] Reprinted by permission from R. P. Kane, *J. Elast. Plast.* **9**, 416 (1977). Copyright by Technomic Publishing Company, Inc., 265 Post Road West, Westport, Connecticut 06880.

ucts have the advantage of being solid plasticizers available in pellet form. Poly(vinyl chloride) can be added in all proportions to give useful compositions. At low PVC levels, improved performance of the thermoplastic polyurethanes in typical flammability tests is one of the major attributes. At higher levels of PVC, the products resemble typical permanently plasticized PVC compositions. The overall property balance is quite similar to Hytrel–PVC blends. The major factors preventing broader commercial acceptance of both Hytrel and the thermoplastic polyurethanes are the low efficiency of the plasticization of PVC (relative to monomeric or oligomeric plasticizers) and economics; Hytrel or the polyester thermoplastic polyurethanes cost approximately five times that of PVC. The poor efficiency of these high molecular weight plasticizers is due to the hard-block content, resulting in a higher T_g than typical PVC plasticizers. With sufficient hard-block content to yield a solid, nonblocking, pellet-form material, the T_g is sufficiently raised such that the efficiency of plasticization is lowered.

Another approach to raise the efficiency as well as lower the raw material cost has been to extend dihydroxy-terminated polyester oligomers with diisocyanates [51–53]. The primary choice for the diisocyanate is toluene diisocyanate (TDI), unless weatherability is desired; in which case, hexamethylene diisocyanate or hydrogenated MDI (H_{12}MDI) can be substituted. A series of variations of the high molecular weight diisocyanate-extended polyesters has been mentioned in the above-cited patents as permanent plasticizers for PVC. These variations, however, are uncrosslinked, very viscous liquids with problems similar to gumstock materials.

7.1.3 Polystyrene–PPO (Noryl)

A commercial venture involving the miscible polymer blend of polystyrene and poly(2,6-dimethyl-1,4-phenylene oxide) (PPO) has demonstrated the basic price/performance advantages and versatility of miscible polymer blends in which the components have widely differing properties. PPO was initially introduced in 1964 by General Electric Co. as a high-temperature polymer designed to compete in the engineering thermoplastic market area [54]. With a T_g of 210°C, the temperatures required for fabrication seriously limited the injection molding latitude. The pendant methyl groups were subject to oxidative attack, lowering its elevated temperature stability.

As combinations of PPO and polystyrene yield miscible blends with single T_g's, any heat distortion temperature between that of polystyrene (85°C) and PPO (195°C) can be obtained. With the addition of polystyrene, PPO can be processed at much lower temperatures, and thus the injection molding deficiencies are virtually eliminated. The much lower selling price of polystyrene greatly enhances the commercial acceptance of the blend relative to unmodified PPO. The addition of PPO to polystyrene produces load-bearing capabilities at higher temperatures, improved mechanical properties (PPO is tougher and stronger than polystyrene), and improved resistance to flammability as defined by existing flammability tests (e.g., Underwriter's Laboratories Test UL-94). The ability to achieve improved flammability ratings has been one of the major reasons for the commercial acceptance of the PPO–polystyrene blends. The addition of PPO even at low concentrations to polystyrene allows the use of low-cost phosphorus-containing additives (e.g., triphenyl phosphate) [55] at relatively low concentrations to achieve V-1 ratings [56] as defined by Test UL-94. The basic patent concerning PPO–polystyrene blends is listed in Cizek [57].

In the commercial products marketed by General Electric under the trade name Noryl, rubber-modified polystyrene is used to achieve impact-resistant compositions. Until 1977, commercial Noryl products contained up to 50% PPO. Noryl 731, which was the product with the highest heat distortion temperature (unreinforced) listed in General Electric product literature prior to 1977, is believed to contain approximately 50 wt% PPO. Typical properties of Noryl 731 versus those for impact polystyrene and PPO are listed in Table 7.3, illustrating the achievement of average properties of the components. Recently a version of Noryl (Noryl N300) with a higher heat distortion temperature has been introduced, further broadening the property profile [58].

TABLE 7.3

Mechanical Property Data for Polystyrene, PPO, and Commercial Blend of Polystyrene–PPO (Noryl)[a]

	Impact polystyrene ($\sim$10 to 12% rubber)	Noryl 731	PPO[b]
Tensile modulus (psi)	300,000	350,000	370,000
Tensile strength (psi)	4,750	9,500	10,500
Ultimate elongation (%)	30	25	50
Flexural strength (psi)	8,000	13,500	14,000
Flexural modulus (psi)	300,000	360,000	370,000
Heat distortion temperature (°C) (264 psi), 0.25-in. bar	88	129	190–195°
T_g (°C)	100	145	210
Specific gravity	1.04	1.06	1.06
Notched Izod impact strength (ft-lb/in. of notch)	1.2	2 to 4	1.6

[a] Data are from Lee et al. [54], Modern Plastics Encyclopedias, and relevant product literature.

[b] PPO = poly(2,6-dimethyl-1,4-phenylene oxide).

7.1. Industrial Examples

With the improved heat distortion temperature, toughness, flame resistance, and creep resistance relative to impact polystyrene, Noryl has been able to penetrate markets available to the family of materials generally classified as engineering thermoplastics. With the increased demands on the flame resistance of polymers used in the electrical/electronic and appliance applications, the fact that Noryl satisfies the requirements of Underwriter's Laboratories Test UL-94 for a V-1 rating (i.e., Noryl SE-100 and Noryl SE-1) makes it one of the prime candidates for these market areas.

Specific applications cited by General Electric trade literature and plastic journal articles [59–61] include examples in the following list, with specific attributes noted.

(1) Swimming pool pumps: hydrolytic resistance and low creep
(2) Dishwasher parts: hydrolytic resistance and low creep
(3) Computer terminal housings: flammability rating
(4) Hospital furniture: flammability rating
(5) Hair dryer components: heat distortion temperature and flammability rating
(6) Humidifier parts: hydrolytic resistance and flammability rating
(7) Curling irons: heat distortion temperature and flammability rating
(8) Vacuum cleaner housings: flammability rating
(9) Coffeemaker components: heat distortion temperature and flammability rating
(10) Automotive dashboards: heat distortion temperature and impact strength
(11) Automotive grills and trim: ability to be plated and heat distortion temperature
(12) Plumbing fixtures: heat distortion temperature and ability to be plated
(13) Tuner and deflection yoke for televisions: heat distortion temperature and flammability rating
(14) Business machine housing: impact strength and flammability rating

7.1.4 Block Copolymers

Several commercial block copolymers are available in which the component blocks exhibit certain characteristics of amorphous phase miscibility. These materials include thermoplastic polyurethanes and poly(butylene terephthalate)–poly(tetrahydrofuran) $(AB)_n$ block copolymers. With the thermoplastic polyurethanes, the miscibility of the amorphous phase is not universally accepted, although data exist to support at least partial miscibility, as will be discussed. Both of these cases have crystallizable block segments. With block copolymers containing amorphous blocks, the miscibility of the component blocks will yield properties quite similar to those of random copolymers of the monomeric constituents. To achieve the desired elastomeric properties typical of block copolymers with a low T_g block, phase separation is essential. Elastomeric properties can be obtained with block copolymers exhibiting miscibility if one of the blocks is crystallizable with a melting point above the test temperature. Another block of the copolymer must have a T_g below the test temperature.

The structure of the polyester- or polyether-based thermoplastic polyurethanes has been the subject of many investigations, with varying conclusions [62–67]. With thermoplastic polyurethanes, the soft block consists of segments of 1000 to 3000 M_n of a low T_g polyester [e.g., poly(ε-caprolactone)] or a polyether [e.g., poly(tetrahydrofuran)]. These segments can be extended to higher molecular weight by methylene–bis(4)phenyl isocyanate (MDI) or short blocks of MDI–butane diol. The hard blocks of commercial thermoplastic polyurethanes consist of MDI–butane diol units.

(Thermoplastic polyurethane hard block)

The modulus–temperature behavior of the thermoplastic polyurethanes clearly indicates a two-phase structure. Initial studies attributed the higher temperature transition to the association of hard, glassy segments of the urethane block. However, more recent studies have concluded that the hard block is crystalline. It is interesting to note that toluene diisocyanate (mixed 2,4 and 2,6 isomers) does not yield useful thermoplastic polyurethanes when substituted for MDI, presumably due to the inability of TDI to yield crystalline domains [68]. The 2,6 isomer of TDI, however, has been shown to develop a modulus plateau above the T_g, resulting in behavior similar to MDI-based materials, presumably due to the ability to form crystalline segments [69]. Increasing the hard block/soft block ratio results in an increase in the T_g, indicating at least a certain level of miscibility in the residual amorphous phase [70]. These results, combined with the crystalline nature of the hard block, lead to the conclusion that the amorphous phase of the thermoplastic polyurethanes exhibits a level of mixing between the uncrystallized polyurethane segments and the polyester or polyether blocks.

While the low-temperature transition of thermoplastic polyurethanes has generally been ascribed to the soft polyester or polyether phase, the values

7.1. Industrial Examples

are significantly higher than the pure-component T_g's, thus indicating at least partial miscibility. Mixtures of the polyester-based thermoplastic polyurethanes with the poly(hydroxy ether) of bisphenol A (Phenoxy) [71] or PVC [50] yield single-phase blends with an apparent depression of the polyurethane crystallinity, based on modulus–temperature data. These blends exhibit a sharp, single T_g between those observed for the constituents; the absence of a separate transition for the urethane blocks indicates that the block copolymer behaves as a single-phase system in the amorphous state. Kaplan [72] recently concluded miscibility of the amorphous phase of thermoplastic polyurethanes for the block molecular weights found in commercial products. At higher molecular weights, it was noted that two glass transitions could be resolved. Although the amorphous phase structure of thermoplastic polyurethanes is still in doubt (based on a composite view of the literature), we have included the thermoplastic polyurethanes in our discussion of industrial utility, because the overall evidence appears to favor at least partial miscibility for the commercially available products.

The combination of flexibility, high tensile strength, high tear strength, excellent wear and abrasion resistance, good oil resistance, and processability by thermoplastic fabrication techniques yields a property balance vastly superior to conventional vulcanized elastomers in many application areas. Widespread use has resulted in a significant number of commercial applications. A typical set of thermoplastic polyurethane properties is listed in Table 7.4. The excellent ultimate properties of the polyester or polyether thermoplastic polyurethanes are the result of the ability of the polyester or polyether blocks to stress-crystallize. Noncrystallizable segments [e.g., poly-(propylene glycol)] substituted for these blocks yield inferior properties.

Thermoplastic polyurethanes were initially introduced in the early 1960s by B. F. Goodrich under the trade name Estane. This technology, subject of U. S. Patent No. 2,871,218 by Schollenberger, involved the polyester-based urethane elastomers. Many other commercial polyurethanes have been introduced since this initial entry, with examples listed in the following tabulation:

Producer	Trade name [73]	Producer	Trade name [73]
Upjohn Chemical Co.	Pellethane	Mobay Chemical Co.	Texin
Uniroyal Chemical Co.	Roylar	K. J. Quinn Co.	Q. E.
Hooker Chemical Co.	Rucothane		

Applications of the many thermoplastic polyurethane variations are numerous, with specific examples listed below:

(1) Automotive: fascia, fender extenders, filler panels: broad temperature utility, flexibility, durability

TABLE 7.4

Property Profile of Commercial Thermoplastic Polyurethanes (Pellethane)[a]

	Grade				
	2102-80A	2102-90A	2102-55D	2103-80A	2103-55
Soft block type	Polyester[b]	Polyester	Polyester	Polyether[c]	Polyether
Specific gravity	1.18	1.10	1.22	1.13	1.15
Hardness, shore	80A	90A	55D	80A	55D
Tensile strength (psi)	7500	7500	7500	7000	7000
100% Modulus (psi)	800	1600	2600	1000	2600
300% Modulus (psi)	1700	3800	5000	2300	5000
Elongation at break (%)	600	475	450	550	375
Elongation set (%)	15	25	25	15	25
Die "C" tear, pli	600	850	1200	580	1200
Compression set (%) (D395 Method B)	25–30	25–30	25–30	25–30	25–30

[a] Reprinted by permission from H. W. Bonk, A. A. Sardanopoli, H. Ulrich, and A. A. R. Sayigh, *J. Elastoplast.* **3**, 157 (1971). Copyright by Technomic Publishing Co., Inc., 265 Post Road West, Westport, Connecticut 06880.
[b] Polyester = poly(ε-caprolactone).
[c] Polyether = poly(tetramethylene ether).

(2) Identification tags for cattle: wear resistance, toughness
(3) Jacketing for wire and cable: oil resistance, flexibility, abrasion resistance
(4) Leather finishes: improved durability
(5) Permanent plasticizer for PVC: polyester-type
(6) Coated fabrics: durable leather substitute
(7) Conveyor belts: flexible, abrasion resistant
(8) Adhesives: strong adhesion to various substrates
(9) High-performance footwear: flexible, tough, and abrasion resistant
(10) Vibration insulators: damping characteristics and fatigue resistance
(11) Oil and grease pouches: flexibility and oil resistance

Poly(butylene terephthalate)–poly(tetrahydrofuran) $(AB)_n$ block copolymers have many similarities to thermoplastic polyurethanes in both properties and structure. Poly(butylene terephthalate) provides the crystallizable segment, whereas poly(tetrahydrofuran) yields the low T_g necessary for flexibility. The poly(tetrahydrofuran) segments are also presumed to stress-crystallize, yielding excellent ultimate properties. The calorimetric

7.1. Industrial Examples

TABLE 7.5

Property Profile of Poly(butylene terephthalate)–Poly(tetrahydrofuran) Block Copolymers [76][a]

Poly(butylene terephthalate) (wt %)	ASTM Test	33	58	76
Property				
T_m (°C), DSC		176	202	212
Specific gravity		1.15	1.20	1.22
Durometer hardness	D2240	92A	55D	63D
Tensile strength (psi)	D412	5,700	6,400	6,900
Elongation at break (%)	D412	810	760	510
10% Modulus (psi)	D638	520	1,450	2,450
100% Modulus (psi)	D412	1,150	2,150	2,860
Bashore resilience (%)	D2632	60	53	40
Flexural modulus (psi)	D797	6,500	30,000	72,000
Brittle point (°C)	D746	<-70	<-70	<-70
Izod impact strength (ft-lb/in. of notch)	D256			
22°C		>20	>20	20
−40°C		>20	>20	1.0
Oil swell (% volume increase ASTM No. 3 oil in 7 days/100°C)	D471	22.0	12.2	6.6

[a] Reprinted by permission from G. K. Hoeschele, *Polym. Eng. Sci.* **14**, 848 (1974). Copyright by the Society of Plastics Engineers, Inc.

and dynamic mechanical results reported by Lilaonitkul et al. [74] indicate a single T_g for the amorphous phase which increases with increasing poly(butylene terephthalate) content. This indicates a definite level of miscibility of the components in the amorphous phase. Other investigators have reached similar conclusions [75]. The property profiles of varying poly(butylene terephthalate) content for the commercial products (Hytrel, duPont) are listed in Table 7.5 [76]. Comparison with the thermoplastic polyurethane data in Table 7.4 reveals noticeable similarity with the lower modulus Hytrel products. Versions of Hytrel are available with higher modulus than commercial thermoplastic polyurethanes. The higher modulus Hytrel products have properties comparable with those of plasticized nylon 11 and nylon 12. Applications cited for the poly(butylene terephthalate)–poly(tetrahydrofuran) $(AB)_n$ block copolymers are listed below:

(1) Wire and cable insulation: flexibility and abrasion resistance
(2) Extruded hose: chemical and abrasion resistance, flexibility
(3) Snowmobile tracks: low-temperature flexibility, abrasion resistance
(4) O-rings, gaskets, and seals: chemical resistance

(5) Power transmission belts: flex fatigue resistance, high coefficient of friction
(6) Gears: durability, low noise level
(7) Tires (low speed, off-the-road): flexibility, weatherability
(8) Fuel tanks: ease of rotational molding, chemical resistance, impact resistance

7.1.5 Polyelectrolyte Complexes

The commercial utility of polyelectrolyte complexes has been most thoroughly investigated by Michaels *et al.* at Amicon Corp. (Lexington, Mass.) [77-80]. Ultrafiltration membranes based on polyelectrolyte complexes marketed under the trade name Diaflo have been promoted for use in filtration of macromolecules (e.g., proteins and polysaccharides) from water solutions. These membranes are of particular interest in biochemical purification or solution concentration. The salt-rejection aspects of these membranes are, however, inferior to cellulose acetate (i.e., 50-60% versus 97-98% rejection, respectively) [77]. For high molecular weight substances, the polyelectrolyte complexes are clearly superior to cellulose acetate due to a 10- to 50-fold higher water permeability rate. Polyelectrolyte complex membranes have been proposed for hemodialyzers (artificial kidneys), hemooxygenators (artificial lungs), battery separators, and fuel-cell membranes [78].

Michaels [78] proposed the utility of Ioplex resin [poly(vinyl benzyl trimethyl ammonium chloride)-sodium poly(styrene sulfonate)] as an additive to plastic films (i.e., thermoplastic polyurethanes or plasticized PVC) to improve the breathability (increased moisture vapor transmission) of upholstery, wearing apparel, and leather substitutes. Electrolyte-containing polyelectrolyte complexes are electrically conductive and thus were proposed as electrically conductive and antistatic coatings.

A method to make photoresistant sheets from polyelectrolyte complexes was described by Taylor [81]. This *in situ* polymerization technique involved a water solution of a strong polyelectrolyte with a monomer of an oppositely charged electrolyte capable of polymerizing with exposure to ultraviolet light. Exposed areas will consist of an insoluble complex, thus resisting removal via washing.

Various biomedical applications have been tested using polyelectrolyte-complex hydrogels. The high water sorption yields certain characteristics similar to body tissue. Extensive tests have been conducted on the utility of polyelectrolyte complexes as antithrombogenic surfaces. Using the *in vivo* vena cava Gott Ring test (*in vivo* vena cava ring implanted in dogs) as a criterion for thromboresistance, the form of Ioplex 101 containing 0.5 mEq

7.1. Industrial Examples

of excess polyanion exhibited superior thromboresistance, with promising results up to 2 weeks postimplantation [82, 83]. As the excess-polyanion polyelectrolyte complexes contain a high sulfonate ion density, the authors pointed out that heparin (an anticoagulant) also has a high sulfonate ion content. The results of the *in vivo* vena cava test are illustrated in Fig. 7.1 at different polyanion:polycation ratios. These results were considered to be a significant advance in the development of antithromboresistant materials and are described in two U. S. patents [84, 85].

Microencapsulation systems based on complexes from naturally occurring polyelectrolytes (gelatin:polycation–gum arabic:polyanion) were investigated by Green [86] at the National Cash Register Co. Other potential uses for polyelectrolyte complexes are [78, 87]

(1) wound and burn dressings
(2) soft contact lenses and cornea substitutes

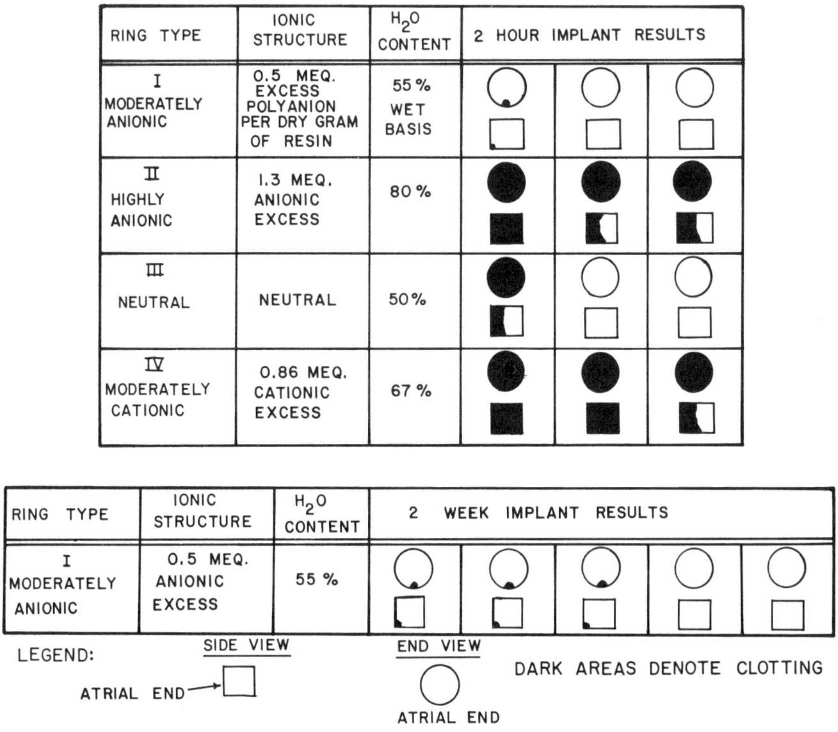

RING TYPE	IONIC STRUCTURE	H_2O CONTENT	2 HOUR IMPLANT RESULTS
I MODERATELY ANIONIC	0.5 MEQ. EXCESS POLYANION PER DRY GRAM OF RESIN	55% WET BASIS	
II HIGHLY ANIONIC	1.3 MEQ. ANIONIC EXCESS	80%	
III NEUTRAL	NEUTRAL	50%	
IV MODERATELY CATIONIC	0.86 MEQ. CATIONIC EXCESS	67%	

RING TYPE	IONIC STRUCTURE	H_2O CONTENT	2 WEEK IMPLANT RESULTS
I MODERATELY ANIONIC	0.5 MEQ. ANIONIC EXCESS	55%	

LEGEND: SIDE VIEW — ATRIAL END □   END VIEW — ATRIAL END ○   DARK AREAS DENOTE CLOTTING

Fig. 7.1. Thromboresistance studies involving polyion hydrogels (Gott ring test). [Reprinted by permission from M. K. Vogel, R. A. Cross, H. J. Bixler, and R. J. Guzman, *J. Macromol. Sci., Chem.* **4** (3), 675 (1970). Copyright by Marcel Dekker, Inc.]

(3) environmental sensors and chemical detectors
(4) coating for electrostatic photocopy applications
(5) dielectric filler for dielectrically heated plastics
(6) nonfogging, transparent coating for windows
(7) matrix for slow release of implantable drugs

7.1.6 Miscellaneous Miscible Polymer Uses

High-heat ABS involves substitution of α-methylstyrene for styrene to increase the T_g of the glassy matrix (styrene/acrylonitrile copolymer). Two basic approaches exist to achieve this T_g increase; one involves the miscible polymer approach. One obvious approach is the direct substitution of α-methylstyrene in the polymerization for styrene, yielding a random copolymer with a higher T_g. A less obvious approach, believed to be commercially utilized, consists of mixing a copolymer of α-methylstyrene/acrylonitrile with ABS. The azeotropic composition of this copolymer (69 wt% α-methylstyrene) and the azeotropic composition of styrene/acrylonitrile (76 wt% styrene) were observed to yield single-phase blends by Slocombe [88]. The patent literature also cites this mixture as a method of enhancing the heat distortion temperature of ABS [89]. High-heat ABS is used in applications requiring lower creep and higher heat distortion temperatures than standard ABS. These applications include electrical appliances, short-term boiling water exposure, or parts requiring higher temperatures in finishing operations, such as a paint drying cycle.

Rigid poly(vinyl chloride) compositions exhibit a price/performance balance that allows penetration into large commodity applications such as pipe and house siding. The T_g of rigid PVC (maximum of 85°C) results in heat distortion temperatures which are borderline in many applications. The utilization of the miscible polymer additive approach is recognized as a viable method to alleviate this problem. The patent literature cites several compositions useful as heat distortion builders for PVC, such as α-methylstyrene/acrylonitrile copolymers and α-methylstyrene/methyl methacrylate/acrylonitrile terpolymers [90–93]. A compound product (Blendex 586, Borg-Warner Co.) claimed to be a terpolymer of α-methylstyrene/acrylonitrile/styrene is cited as a heat distortion temperature enhancer for PVC in the product literature. The potential applications of higher T_g compositions of PVC include

(1) appliances requiring UL flammability rating
(2) pipe for hot fluid transport
(3) house siding in warm climates or with darker colors in normal climates

7.1. Industrial Examples

(4) containers filled with hot fluids
(5) thermoformed components for recreational equipment

Polycarbonates of the general structure

$$\left(-O-\underset{H_3C}{\overset{H_3C}{\bigcirc}}-X-\underset{CH_3}{\overset{CH_3}{\bigcirc}}-O-\overset{O}{\underset{\|}{C}}-\right)_n$$

have been cited in several patents [94, 95] assigned to Bayer as heat distortion temperature enhancers for PVC. Good transparency is also claimed, hinting at possible miscibility.

Polymeric processing aids are commonly added to PVC in the range of 1–5 wt% to improve the processability of PVC without changing the physical properties. Zelinger *et al.* [96] investigated the miscibility of these processing aids with PVC at higher concentrations than normally used commercially. They observed that the methyl methacrylate-based materials (Paraloid K-120N, 95/5 methyl methacrylate/ethyl acrylate, Rohm & Haas; Degalan V26, 90/10 methyl methacrylate/butyl acrylate, Degussa A. G.) were miscible with PVC. It was hypothesized that the miscible modifiers were able to break down the structure of the initial PVC particles, improving the deformability of the PVC phase during processing.

The miscibility of poly(methyl methacrylate)–poly(vinylidene fluoride) (PMMA–PVF_2) blends has been studied in detail [97–99]. PMMA can be added to PVF_2 in low amounts to improve the processing [100]. PVF_2 can be added to PMMA in low concentrations to act as a polymeric plasticizer [101]. Noland *et al.* [97] noted that the ultraviolet (uv) degradation of the PMMA–PVF_2 blends was consistent with that of the two components considered separately. As both are highly resistant to uv degradation, weatherable film applications of the blend are therefore possible.

The poly(hydroxy ether) of bisphenol A (Phenoxy) exhibits excellent adhesion to inorganic substances. This utility was cited in a patent [102] utilizing Phenoxy as an additive to filled poly(butylene terephthalate) blends. In a similar case, Horn [103] cited the utility of bisphenol A polycarbonate in blends with filled poly(butylene terephthalate). As polybutylene terephthalate) is emerging as an important engineering thermoplastic, these blends may have future interest. Polycarbonate–poly(butylene terephthalate) blends have been shown to have only limited miscibility [104].

The coatings industry has employed miscible polymer blends for years, especially to improve the film properties of nitrocellulose [105] as well as other modified cellulosics, such as cellulose acetate butyrate. Polymer

TABLE 7.6

Additional Examples of Potential Utility of Miscible Polymer Blends

Miscible polymer blend	Utility	Reference
Bisphenol A polycarbonate–poly-(ε-caprolactone)	Improved moldability	108
Poly(vinylidene fluoride)–methyl methacrylate/vinyl fluoride (68/32) copolymer	Excellent transparency	109
Bisphenol A polycarbonate–poly(ethylene terephthalate)	Clear film and sheeting for packaging applications	110
Poly(hydroxy ether) of bisphenol A–thermoplastic polyurethane	Binder composition for magnetic tape	111
Poly(vinyl butyrate)–poly(n-propyl acrylate)		112
Epoxy–poly(butylene terephthalate)/poly(tetrahydrofuran) $(AB)_n$ block copolymers	Improved toughness	113
Poly(ε-caprolactone)–poly(hydroxy ether) of bisphenol A	Improved toughness; improved stress crack resistance	114
Ethylene/N-methyl-N-vinyl acetamide or ethylene/N,N-dimethyl acrylamide–poly(vinyl chloride)*	Improved toughness and flexibility for PVC	115
Poly(ε-caprolactone)-g-polystyrene–poly(vinyl chloride)	Permanently plasticized PVC	116–118
Poly(ε-caprolactone)/diene block copolymer–styrene/acrylonitrile copolymers	Improved toughness for styrene/acrylonitrile copolymers	119
ABS–thermoplastic polyurethane		58
Bisphenol A polycarbonate–poly(ethylene terephthalate)	Improved flow in fiberglass filled with compositions for PET addition to polycarbonate	120
Poly(vinyl chloride)–poly(butylene terephthalate)–poly(tetrahydrofuran) $(AB)_n$ block copolymer	Acoustical damping	44
Cellulose nitrate–poly(methyl vinyl ether) or poly(ethyl vinyl ether)	Plasticized composition of cellulose nitrate	121
PVC–α-methylstyrene/methyl methacrylate/acrylonitrile (36/55.5/8.5)	Transparent impact-resistant lens (graft copolymer incorporated with matched refractive index with miscible polymer matrix)	122

modifications were generally lower molecular weight oligomers, although high molecular weight polymers such as poly(ε-caprolactone) have been proposed as excellent modifiers [106].

As with coatings, adhesive applications have utilized miscible polymer systems in specific cases. Low molecular weight resins in neoprene adhesive

7.2. Mechanical Compatibility versus Miscibility in Polymer Blends

formulations yield improved tack or promote adhesion specifically for certain substrates. These additives include phenolics, terpene–phenolics resins, and coumarone–indene resins [107]. Nitrile rubber adhesives often contain low molecular weight additives, including phenolic resins and lower molecular weight vinyl chloride co- and terpolymers [107].

Although the patent literature was not thoroughly searched for this treatise, several examples of potentially useful blends of miscible polymers were uncovered. These previously uncited examples are summarized in Table 7.6 [44, 58, 108–122].

7.2 MECHANICAL COMPATIBILITY VERSUS MISCIBILITY IN POLYMER BLENDS

7.2.1 Important Commercial Examples of Two-Phase Blends

The utility of polymer blends obviously does not require achieving miscibility. In fact, most of the multicomponent polymer systems commercially utilized are two-phase blends. This is somewhat a consequence of the limited number of interesting miscible polymer blends. However, there are many cases where a two-phase system offers a specific advantage over that expected for a single-phase blend of the same constituents.

Important examples of commercial two-phase blends are impact polystyrene and ABS. The initial commercial impact polystyrenes were melt blends of polystyrene and rubber (e.g., styrene/butadiene rubber). However, the rubber efficiency was generally poor, primarily due to the inability to achieve and maintain an optimum rubber particle size, which has emerged as a major variable in detailed studies [123]. The *in situ* polymerization under specific conditions (i.e., optimum shear rate during phase inversion) of impact polystyrene from a solution of styrene and rubber solved the morphological problems inherent with melt blends. Optimum particle size of the rubber, ability to maintain these dimensions in a shear field due to crosslinking, improved compatibility of the rubber with the matrix due to grafting, and occlusions of polystyrene within the rubber particle to improve the rubber efficiency are all inherent characteristics of the *in situ* polymerization technique utilized commercially for impact polystyrene [123–125].

The impact improvement of polystyrene by the incorporation of a low modulus rubber phase is dramatic due to the ability of the polystyrene matrix to undergo a myriad of craze formation between the rubber particles, thus adsorbing a large amount of energy before failure can occur. The toughening of ABS (styrene/acrylonitrile matrix with polybutadiene or styrene/butadiene rubber particles grafted with matrix SAN and lightly cross-linked) is analogous to impact polystyrene.

Poly(vinyl chloride) and poly(methyl methacrylate) are commonly impact modified by the incorporation of grafted, lightly cross-linked rubber particles. With poly(vinyl chloride), the grafted polymer on the rubber phase is chosen to exhibit mechanical compatibility with PVC. Typical PVC impact modifiers consist of styrene/acrylonitrile copolymers or poly(methyl methacrylate) grafted onto butadiene-based rubber [126].

Poly(vinyl chloride)–impact poly(methyl methacrylate) blends [available from duPont (DKE), Rohm & Haas (Kydex), and Shulman (Polydene)] offer an excellent compromise of the specific attributes of the constituents. Impact PMMA offers improved heat distortion temperatures, while PVC allows the blend to achieve important UL-94 V-0 flammability ratings. The addition of impact PMMA to PVC also improves the thermoforming characteristics. The notched impact strength is higher than either of the components. Applications include extruded paneling, seat backs for mass transit, machine housing, aircraft interior components, fume ducting, and industrial wall paneling [58, 127]. Note that atactic PMMA–PVC blends were cited to be two-phase systems in Chapter 5, whereas syndiotactic PMMA–PVC blends exhibited miscibility. The proximity of miscibility of the atactic PMMA with PVC results in excellent mechanical compatibility of the PMMA–PVC blends.

ABS–PVC blends have quite similar property profiles to the impact PMMA–PVC blends discussed above. These blends are commercially available from Borg-Warner (Cycovin), Abtec (Abson), and Shulman (Polyman). ABS offers improved heat distortion temperatures and processability, whereas PVC offers flame-retardant properties. The blend can have higher notched impact strength than either of the components. Applications include power-tool handles, sanitary ware, communication relays, electrical terminal blocks, and electronic housings [120, 127]. Note that Shur and Ranby concluded that the SAN matrix of ABS was miscible with PVC [128]. However, other investigators have reported two-phase behavior of styrene/acrylonitrile (in the range of acrylonitrile content used in ABS) and PVC [129]. High-heat ABS based on the incorporation of α-methylstyrene may indeed be miscible with PVC, as discussed earlier in this chapter.

Bisphenol A polycarbonate exhibits the highest notched impact strength of unmodified rigid commercial polymers. The susceptibility to stress cracking in the presence of many organic chemical environments is a deficiency. Blends with ABS retain the high notched impact strength, lower the cost relative to polycarbonate, and improve the resistance to environmental stress cracking. Commercial polycarbonate–ABS blends are available under the trade names Cycoloy (Borg-Warner) and Bayblend (Mobay) [1, 58, 127].

Polyolefin blends have been utilized in many forms to achieve modifications yielding environmental stress rupture resistance, improved impact

7.2. Mechanical Compatibility versus Miscibility in Polymer Blends

strength, flexibility, and filler acceptance [130–133]. The addition of ethylene copolymers (e.g., ethylene/vinyl acetate or ethylene/ethyl acrylate), ethylene/propylene rubber, or polyisobutylene to either low density or high density polyethylene or polypropylene has been cited as providing the above property improvements. The addition of ethylene/propylene rubber (EPR) or blends of ethylene/propylene rubber and high density polyethylene to polypropylene has been specifically utilized for improving the low-temperature impact strength [134]. Recently, blends of $>50\%$ EPR in polypropylene have yielded a new family of low modulus materials. Some of these materials are true blends, whereas others contain grafted polypropylene on EPR as well as the ungrafted constituents. These products include those under the trade names TPR, Somel, and Telcar [135, 136]. Applications cited for these materials include replacements for plasticized PVC, hose, wire and cable insulation, and automotive bumpers and fascia.

The use of elastomer blends has been the subject of several reviews [6, 7, 137] in which commercial examples and property advantages of immiscible elastomer blends are cited. Blends of natural rubber and polybutadiene have shown various advantages, including heat stability, improved elasticity, and abrasion resistance [7]. EPDM (ethylene/propylene/diene monomer) blended with SBR (styrene/butadiene rubber) has shown improvements in ozone and chemical resistance with better compression set properties [138]. Blends of EPR and nitrile rubber have been cited as a compromise for obtaining moderate oil and ozone resistance with improved low-temperature properties [7]. Neoprene–polybutadiene blends offer improved low-temperature properties and abrasion resistance with better processing characteristics [139].

Ethylene/vinyl acetate copolymers (at vinyl acetate levels lower than that yielding miscibility) have been cited for their potential as impact modifiers for PVC [140]. Chlorinated polyethylene added to PVC enhances the impact resistance of rigid formulations and can be used as an additive in plasticized formulations to yield improved flammability test ratings [141].

Blends of nylon 6 with low density polyethylene provide lower water sorption and improved dimensional stability. Scrap fiber-grade nylon 6,6 has been reported to be a useful injection molding compound when blended with 10% of an ethylene/vinyl acrylate copolymer [142].

Immiscible blends will generally be opaque except when the components have matched refractive indexes (transparent) or similar refractive indexes (translucent). In various cases, blends will exhibit a pearlescent appearance, which has been used for decorative effects [e.g., polystyrene–poly(methyl methacrylate) blends].

AB or ABA block copolymers based on polystyrene and polybutadiene or polyisoprene have been commercial for over a decade. The commercial products exhibit two-phase behavior with ideal mechanical compatibility, as expected from the covalent bonding of the phases. The AB styrene–butadiene block copolymer has been utilized primarily in blends with other vulcanizable elastomers to improve processing and low-temperature flexibility [143]. The ABA block copolymers, with greatly superior ultimate properties compared to the AB counterparts, have commercial utility in pharmaceutical and food contact applications, footwear, hot melt and solution adhesives, protective coatings, and general-purpose elastomer applications. Hydrogenation of the center polybutadiene block of an ABA block copolymer to yield a polystyrene–ethylene/butylene–polystyrene block copolymer has been recently commercialized by Shell Chem. Co. under the classification Kraton G [12]. Improved strength, weatherability, and oxidative and ozone resistance are performance parameters of Kraton G that are expected to extend the applications of the ABA block copolymers. While the styrene–diene block copolymers and their hydrogenated versions constitute an important class of immiscible polymer systems, blends of these block copolymers with polystyrene, low density polyethylene, and polypropylene have been cited by Bull and Holden [144] as offering commercial potential.

The synthesis of radial styrene–butadiene block copolymers is also a recent advance offering improved mechanical properties and processability over the linear ABA counterparts, thus extending the utility of these block copolymers [145, 146]. A high styrene content radial block copolymer introduced by Phillips (K-resin) offers the rigidity, transparency, and good impact strength needed for packaging and thermoforming applications [147]. High butadiene content radial block copolymers have been used in hot melt and pressure-sensitive adhesives [148, 149].

Fibrous or particulate poly(tetrafluoroethylene) is commonly added to polymers to improve the multiple traversal wear resistance and to impart lubricity [150]. An example of this is the blend of poly(phenylene sulfide) and poly(tetrafluoroethylene) used as a cookware coating with nonstick properties [151].

To improve the surface characteristics of reinforced unsaturated-polyester composites, as well as to prevent shrinkage, polymers termed low profile additives are commonly added to the unreacted styrene–polyester mixture. These polymers are generally soluble in the styrene monomer but usually-phase separate as the styrene polymerization proceeds. Examples of low profile additives include modified cellulosics, acrylic polymers, poly(ε-caprolactone), and poly(vinyl acetate).

The toughness of epoxy resins can be enhanced by the addition of reactive low T_g oligomers. These include carboxy-terminated butadiene/acrylonitrile

7.2. Mechanical Compatibility versus Miscibility in Polymer Blends

TABLE 7.7

Additional Examples of Present or Potentially Useful Two-Phase Blends

Two-phase blend	Utility	Reference
Poly(phenylene sulfide)–bisphenol A polycarbonate	Improved flame rating resistance of bisphenol A polycarbonate	155
Poly(phenylene sulfide)–polyimide	Electrical/electronic applications and high temperature bearings	58, 156, 157
Poly(sulfone)–ABS	Food trays; electroplatable applications	156
Nylon 6,6–ABS	Fiberglass-reinforced injection molded parts	58
Poly(oxybenzoyl)–poly(tetrafluoroethylene)	Improved wear resistance over unmodified PTFE	158
Poly(phenylene sulfide)–poly(aryl sulfone)	Improved strength and toughness over unmodified poly(phenylene sulfide)	159
Poly(butylene terephthalate)–bisphenol A/tetrabromobisphenol A polycarbonate	Improved flammability rating for poly(butylene terephthalate)	160
Ethylene/vinyl acetate copolymer–g-poly(vinyl chloride)–poly(vinyl chloride)	Impact modification of PVC	161
Isopropylidene bis(2,6-dichloro-p-phenylene) bis(polyhalophenyl) carbonate–polystyrene or ABS	Improved flammability rating for polystyrene or ABS	162
Impact-modified maleic anhydride/styrene copolymer–styrene/butadiene/styrene ABA block copolymer	Improved impact strength	163
Poly(ethylene oxide)–silicone rubber block copolymers	Nonionic surfactant	164
Poly(ethylene oxide)–poly(propylene oxide) block copolymers	Surfactants	165
Polyethylene–polystyrene	Biaxially oriented films for synthetic paper	166

copolymers [152], hydroxy-terminated poly(ε-caprolactone) [153], hydroxy-terminated poly(propylene oxide) [153], and powdered high-rubber ABS. Drake and Siebert [154] have reviewed elastomer-modified epoxies, particularily those with nitrile rubber addition. Phenolics can be impact modified with butadiene/acrylonitrile rubber and used as structural adhesives.

The above examples cover the more common two-phase polymer mixtures utilized commercially. There are other interesting examples, which are given in Table 7.7 [58, 153, 156–166].

7.2.2 Specific Advantages of Two-Phase Behavior

The primary utility of two-phase behavior in polymer blends, as illustrated by commercial prominence, lies in the ability to improve the impact strength of brittle, glassy polymers. The important variables of inclusion of rubber particles into a glassy or highly crystalline polymer matrix include particle size, degree of cross-linking, and mechanical compatibility of the rubber phase with the matrix (usually achieved by grafting the matrix polymer onto the rubber backbone). Discussions of these important variables can be found elsewhere [123–125, 167–169].

In block copolymers sold as thermoplastic elastomers, two-phase behavior is the key to elastomeric properties at normal-use temperatures combined with thermoplastic characteristics at temperatures suitable for conventional thermoplastic fabrication. To achieve this, the predominant phase (continuous phase) must be amorphous with a T_g below the normal-use temperature, whereas the dispersed phase (discontinuous phase) must have a T_g or T_m above the normal-use temperature range. The dispersed phase physically restricts the soft-block chain ends to a specific boundary (interface) and therefore presents a situation similar to cross-linking. The dispersed phase is also a reinforcing material similar to carbon black in conventional thermoset elastomers. While achieving the two-phase behavior in block copolymers is not a difficult task due to the general immiscibility of polymers, it has been clearly established that blocks of widely dissimilar polymers are not desired because the thermoplastic character can be hindered or even eliminated. This is due to the retention of phase separation above the major transitions for both block copolymer components. In order to break down this structure, the domains must be capable of being somewhat dispersed into the continuous phase after application of an applied external force during fabrication. Widely dissimilar blocks will resist this breakdown, as discussed by McGrath and Matzner [170, 171]. A general relationship was observed in which the solubility parameter difference between the block constituents for a series of silicone rubber block copolymers was related to the moldability. Low values of $\Delta\delta_p$ (solubility parameter difference between the blocks) yielded moldable materials, whereas high $\Delta\delta_p$ values resulted in intractable materials quite similar to thermosets except that they were soluble in a series of common solvents.

Amorphous, glassy thermoplastics (e.g., polystyrene, bis A polycarbonate) suffer from solvent-induced stress rupture at stress levels considerably lower than their normal tensile strengths. Two basic explanations have been promoted to account for this behavior: a decrease in the surface energy and plasticization of the polymer by the environment [172, 173]. The diffusion of a penetrant into a glassy polymer can create high internal stresses even up

7.2. Mechanical Compatibility versus Miscibility in Polymer Blends 345

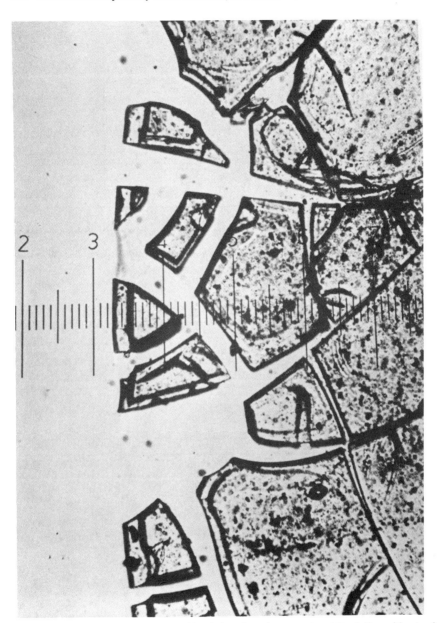

Fig. 7.2. Dissolution by stress cracking below the gel temperature (dissolution without gel layer). [Reprinted by permission from K. Ueberreiter, *in* "Diffusion in Polymers" (J. Crank and G. S. Park, eds.), p. 219. Academic Press, New York, 1968.]

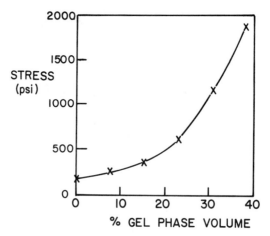

Fig. 7.3. Stress required for 1-min rupture in *n*-heptane for impact polystyrene at various gel phase volume levels. [Reprinted by permission from L. M. Robeson, *in* "Problem Solving with Plastics" (L. S. VanDelinder, ed.), p. 89. Copyright 1971 by the National Association of Corrosion Engineers, Houston, Texas.]

to or exceeding the tensile strength. Applied stress or molded-in-stress will be additive to the created internal stress, thus resulting in microcrack formation, which then can propagate yielding material failure. Ueberreiter showed that stress cracking could occur with glassy polymers to an extent that dissolution by stress cracking below the gel temperature was possible [174]. This is illustrated in Fig. 7.2 for the case of poly(methyl methacrylate)–dimethyl phthalate.

Addition of rubber inclusions has been shown to yield definite improvements in the environmental stress rupture resistance. This improvement is believed to be analogous to impact strength enhancement in that the cracks or crazes formed by solvent can be arrested with low modulus inclusions [175]. This is the case with impact polystyrene, as illustrated in Fig. 7.3. ABS is similar in that improvements over the SAN matrix are observed. The addition of ABS to bis A polycarbonate also improves the resistance to stress cracking in various environments.

Crystalline polymers exhibit a vastly improved environmental stress rupture resistance over their amorphous counterparts. The addition of crystalline polymers to modify an amorphous polymer will therefore improve the solvent-induced stress rupture resistance. As the level of improvement is related to the continuous phase structure, a certain level of continuity of the crystalline polymer will be required to yield noticeable improvements. This is shown in Fig. 7.4 for polysulfone blends with a block copolymer of polysulfone and nylon 6. The stress rupture resistance clearly denotes the

7.2. Mechanical Compatibility versus Miscibility in Polymer Blends

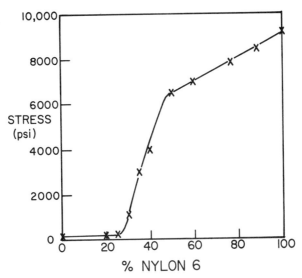

Fig. 7.4. Stress level required for 1-min rupture in acetone for polysulfone–nylon-6 block copolymer blends versus nylon-6 composition. [Reprinted by permission from J. E. McGrath, L. M. Robeson, and M. Matzner, *in* "Recent Advances in Polymer Blends, Grafts, and Blocks" (L. H. Sperling, ed.), p. 195. Copyright 1974 by Plenum Press.]

continuous phase structure with the sigmoidal behavior expected with immiscible polymer blends [176].

In certain cases, extreme levels of immiscibility may be desired to yield the desired surface properties brought about by the migration of a certain polymer during processing. Anti-slip, anti-block, and low coefficient of friction are properties attainable with appropriate immiscible polymer blends. Fibrous or particulate poly(tetrafluoroethylene) (PTFE) is commonly employed in polymer blends to reduce the frictional heating and lower the multiple traversal wear. In many cases, the polymer blend is fabricated below the melting point of PTFE, and thus PTFE is an inert filler retaining the basic physical characteristics and particle size prior to incorporation.

The preparation of ultra-thin fibers ($\sim 1\ \mu$m) is an extremely difficult task using conventional fiber spinning techniques. Merriam and Miller [177] showed that immiscible polymer blends allow for a facile solution to this problem. By blending two immiscible polymers followed by uniaxial orientation into a tape and solvent leaching of one component, very fine fibers of about 1-μm diameter can be obtained. An example of this technique was used in preparing fiber webs for cell culture growth as part of the NIH-sponsored study involving materials for artificial hearts [178]. Blends of an

TABLE 7.8

Additional Examples of Property Attributes of Immiscible Polymer Blends

Two-phase blend	Property enhancement	Reference
Poly(isobutylene)–ethylene/propylene rubber	Reduced compression set and superior physical properties at 100°C	179
Natural rubber–halobutyl rubber–ethylene/propylene/diene (EPDM) rubber (three-phase blend)	Improved ozone resistance	180
Nitrile rubber–ethylene/propylene/diene rubber	Excellent property compromise of oil and ozone resistance; low-temperature flexibility	181
Poly(butadiene)–natural rubber	Improved crack growth resistance	182
Styrene/butadiene–EPDM rubber	Paintable elastomer for automotive use	183
Neoprene–natural rubber	10–30% natural rubber improves tear strength	184
Neoprene–styrene/butadiene rubber	Improved resistance to sunlight discoloration; better low-temperature toughness	184
Polsulfide rubber–Neoprene	Neoprene addition improves processing	184
Polyisoprene–polychloroprene	Improved tensile strength and fatigue life with addition of 5–15% polychloroprene to polyisoprene	185
Poly(acrylic acid) grafted on polyethylene	Improved filler acceptance	186
Poly(ε-caprolactone)–poly(vinyl alkyl ethers)	Improved adhesion compositions	187
Poly(ε-caprolactone)–various elastomers	Improved processability, additive acceptance, and strength	188

ethylene/acrylic acid ionomer and polypropylene were prepared and oriented, followed by water extraction of the ethylene/acrylic acid ionomer. Drafting the resultant tape perpendicular to the fiber orientation yielded a nonwoven structure with pore dimensions similar to cell dimensions.

Elastomer blends of immiscible rubber components have been cited as improving almost every property imaginable. These blends, in reality, are generally considered as a compromise in properties to alleviate a particular deficiency of the major component. Property enhancements quoted include the examples given in Table 7.8 [179–188]. In addition to the performance improvements cited for elastomer blends, other cases worthy of mention are also given in Table 7.8.

7.2. Mechanical Compatibility versus Miscibility in Polymer Blends

7.2.3 Importance of the Interface

For phase-separated systems, the interfacial adhesion between the respective phases governs the ultimate mechanical properties. For polymeric constituents having limited affinity for each other, the interface represents a flaw yielding ultimate properties significantly lower than that expected from constituent values.

Precise physicochemical characterization of solid surfaces and interfaces is an extremely difficult task, and interpretations are generally not straightforward. For liquids, however, the situation is much more amenable to sound theoretical and practical interpretations inasmuch as surface and interfacial tensions are the fundamental parameters involved. They play an important role in the areas of adhesion, wettability, miscibility, and dispersibility. This is not the case in pure materials, which contain no appreciable amounts of surface-active species. The molecular forces pertaining to surface molecules are similar in nature to those pertaining to the molecules in the bulk of the system. In emulsions, suspensions, and foams, the system is thermodynamically unstable; it can be kinetically stabilized if and only if the interfacial free energy is minimized. This is why surfactants are commonly used to stabilize emulsions and foams.

In the field of polymer–polymer miscibility, the existence of a finite interfacial tension is usually taken as an indication of immiscibility. Attempts have been made to develop a multidimensional surface tension scheme which could guide the attainment of desirable miscible polymers in a rapid, methodical manner [189, 190]. A typical two-dimensional surface tension method is the geometric–harmonic-mean relationship:

$$\gamma_{12} = \gamma_1 + \gamma_2 - 2(\gamma_1^d \gamma_2^d)^{0.5} - [4\gamma_1^p \gamma_2^p / (\gamma_1^p + \gamma_2^p)]$$

where γ is the surface tension, subscript 1 is component 1, subscript 2 is component 2, superscript d is the dispersive contribution to surface tension, and superscript p is the polar contribution to surface tension.

In practice, this type of quantitative approach has not been used in the design of miscible polymer mixtures; rather, it has found application in characterizing the interface between polymers known to be miscible or immiscible. From measured or reported surface and interfacial tensions of a brittle and a rubbery plastic, one could infer the level of adhesion at the interface between the pair and hence the level of impact resistance. For example, it is known that if the interfacial tension between the rubber and the matrix is negligible and the polymers are miscible, the rubber will completely dissolve in the rigid material, and no impact modification occurs. If, on the other hand, the interfacial tension is large and the system is highly immiscible, no adhesion, and hence no reinforcement, is achieved between the two polymers. That is, immiscibility goes hand in hand with finite interfacial tension and is accompanied by diminished adhesion.

For phase-separated systems, the characteristics of the interfacial regions generally govern the ultimate mechanical properties (e.g., strength and toughness) attainable with the blend. As is generally the case with two-phase blends, poor adhesion leads to a myriad of flaws located at the interfacial regions, resulting in much lower ultimate properties than expected from averaged properties. The generalizations used to assess the potential miscibility of polymer blends can also be utilized to predict the adhesion of one polymer to another. If indeed the polymers are miscible, excellent adhesion is expected, providing sufficient temperature and pressure are employed to allow for molecular mixing at the interface. Polymers which have similar dispersive and polar forces will exhibit improved interfacial adhesion over those cases widely differing in molecular forces. Polycarbonate–ABS blends, for example, exhibit superior properties over polycarbonate–impact polystyrene blends even when the property differences of ABS and impact polystyrene are appropriately accounted for.

An excellent example of the importance of good interfacial adhesion is provided by the work of Longworth [191] in his novel approach to improving the mechanical compatibility of polyolefin blends (primarily polyethylene and polypropylene). By grafting 0.1–0.5 wt% of an acidic monomer to one polymer and a basic monomer to the other, the blends were observed to retain the ductility of the unblended constituents, whereas the blends of the unmodified components were brittle.

In theory, some intermixing of the blend components at the interfacial regions should occur; however, in extreme cases of immiscibility, the degree predicted by phase diagrams would be negligible. For cases where limited miscibility occurs, interpenetration of one molecular species into and beyond the interfacial region would be expected to yield improved mechanical compatibility. These cases present a gradient of composition across the interfacial regions and reduce the magnitude of structural flaws imposed by the interface.

A successfully impact-modified rigid polymer contains an immiscible dispersed phase composed of rubber particles with excellent adhesion to the matrix. Impact-modified polymers generally consist of blending rigid, brittle polymers with elastomeric graft (or block) copolymers. One constituent of the graft (or block) copolymer exhibits excellent adhesion to the matrix polymer and in many cases is the matrix polymer. Impact polystyrene, impact-modified PMMA, ABS, and impact-modified PVC are important commercial examples of this principle. Note that the block or graft copolymers act essentially like surface-active agents which couple the components

7.2. Mechanical Compatibility versus Miscibility in Polymer Blends

in the mixture and maintain the elastomeric polymer as if it were a colloidal dispersion in the plastic matrix. That is, the interfacial free energy is minimized and the thermodynamically unstable system is made kinetically stable.

Block copolymer technology can be used successfully to attain useful properties of highly mechanically incompatible polymer blends. Polysulfone–silicone rubber blends are grossly incompatible with properties best described as "cheesy." Block copolymers of these constituents offer an excellent compromise of properties [192]. Polystyrene–polybutadiene (30 wt% polystyrene) blends offer no useful elastomeric properties (unvulcanized); however, block copolymers (i.e., ABA type) are important commercial thermoplastic elastomers. The adhesion at the interface for the block copolymer is excellent because covalent bonds must be ruptured for interfacial failure, as opposed to physical bonds for simple blends.

7.2.4 Specific Advantages of One-Phase Behavior in Polymer Blends

The primary advantage realized with miscible polymer blends compared to the immiscible blend counterparts is the assurance of mechanical compatibility. The simplified average of ultimate mechanical properties is not always the observed or expected case, due to the importance of the position of the T_g and of brittle–ductile behavior. The blend can indeed attain a tensile strength higher than the constituents in the case of mixing a lower modulus ductile polymer with a high modulus, brittle polymer, as shown for α-methylstyrene/methacrylonitrile–poly(vinyl chloride) [193] and PPO–polystyrene blends [194].

In fabrication, miscible blends will also offer advantages over immiscible blends in regard to surface characteristics and weld-line strength. These processing variables are more commonly encountered in injection molding (due to the higher shear rates), as opposed to extrusion or compression molding. During the injection molding process, very high shear rates are present during the mold-injection step. This can result in migration of the lower viscosity phase to the surface, thus altering both the surface appearance (i.e., uniformity) and the overall properties. Also in the molding process, the use of multiple gating or complex shapes can result in cooled surfaces meeting in the mold to form what is commonly referred to as a weld line. In polymer mixtures exhibiting two-phase behavior, the weld line strengths are generally inferior to single-phase mixtures. With impact polystyrene and ABS, this is less of a problem because the dispersed higher viscosity phase does not constitute a significant portion of the contacting surfaces forming weld lines.

However, with equal volume fraction of the components, as with certain commercial bis A polycarbonate–ABS blends, weld line strength problems can seriously limit broad utility.

In two-phase elastomer blends, segregation of the cross-linking agents (e.g., sulfur or peroxides), carbon black, or antioxidants preferentially into one phase can result in failure to attain optimum properties. Various examples of this problem were cited by Cornish in a review of elastomer blends in which the soluble and insoluble compounding ingredients were found to be preferentially concentrated in one phase [7]. For curing agents, this concentration gradient could arise during the curing process due to diffusion to the phase exhibiting higher curing rates. The balance of optimum curing of both phases, therefore, presents a difficult problem. Single-phase elastomer blends would, of course, not be affected by this dilemma.

For polymers commonly plasticized with low molecular weight soluble additives, the miscible polymer approach offers a viable means of permanent plasticization. The plasticization of PVC is basically the result of lowering the T_g; thus, an immiscible polymer additive cannot realistically be considered for a permanent plasticizer. The reverse situation, raising the T_g or heat distortion temperature, is also a situation where miscibility is required. In principle, the heat distortion temperature of a specific polymer can be raised by the addition of a higher T_g immiscible polymer; however, this approach is only feasible where the higher T_g component is primarily the continuous phase. As the T_g is the basic property of a polymer affecting the overall physical properties, the raising or lowering of the T_g of a specific polymer by the incorporation of another miscible polymer is presently the major reason for commercial interest in polymer–polymer miscibility.

Complex formation of miscible water-soluble polymers offers a method similar to cross-linking for achieving a water-swellable gel. These gels, unlike the cross-linked analogs, would offer reversibility as a function of solution pH. Stronger polyelectrolyte complexes, however, are less affected by pH and are generally soluble only in specific ternary solvent mixtures. Applications for this unique class of miscible polymers have been covered in Section 7.1.5.

Generally, two-phase mixtures of polymers have utility only when one phase is the major continuous phase. With miscible polymer blends, useful and commercially viable properties are often possible over the entire composition range. The versatility of matching the price/performance requirements of specific applications allows for a myriad of different products from combinations of the two miscible polymers. This characteristic is particularly apparent with the combination of PPO and polystyrene.

Several important commercial examples of miscible polymer blends have been discussed in detail in this treatise. The utility of immiscible multicomponent polymer blends is of far greater commercial importance now. This is at least partially due to the relatively low number of known miscible polymer blends, combined with only a superficial understanding of the nature of polymer–polymer miscibility. Technology relating to polymer–polymer miscibility is an emerging field of future importance, and, as further significant developments in this technology occur, the commercial importance of polymer–polymer miscibility will expand accordingly. We hope that this treatise will help provide a base point for these future developments.

REFERENCES

1. R. D. Deanin and R. R. Geoffroy, *Am. Chem. Soc., Div. Org. Coat. Plast. Chem., Pap.* **37** (1), 257 (1977).
2. H. L. Morris, *J. Elast. Plast.* **6**, 119 (1974).
3. *Rubber World* **167** (5), 49 (1973).
4. G. Bier, *Kunststoffe* **55**, 694 (1965).
5. R. G. Alsup, *Tech. Pap., Reg. Tech. Conf., Soc. Plast. Eng., Ohio Sect.* p. 66 (1976).
6. J. R. Dunn, *Rubber Chem. Technol.* **49**, 978 (1976).
7. P. J. Corish and B. D. W. Powell, *Rubber Chem. Technol.* **47**, 481 (1974).
8. G. Holden, E. T. Bishop, and N. R. Legge, *J. Polym. Sci.* **26**, 37 (1969).
9. H. A. Vaughn, *J. Polym. Sci., Part B* **7**, 569 (1969).
10. A. C. Soldatos and A. S. Burhans, *Ind. Eng. Chem., Prod. Res. Dev.* **9**, 296 (1970); *Adv. Chem. Ser.* **97**, 531 (1970).
11. G. Kraus and H. E. Railsback, *in* "Recent Advances in Polymer Blends, Grafts, and Blocks" (L. H. Sperling, ed.), p. 245. Plenum, New York, 1974.
12. G. Holden, *in* "Recent Advances in Polymer Blends, Grafts, and Blocks" (L. H. Sperling, ed.), p. 269. Plenum, New York, 1974.
13. Badum, U. S. Patent 2,297,194 (1942).
14. R. A. Emmett, *Ind. Eng. Chem.* **36**, 730 (1944).
15. J. E. Pittenger and C. F. Cohan, *Mod. Plast.* **25** (9), 81 (1947).
16. M. C. Reed, *Mod. Plast.* **27** (12), 117 (1949).
17. C. F. Hammer, *Am. Chem. Soc., Div. Org. Coat. Plast. Chem., Pap.* **37** (1), 234 (1977).
18. H. F. Schwarz and W. S. Edwards, *Appl. Polym. Symp.* **25**, 243 (1974).
19. V. R. Landi, *Appl. Polym. Symp.* **25**, 223 (1974).
20. R. D. Deanin, R. O. Normandin, and I. T. Patel, *Am. Chem. Soc., Div. Org. Coat. Plast. Chem., Pap.* **36** (1), 304 (1976).
21. E. F. Jordan, Jr., B. Artymyshyn, G. R. Riser, and A. N. Wrigley, *J. Appl. Polym. Sci.* **20**, 2715 (1976).
22. E. F. Jordan, Jr., B. Artymyshyn, G. R. Riser, and A. N. Wrigley, *J. Appl. Polym. Sci.* **20**, 2737 (1976).
23. E. F. Jordan, Jr., B. Artymyshyn, and G. R. Riser, *J. Appl. Polym. Sci.* **20**, 2757 (1976).
24. R. D. DeMarco, M. E. Woods, and L. F. Arnold, *Rubber Chem. Technol.* **45**, 1111 (1972).
25. M. E. Woods and D. G. Frazer, *Soc. Plast. Eng., Tech. Pap.* **32**, 426 (1974).
26. F. L. Mayer, D. L. Stalling, and J. L. Johnson, *Nature (London)* **238**, 411 (1972).
27. F. R. Williams and R. D. Aylesworth, U. S. Patent 3,972,962 (1976) (assigned to Emery Industries Inc.).

28. R. L. Adelman, U. S. Patent 3,723,570 (1973) (assigned to E. I. duPont de Nemours and Co.).
29. Great Britain Patent 1,455,390 (1976) (assigned to Ciba-Geigy).
30. P. Penczek and G. Cynkowska, *Int. Polym. Sci. Technol.* **4** (5), 19 (1977).
31. G. Matthews, "Vinyl and Allied Polymers," Vol. 2, p. 109. CRC Press, Cleveland, Ohio, 1972.
32. D. Newton and J. Cronin, *Br. Plast.* **31,** 426 (1958).
33. W. S. Penn, "PVC Technology," 3rd ed. Wiley (Interscience), New York, 1972.
34. J. Karoly, *Ind. Eng. Chem.* **45,** 1060 (1953).
35. C. F. Hammer, *Macromolecules* **4,** 69 (1971).
36. C. F. Hammer, U. S. Patent 3,684,778 (1972) (assigned to E. I. duPont de Nemours and Co.).
37. J. J. Hickman and R. M. Ikeda, *J. Polym. Sci., Polym. Phys. Ed.* **11,** 1713 (1973).
38. L. M. Robeson and J. E. McGrath, *Pap., 82nd Nat. Meet.* AIChE, *1976.*
39. H. E. Bair, D. Williams, T. K. Kwei, and F. J. Padden, Jr., *Am. Chem. Soc., Div. Org. Coat. Plast. Chem., Pap.* **37** (1), 240 (1977).
40. C. F. Hammer, German Offen. 2,238,555 (1973) (assigned to E. I. duPont de Nemours and Co.).
41. C. F. Hammer, U. S. Patent 3,780,140 (1973) (assigned to E. I. duPont de Nemours).
42. T. Nishi and T. K. Kwei, *J. Appl. Polym. Sci.* **20,** 1331 (1976).
43. T. Nishi, T. K. Kwei, and T. T. Wang, *J. Appl. Phys.* **46,** 4157 (1975).
44. D. J. Hourston and I. D. Hughes, *J. Appl. Polym. Sci.* **21,** 3093 (1977).
45. R. W. Crawford and W. K. Witsiepe, U. S. Patent 3,718,715 (1973) (assigned to E. I. duPont de Nemours and Co.).
46. R. P. Kane, *J. Elast. Plast.* **9,** 416 (1977).
47. A. Reischl, W. Gobel, and K. L. Schmidt, U. S. Patent 3,444,266 (1969) (assigned to Farbenfabriken Bayer.
48. W. Keberle and W. Gobel, U. S. Patent 3,637,553 (1972) (assigned to Farbenfabriken Bayer).
49. A. B. Magnusson and J. A. Parker, U. S. Patent 3,366,707 (1968).
50. H. M. Bonk, A. A. Sardanopoli, H. Ulrich, and A. A. R. Sayigh, *J. Elastoplast.* **3,** 157 (1971).
51. R. P. Carter, Jr., U. S. Patent 3,487,126 (1969) (assigned to Goodyear Tire and Rubber Co.).
52. L. M. Higashi and R. McEvers, U. S. Patent 3,650,828 (1972) (assigned to Karex, Inc.).
53. L. M. Robeson, W. J. Saunders, S. W. Chow, and M. Matzner, "Development of Improved Materials for Extraoral Maxillofacial Prothesis," Compr. Final Tech. Rep., NIH-NIDR Contract No. N01-DE-42436. Union Carbide Corp., 1976.
54. H. Lee, D. Stoffey, and K. Neville, "New Linear Polymers," p. 63. McGraw-Hill, New York, 1967.
55. A. Katchman and G. F. Lee, Jr., Great Britain Patent 1,372,634 (1974) (assigned to General Electric Co.).
56. *Mod. Plast.* **53** (10A), 1976.
57. E. P. Cizek, U. S. Patent 3,383,435 (1968) (assigned to General Electric Co.).
58. G. R. Forger, *Mater. Eng.* **85** (8), 44 (1977).
59. R. P. Belanger, *Mod. Plast.* **52** (10A), 46 (1975).
60. *Mod. Plast.* **52** (2), 34 (1975); *Plast. Technol.* **18** (2), 19 (1972).
61. "Noryl Thermoplastic Resins," General Electric Product Literature.
62. G. M. Estes, S. L. Cooper, and A. V. Tobolsky, *J. Macromol. Sci., Rev. Macromol. Chem.* **4** (2), 313 (1970).
63. J. L. Illinger, N. S. Schneider, and F. E. Karasz, *Polym. Eng. Sci.* **12,** 25 (1972).

References

64. S. L. Samuels and G. L. Wilkes, *J. Polym. Sci., Polym. Phys. Ed.* **11,** 807 (1970).
65. J. B. Clough, N. S. Schneider, and A. O. King, *J. Macromol. Sci., Phys.* **2** (4), 641 (1968).
66. J. A. Koutsky, N. V. Hien, and S. L. Cooper, *J. Polym. Sci., Part B* **8,** 353 (1970).
67. G. M. Palyutkin, A. R. Sokolov, B. V. Vasil'ev, and O. G. Tarakanov, *Vysokomol. Soedin., Ser. A* **13** (10), 2286 (1971).
68. W. J. MacKnight, M. Yang, and T. Kajiyama, *Anal. Calorimetry Proc. Am. Chem. Soc. Symp., 1968* p. 99 (1968).
69. C. S. P. Sung, N. S. Scheider, R. W. Matton, and J. L. Illinger, *Polym. Prepr., Am. Chem. Soc., Div. Polym. Chem.* **15** (1), 620 (1974).
70. D. S. Huk and S. L. Cooper, *Polym. Eng. Sci.* **11,** 369 (1971).
71. C. G. Seefried, Jr., J. V. Koleske, and F. E. Critchfield, *Polym. Eng. Sci.* **16,** 771 (1976).
72. D. S. Kaplan, *J. Appl. Polym. Sci.* **20,** 261 (1976).
73. S. C. Wells, *J. Elastoplast.* **5,** 102 (1973).
74. A. Lilaonitkul, J. C. West, and S. L. Cooper, *J. Macromol. Sci., Phys.* **12** (4), 563 (1976).
75. N. K. Kalfoglou, *J. Appl. Polym. Sci.* **21,** 543 (1977).
76. G. K. Hoeschele, *Polym. Eng. Sci.* **14,** 848 (1974).
77. "Polyelectrolyte Complex Films as Reverse-Osmosis Desalination Membranes," Final Report, Contract No. 14-01-0001-315. U. S. Dept of Interior, Office of Saline Water Conversion, Washington, D.C. Cited in Reference 78.
78. A. S. Michaels, *Ind. Eng. Chem.* **57** (10), 32 (1965).
79. A. S. Michaels, U. S. Patents 3,419,430 and 3,419,431 (1968) (assigned to Amicon Corp.).
80. A. S. Michaels, U. S. Patent 3,467,604 (1969) (assigned to Amicon Corp.).
81. L. D. Taylor, U. S. Patent 3,578,458 (1971) (assigned to Polaroid Corp.).
82. H. J. Bixler, R. A. Cross, and D. W. Marshall, *in* "Artificial Heart Program Conference Proceedings" (R. J. Hegyeli, ed.), p. 79. US Govt. Printing Office, Washington, D.C., 1969.
83. M. K. Vogel, R. A. Cross, H. J. Bixler, and R. J. Guzman, *J. Macromol. Sci., Chem.* **4** (3), 675 (1970).
84. H. J. Bixler, U. S. Patent 3,475,358 (1969) (assigned to Amicon Corp.).
85. H. J. Bixler, U. S. Patent 3,514,438 (1970) (assigned to Amicon Corp.).
86. B. K. Green, U. S. Patent 2,800,457 (1957) (assigned to National Cash Register Co.).
87. M. J. Lysaght, *in* "Ionic Polymers" (*L. Holliday, ed.*), p. 281. Wiley, 1975.
88. R. J. Slocombe, *J. Polym. Sci.* **26,** 9 (1957).
89. H. H. Irvin, U. S. Patent 3,010,936 (1961) (assigned to Borg-Warner Corp.).
90. Y. C. Lee and G. A. Trementozzi, U. S. Patent 3,644,577 (1972) (assigned to Monsanto Co.).
91. K. Sugimoto, S. Tanaka, and H. Fujita, U. S. Patent 3,520,953 (1970) (assigned to the Japanese Geon Co., Ltd.).
92. K. Saito, M. Yoshino, and S. Yoshioka, U. S. Patent 3,287,443 (1966) (assigned to Kanegafuchi Chem. Ind. Co., Ltd.).
93. L. Scarso, E. Cerri, and G. Pezzin, U. S. Patent 3,772,409 (1973) (assigned to Montecatini Edison Sp. A.).
94. Belgian Patent 824,397 (1975) (assigned to Bayer AG).
95. Belgian Patent 824,398 (1975) (assigned to Bayer AG).
96. J. Zelinger, E. Volfova, H. Zahradnikova, and Z. Pelzbauer, *Int. J. Polym. Mater.* **5,** 99 (1976).
97. J. S. Noland, N. N. C. Hsu, R. Saxon, and J. M. Schmitt, *Adv. Chem. Ser.* **99,** 15 (1971).
98. T. Nishi and T. T. Wang, *Macromolecules* **8,** 909 (1975).
99. D. R. Paul and J. O. Altamirano, *Adv. Chem. Ser.* **142,** 371 (1975).
100. F. F. Koblitz, R. G. Petrella, A. A. Dukert, and A. Christofas, U. S. Patent 3,253,060 (1966) (assigned to Pennsalt Chemicals Corp.).
101. C. H. Miller, Jr., U. S. Patent 3,458,391 (1969) (assigned to American Cyanamid Co.).

102. J. S. Gall, U. S. Patent 4,008,199 (1977) (assigned to Celanese Corp.).
103. P. Horn and S. Wolfgang, German Offen. 2,147,002 (1975) (assigned to BASF AG).
104. D. R. Paul, J. W. Barlow, C. A. Cruz, R. N. Mohn, T. R. Nassar, and D. C. Wahrmund, *Am. Chem. Soc., Div. Org. Coat. Plast. Chem., Pap.* **37** (1), 130 (1977).
105. "Nitrocellulose: Properties and Uses." Publication of Hercules Powder Co., 1955.
106. G. V. Olhoft, N. R. Eldred, and J. V. Koleske, U. S. Patent 3,642,507 (1972) (assigned to Union Carbide Corp.).
107. I. Skeist, ed., "Handbook of Adhesives," Van Nostrand-Reinhold, Princeton, New Jersey, 1977.
108. A. L. Baron and P. Sivaramakrishnan, German Offen. 2,622,412 (1976) (assigned to Mobay Chem. Co.); also German Offen. 2,622,413 (1976).
109. K. Kidoh, Y. Kudo, and F. Suzuki, German Offen. 2,622,498 (1976) (assigned to Kureha Chem. Ind. Co., Ltd.).
110. P. S. Bollen, U. S. Patent 4,029,631 (1977) (assigned to Allied Chem. Corp.).
111. R. B. Navidad, U. S. Patent 3,911,196 (1975) (assigned to Ampex Corp.).
112. R. J. Kern, U. S. Patent 2,806,015 (1957) (assigned to Monsanto Chem Co.).
113. G. K. Hoeschele, U. S. Patent 3,723,569 (1973) (assigned to E. I. duPont de Nemours and Co.).
114. J. V. Koleske, C. J. Whitworth, Jr., and R. D. Lundberg, U. S. Patent 3,925,504 (1975) (assigned to Union Carbide Corp.).
115. J. E. McGrath and M. Matzner, U. S. Patent 3,798,289 (1974) (assigned to Union Carbide Corp.).
116. E. B. Harris and D. B. Braun, U. S. Patent 3,855,357 (1974) (assigned to Union Carbide Corp.).
117. L. A. Pilato, J. V. Koleske, B. L. Joesten, and L. M. Robeson, *Polym. Prepr., Am. Chem. Soc., Div. Polym. Chem.* **17** (2), 824 (1976).
118. F. E. Critchfield and J. V. Koleske, U. S. Patent 3,864,434 (1975) (assigned to Union Carbide Corp.).
119. C. W. Childers and E. Clark, U. S. Patent 3,649,716 (1972) (assigned to Phillips Petroleum Co.).
120. M. Grundmeier, R. Binsack, and H. Vernaleken, U. S. Patent 4,056,504 (1977) (assigned to Bayer).
121. B. D. Gesner, *Encycl. Polym. Sci. Technol.* **10**, 694 (1969).
122. R. Casper, W. Notte, and H. Braese, German Offen. 2,613,121 (1977) (assigned to Bayer).
123. E. R. Wagner and L. M. Robeson, *Rubber Chem. Technol.* **43**, 1129 (1970).
124. G. E. Molau and H. Keskkula, *J. Polym. Sci., Part A-1* **4**, 1595 (1966).
125. L. Bohn, *Adv. Chem. Ser.* **142**, 66 (1975).
126. R. D. Deanin and A. M. Crugnola, eds., "Toughness and Brittleness of Plastics," Adv. Chem. Ser. No. 154. Am. Chem. Soc., Washington, D. C., 1976.
127. *Plast. World*, **35** (11), 56 (1977).
128. Y. J. Shur and B. G. Ranby, *J. Appl. Polym. Sci.* **20**, 3121 (1976).
129. M. T. Shaw, *J. Appl. Polym. Sci.* **18**, 449 (1974).
130. "Exxon Elastomers for Polyolefin Modification," Product literature of Exxon Chem. Co., 1975.
131. W. M. Speri and G. R. Patrick, *Polym. Eng. Sci.* **15**, 668 (1975).
132. K. J. Kumbhani, *Soc. Plast. Eng., Tech. Pap.* **35**, 23 (1977).
133. O. F. Noel, III and J. F. Carley, *Polym. Eng. Sci.* **15**, 117 (1975).
134. R. C. Thamm, *Rubber Chem. Technol.* **50**, 24 (1977).
135. H. L. Morris, *J. Elast. Plast.* **6**, 1 (1974).
136. W. K. Fischer, *Mod. Plast.* **51** (10A), 116 (1974).
137. P. J. Corish, *Rubber Chem. Technol.* **40**, 324 (1967).

138. M. S. Sutton, *Rubber World* **149** (5), 62 (1964).
139. D. E. Wingrove, *Rubber Age* **102** (4), 74 (1976).
140. D. Hardt, *Br. Polym. J.* **1**, 225 (1969).
141. D. Fleischer, H. Scherer, and J. Brandrup, *Angew. Makromol. Chem.* **58/59**, 121 (1977).
142. R. L. Jalbert and J. P. Smejkal, *Mod. Plast.* **53** (10A), 108 (1976).
143. H. E. Railsback and G. Porta, *Mater. Plast. Elast.* **35**, 63 (1969).
144. A. L. Bull and G. Holden, *J. Elast. Plast.* **9**, 281 (1977).
145. L. K. Bi and L. J. Fetters, *Macromolecules* **8**, 98 (1975).
146. L. K. Bi and L. J. Fetters, *Macromolecules* **9**, 732 (1976).
147. L. M. Foclor, A. G. Kitchen, and C. C. Baird, *Am. Chem. Soc., Div. Org. Coat. Plast. Chem., Pap.* **34** (1), 130 (1974).
148. O. L. Marrs, F. E. Naylor, and L. O. Edmonds, *J. Adhes.* **4**, 211 (1972).
149. O. L. Marrs and L. O. Edmonds, *Adhes. Age* **14** (12), 15 (1971).
150. B. Arkles, J. Thebarge, and M. Schireson, *J. Am. Soc. Lub. Eng.* **33** (1), 33 (1976).
151. R. B. Seymour, *Mod. Plast.* **53** (10A), 198 (1976).
152. J. N. Sultan, R. C. Laible, and R. G. McGarry, *Appl. Polym. Symp.* **16**, 127 (1971).
153. A. Noshay and L. M. Robeson, *J. Polym. Sci., Polym. Chem. Ed.* **12**, 689 (1974).
154. R. Drake and A. Siebert, *SAMPE Q.* **6** (4), 11 (1975).
155. S. Adelmann, D. Margotte, J. Merten, and H. Vermaleken, Great Britain Patent 1,477,994 (1977) (assigned to Bayer).
156. *Mod. Plast.*, **54** (11), 42 (**1977**).
157. R. T. Alvarez, U. S. Patent 4,017,555 (1977).
158. S. G. Cottis and B. E. Nowak, *Mod. Plast.* **52** (10A), 56 (1975).
159. F. W. Bailey, U. S. Patent 4,021,596 (1977) (assigned to Phillips Petroleum Co.).
160. T. J. Dolce, German Offen. 2,654,840 (1977) (assigned to General Electric Co.).
161. J. Zelinger, V. Heidingsfeld, and V. Altmann, *Plasty Kauc.* **13** (5), 129 (1976).
162. J. A. Gunsher and R. G. Pews, U. S. Patent 3,846,469 (1974) (assigned to Dow. Chem. Co.).
163. G. F. Lee, Jr., German Offen. 2,713,455 (1977) (assigned to General Electric Co.).
164. M. Matzner, L. M. Robeson, A. Noshay, and J. E. McGrath, *Encycl. Polym. Sci. Technol., Suppl.* **2**, 129 (1977).
165. A. Noshay and J. E. McGrath, "Block Copolymers; Overview and Critical Survey," p. 243. Academic Press, New York, 1977.
166. J. A. Manson and L. H. Sperling, "Polymer Blends and Composites," p. 279. Plenum, New York, 1976.
167. C. B. Bucknall, *Br. Plast.* **40** (11), 118 (1967).
168. M. Matsuo, T. T. Wang, and T. K. Kwei, *J. Polym. Sci., Part A-2* **10**, 1085 (1972).
169. C. B. Bucknall, I. C. Drinkwater, and W. Keast, *Polymer* **13**, 115 (1972).
170. M. Matzner, A. Noshay, and J. E. McGrath, *Polym. Prepr., Am. Chem. Soc., Div. Polym. Chem.* **14** (1), 68 (1973).
171. M. Matzner, A. Noshay, and J. E. McGrath, *Trans. Soc. Rheol.* **21**, 273 (1977).
172. H. Stuart, G. Markowski, and D. Jeshke, *Kunststoffe* **54**, 618 (1964).
173. G. A. Bernier and R. P. Kambour, *Macromolecules* **1**, 393 (1968).
174. K. Ueberreiter, *in* "Diffusion in Polymers" (J. Crank and G. S. Park, eds.), p. 209. Academic Press, New York, 1968.
175. L. M. Robeson, *in* "Problem Solving with Plastics" (L. S. VanDelinder, ed.), p. 87. Natl. Assoc. Corrosion Eng., Houston, Texas, 1971.
176. J. E. McGrath, L. M. Robeson, and M. Matzner, *in* "Recent Advances in Polymer Blends, Grafts, and Blocks" (L. H. Sperling, ed.), p. 195. Plenum, New York, 1974.
177. C. N. Merriam and W. A. Miller, U. S. Patent 3,099,067 (1963) (assigned to Union Carbide Corp.).
178. J. S. Byck, S. W. Chow, L. J. Gonsior, W. A. Miller, W. P. Mulvaney, L. M. Robeson, and

M. A. Spivack, *in* "Artificial Heart Program Conference Proceedings" (R. J. Hegyeli, ed.), p. 123. US Printing Office, Washington, D.C., 1969.
179. D. C. Coulthard, K. Ritchie, and J. Walker, *Pap., 110th Meet., Rubber Div., Am. Chem. Soc.*, 1976.
180. A. E. Crepeau, *Pap., 110th Meet., Rubber Div., Am. Chem. Soc.*, 1976.
181. J. M. Mitchell, *Pap., 110th Meet., Rubber Div., Am. Chem. Soc.*, 1976.
182. J. R. Beatty, *Pap., 110th Meet., Rubber Div., Am. Chem. Soc.*, 1976.
183. J. F. O'Mahoney, Jr., U. S. Patent 4,020,038 (1977) (assigned to Goodyear Tire and Rubber Co.).
184. R. M. Murray and D. C. Thompson, "The Neoprenes." E. I. duPont de Nemours and Co., Wilmington, Delaware, 1963.
185. V. A. Shershnev, V. N. Kuleznev, and E. A. El'shevskaya, *Kauch. Rezina* No. 12, 7 (1976); translated in *Int. Polym. Sci. Technol.* **4** (6), T/17 (1977).
186. N. G. Gaylord, *Adv. Chem. Ser.* **142**, 76 (1975).
187. R. D. Lundberg, J. V. Koleske, D. F. Pollart, and W. H. Smarook, U. S. Patent 3,641,204 (1972) (assigned to Union Carbide Corp.).
188. R. D. Lundberg, J. V. Koleske, and E. R. Walter, U. S. Patent 3,637,544 (1972) (assigned to Union Carbide Corp.).
189. S. Wu, *J. Macromol. Sci., Rev. Macromol. Chem.* **10** (1), 1 (1974).
190. R. J. Roe, V. L. Bacchetta, P. M. G. Wong, *J. Phys. Chem.* **71**, 4190 (1967).
191. R. Longworth, U. S. Patent 3,299,176 (1967) (assigned to E. I. duPont de Nemours and Co.).
192. L. M. Robeson, A. Noshay, C. N. Merriam, and M. Matzner, *Angew. Makromol. Chem.* **29/30**, 47 (1973).
193. J. F. Kenney, *in* "Recent Advances in Polymer Blends, Grafts, and Blocks" (L. H. Sperling, ed.), p. 117. Plenum, New York, 1974.
194. A. F. Yee, *Polym. Prepr., Am. Chem. Soc., Div. Polym. Chem.* **17** (1), 145 (1976).

Appendix 1

Nomenclature

Symbol	Description	Typical units
a_i	Activity of component i	—
a_T	Temperature shift factor	—
A	Free energy at constant volume	cal
A_{ij}	Contact area per molecule	cm^2
$\bar{A}$	Avogadro's number	$mole^{-1}$
b_o	Monolayer thickness	cm
c	Total external degrees of freedom	—
c_{ij}	Cross-term correction for degrees of freedom, c	—
C_p	Specific heat at constant pressure	$cal \cdot mole^{-1} \cdot {}^\circ K^{-1}$
d	Diameter	cm
D	Diffusion coefficient	(various)
	Domain size in phase-separated block copolymer systems	Å
E	Tensile modulus	$dyn \cdot cm^{-2}$
E^*	Complex tensile modulus	$dyn \cdot cm^{-2}$
E', E''	Real and imaginary components of E^*	$dyn \cdot cm^{-2}$
E_d	Activation energy of diffusion	cal
f	Frequency (cycles/sec)	sec^{-1}
$g(r)$	Radial distribution function	—
g_{ij}	Koningsveld interaction parameter between i and j components	—
G	Free energy at constant pressure	cal
	Shear modulus	$dyn \cdot cm^{-2}$
	Spherulite growth rate	$cm \cdot sec^{-1}$
G^*	Complex shear modulus	$dyn \cdot cm^{-2}$
G', G''	Real and imaginary components of G^*	$dyn \cdot cm^{-2}$
G'	Free energy density	$cal \cdot cm^{-3}$
G_s	Surface free energy	cal

Symbol	Description	Typical units
H	Enthalpy	cal
	Classical Hamiltonian function	—
$H(\tau)$	Relaxation spectrum function	$dyn \cdot cm^{-2}$
I	Inertial force	dyn
	Ionization potential	electron volts
k	Boltzmann constant	$cal \cdot {}^\circ K^{-1}$
K	Gradient energy coefficient	
l	Range of molecular interaction	Å
m	Mass of one molecule	g
M_n	Number of average molecular weight	$g \cdot mole^{-1}$
M_w	Weight average molecular weight	$g \cdot mole^{-1}$
n	Refractive index	—
n_i	Number of moles of component i	mole
N_i	Number of molecules of component i	—
P	Pressure	$dyn \cdot cm^{-2}$
	Permeability constant	(various)
P_i	Internal pressure	$dyn \cdot cm^{-2}$
p^*	Characteristic pressure	$dyn \cdot cm^{-2}$
$[P]$	Parachor	$erg^{0.25} \cdot cm^{2.5} \cdot mole^{-1}$
q	Electrostatic charge	coulomb
Q	Configurational partition function	—
Q_{ij}	Flory entropy interaction parameter	—
r	Number of segments in a chain molecule	—
	Intermolecular distance	Å
	Volume ratio	—
R	Gas constant	$cal \cdot {}^\circ K^{-1} \cdot mole^{-1}$
$\bar{R}$	Root mean square end-to-end distance	Å
S	Entropy	$cal \cdot {}^\circ K^{-1}$
	Solubility constant	(various)
S_i	Surface area per segment of component i	cm^2
T	Temperature	${}^\circ K, {}^\circ C$
T^*	Characteristic temperature	${}^\circ K$
T_m	Melting point	${}^\circ K, {}^\circ C$
T_m°	Equilibrium melting point	${}^\circ K, {}^\circ C$
T_g	Glass transition temperature	${}^\circ C$
U	Potential energy	cal
U_0	Lattice equilibrium energy	cal
v_i	Volume per mole of component i	$cm^3 \cdot mole^{-1}$
V	Volume	cm^3
V_i	Volume per molecule of component i	cm^3
V_s	Volume per interacting segment	cm^3
w_i	Weight fraction	—
w_A	Weight fraction of amorphous phase	—
w_C	Weight fraction of crystalline phase	—
W_{ij}	Interaction energy for contact ij	cal
x_i	Mole fraction of component i	—
X_{ij}	Exchange energy parameter for ij contact	—
z, Z	Lattice coordination number	—
Z	Canonical partition function	—

Nomenclature

Greek letters	Description	Typical units
α	Polarizability	$\text{coulomb} \cdot \text{cm}^2 \cdot \text{V}^{-1}$
	Expansion ratio of chain coil	—
	Coefficient of thermal expansion	$°\text{K}^{-1}$
β	Compressibility	$\text{cm}^2 \cdot \text{dyn}^{-1}$
	Wave number	cm^{-1}
γ	Surface tension	$\text{dyn} \cdot \text{cm}^{-1}$
	Geometric factor	—
	Thermal pressure coefficient	$°\text{K}^{-1}$
$\dot{\gamma}$	Shear rate	sec^{-1}
δ	Solubility parameter	$\text{cal}^{0.5} \cdot \text{cm}^{-1.5}$
$\delta_p, \delta_d, \delta_h$	Polar, dispersive, and hydrogen bonding components of δ	$\text{cal}^{0.5} \cdot \text{cm}^{-1.5}$
δ_{ij}	Kronecker delta	—
ΔH	Heat of mixing	cal
ΔH_f	Heat of fusion	cal
ΔG	Free energy of mixing	cal
ΔS_{el}	Elastic entropy	$\text{cal} \cdot °\text{K}^{-1}$
ΔS_p	Placement entropy	$\text{cal} \cdot °\text{K}^{-1}$
ΔS_v	Restricted volume entropy difference	$\text{cal} \cdot °\text{K}^{-1}$
ε	Dielectric constant	—
ε_{ij}	Interaction energy per contact ij	cal
ε^*	Complex dielectric constant	—
$\varepsilon', \varepsilon''$	Real and imaginary components of ε^*	—
η	Viscosity	P (poise)
η_0	Zero-shear-rate viscosity	P (poise)
η_{ij}	Interaction energy for contact ij	cal
θ_i	Site (surface) fraction for component i	—
λ	Wavelength	μm
μ	Dipole moment	$\text{coulomb} \cdot \text{cm}$
μ_i	Chemical potential of component i	cal
ν	Poisson's ratio	—
	Molecular volume	cm^3
ν_i	Cross-link density of component i	cm^{-3}
σ	Segment length in block copolymer	—
σ_{ij}	Stress components	$\text{dyn} \cdot \text{cm}^{-2}$
τ	Relaxation time	sec
χ, χ_{ij}	Flory–Huggins interaction parameter between components i and j	—
ϕ_i, Φ_i	Volume fraction, component i	—
ψ_i	Segment fraction, component i	—
ω	Angular velocity, frequency	$\text{radians} \cdot \text{sec}^{-1}$
Ω	Degrees of freedom	—
	Onsager-type phenomenological coefficient	—
Ω_{comb}	Combinatorial part of partition function	—
$\Omega_{free\,vol}$	Free volume part of partition function	—

Above symbols
───────────

| ~ | Dimensionless, reduced |
| ‾ | Number average |

Superscripts
───────────

| M | Mixing, used where needed for clarity |
| * | Characteristic |

Miscellaneous
───────────

lcst	Lower critical solution temperature
ucst	Upper critical solution temperature
DSC	Differential scanning calorimetry
TEM	Transmission electron microscopy
SEM	Scanning electron microscopy

Appendix 2

Abbreviations for Polymer Names

Abbreviation	Structure
AB, ABA, (AB)$_n$	A block copolymer of polymer B with polymer A
ABS	Poly(styrene–co-acrylonitrile) grafted onto polybutadiene, poly(butadiene–co-acrylonitrile), or poly(butadiene–co-styrene) rubber
AN	Acrylonitrile (monomer or unit)
BR	Polybutadiene rubber
CPVC	Chlorinated poly(vinyl chloride)
E	Ethylene (monomer or unit)
EA	Ethyl acetate (monomer or unit)
EPDM, EPT	Poly(ethylene–co-propylene) with a small amount of diene comonomer
EPR	Poly(ethylene–co-propylene) rubber
EVA, EVAc	Poly(ethylene–co-vinyl acetate)
HDPE	High density (linear) polyethylene
IPN	Interpenetrating network of two or more polymers
LDPE	Low density (branched) polyethylene
NBR	Poly(butadiene–co-acrylonitrile) rubber
NR	Natural rubber, poly(*cis*-1,4-isoprene)
PAA	Poly(acrylic acid)
PAN	Polyacrylonitrile
PB, PBu	Polybutadiene
P (bis S phenyl ether)	($\phi SO_2 \phi$—O—)
PBT	Poly(butylene terephthalate)
PC	Polycarbonate ($\phi + \phi CO_3$-)
PCL	Polycaprolactone
P(Cl-S)	Poly(chlorostyrene)

Abbreviation	Structure
PmClS	Poly(*m*-chlorostyrene)
PoClS	Poly(*o*-chlorostyrene)
PpClS	Poly(*p*-chlorostyrene)
P(Cl$_2$-S)	Poly(dichlorostyrene)
PE	Polyethylene
PEA	Poly(ethyl acrylate)
PEMA	Poly(ethyl methacrylate)
PEO	Poly(ethylene oxide)
PET	Poly(ethylene terephthalate)

Phenoxy	($-\phi+\phi-$O$-$CH$_2$CH$-$CH$_2-$O$-$)
	                                 |
	                                OH

PI	Polyisoprene
PIB	Polyisobutylene
PMAN, MAN	Polymethacrylonitrile
PMMA	Poly(methyl methacrylate)
PαMS, αMS	Poly(α-methylstyrene)
PMVK	Poly(methyl vinyl ketone)
PmMeS	Poly(*m*-methylstyrene)
PoMeS	Poly(*o*-methylstyrene)
PpMeS	Poly(*p*-methylstyrene)
PP	Polypropylene
PPO	Poly(2,6-dimethyl-1,4-phenylene oxide)
PS, PST	Polystyrene
PSF	Polysulfone ($-$O$-\phi+\phi-$O$-\phi$SO$_2\phi-$)
PTFE	Poly(tetrafluoroethylene)
Poly(THF)	Poly(tetrahydrofuran)
PVA, PVAc	Poly(vinyl acetate)
PVA, PVAl	Poly(vinyl alcohol)
PVC	Poly(vinyl chloride)
PVF$_2$	Poly(vinylidene fluoride)
PVME	Poly(vinyl methyl ether)
PVN	Poly(vinyl nitrate)
PVP	Poly(vinyl pyridine)
SAN, PSAN	Poly(styrene-co-acrylonitrile)
SB, SBR	Poly(styrene-co-butadiene)
TPE	Polypropylene modified with EPR
VA, VAc	Vinyl acetate (monomer or unit)
-b-	Block
-g-	Graft
a-	Atactic
i-	Isotactic
s-	Syndiotactic

Index

A

AB, ABA, and (AB)$_n$ block copolymers, 9, 105–109, 322, 329–334, 341, *see also* Block copolymers
Abrasion resistance, 253, 287, 323, 326, 331, 341
ABS (acrylonitrile-butadiene-styrene), 10, 227, 322, 338, 339, 340, 341, 346, 348, 350, 351, 352
ABS, high heat, 15, 16, 336
Absorption isotherms, 181, 182, 226, 233, 304
Acid–base interactions, 29, 219, 247, 306
Acrylic polymers, *see* Poly(acrylates)
Activation energy of diffusion, 301–305
Amide polymers, 245–246, *see also* Nylon 6 and Nylon 6,6
Amorphous state, 20, 119
Applications, 329, 331–339
    adhesives, 11, 323, 331, 332, 337–339, 342, 348
    automotive, 15, 326, 329, 331, 334
    biomedical, 334–336, 342–348
    coatings, 16, 17, 323, 331, 336–338, 342
    film, 14, 323, 334, 343
    packaging, 11, 14, 323
    permanent plasticizers, 14, 220–224, 314, 322–327, 352
    miscellaneous, 14, 15, 315
Azeotropic composition, 16, 227, 336

B

Binodal region, 59, 60, 62, 66, 68, 73, 86, 92–95
Biodegradability, 14, 316
Blends
    single-phase, 2, 3, 215–268, 322–339
    two-phase, 2, 3, 106, 287, 339–351
Block copolymers, 2, 9–11, 104–111, 200–203, 211, 223, 254–259, 284, 302, 322, 329–334, 341–344, 351
Brittle-ductile failure envelope, 290, 291
Butadiene/acrylonitrile copolymers, 3, 13, 14, 133, 134, 139, 197, 209, 218, 219, 221, 253, 254, 265, 267, 279, 286, 288, 297, 304, 322–324, 339, 341, 343, 348
Butadiene/styrene copolymers, 110, 139, 204, 205, 253, 254, 280, 282, 292, 297, 341, 348

C

Cahn-Hilliard result, 44, 45
Calorimetric methods, heat of mixing by, 179–182, *see also* Glass transition
Catenane structures, 204, 263
Cellulosics, 231–233, 249, 314, 337
    cellulose acetate, 16, 231–233
    cellulose butyrate, 231–233, 239, 241
    cellulose nitrate, *see* Nitrocellulose
    cellulose proprionate, 231, 232, 239, 241
Charge-transfer complexes, 30, 31, 207, 209, 242, 293
Chlorinated polyether, 240, 241
Chlorinated polyethylene, 322, 341
Cloud-point methods, 6, 97, 140–142, 149, 165, 227, 228, 267
Coatings, *see* Applications
Coefficient of friction, 342, 347
Coefficient of thermal expansion, *see* Expansion coefficient

Cohesive energy density, 48, 49, 55
Complementary dissimilarity, 206, 268
Complex dielectric constant, 128
Complex modulus, 125, 285, 286
Compressibility, 49, 93
Conventional light scattering, 142–146
Copolymers, miscibility of, 196–197
Crazing, 339, 346
Critical interaction parameter, 106
Critical point, 19, 62, 63, 67, 68, 88, 92, 96, 147
Cross-link density, 111, 264
Cross-linking, 204–206, 240, 241, 262–264, 288, 352
Crystallinity, 218, 219, 234, 254, 257, 259, 260–262, 264, 265, 267, 268, 306–314
Crystallization kinetics, 121, 219, 221, 228, 234, 254, 257, 258, 262, 277, 306–314

D

Debye-Bueche theory, 156, 312
Degradation
 biological, 316
 thermal, 314–315
 ultraviolet, 337
Dielectric constant, 128–132
Dielectric loss factor, 128–132, 292
Differential scanning calorimeter, 133–134, 225, 236
Differential thermal analysis, 133–134
Dilatometry, 132, 211, 234–237, 252, 253, 267
Dipole moment, 56
 induced, 26–28
 random, 26–28
Domain, 105, 109, 111, 120–121, 138, 156
Dynamic mechanical characterization, 122–125, 201, 218, 221–223, 225, 228, 229, 232, 236, 237, 241, 246, 258, 259, 261, 263, 282, 283, 287, 292, 293

E

Elasticity entropy, 107
Elastomer blends, 139, 175, 205, 252–254, 341, 348, 352
Electrical conductivity, 209, 243, 247, 293–295, 334
Electrical properties, 243, 247, 292–295
Electron microscopy, 110, 111, 117, 137–140, 218, 262

Emulsions, 235
Environmental stress crack resistance, 9, 287, 344–346
Epoxy resins, 100, 101, 204, 256, 258, 262, 263, 290, 338, 342–343
Equation-of-state theory, 3, 75–104, 168, 226
Equimolar stoichiometric complex, 250, 293, 294
Etching
 electron beam, 140
 solvent, 137
Ethylene copolymers, 122, 124, 130, 131, 133, 139, 146, 197, 203, 209, 210, 218, 219, 223, 224, 238, 245, 246, 260, 261, 265, 266, 281, 288, 292, 303, 304, 306, 325, 341
Ethylene/propylene rubber, 163, 203, 260, 297, 322, 341, 348
Expansion coefficient, 51, 53, 92, 93, 179

F

Fibers, 347
Fillers, 211
Films, *see* Applications
Flammability, 9, 15, 328–329, 340, 343
Flory-Huggins interaction parameter, 26, 46, 65–68, 106, 153, 154, 161, 165, 168, 176, 183, 313, 314
Fox equation, 235, 279, 280
Free energy, 5, 24–26, 32, 33, 84, 108, 166
Free volume, 120, 175
Freeze drying, 210–211, 235, 287
Friction, 9, 342, 347

G

Gas permeability, 278, 300–306
Gas solubility, 301, 304, 305
Gibbs-Dimarzio theory, 120
Glass transition
 general, 4, 6, 7, 119–136, 174
 calorimetric methods, 133–134
 dielectric methods, 128–132
 mechanical methods, 121–127
 radio luminescence, 135–136
 thermo-optical analysis, 134–135
Gordon-Taylor equation, 279
Graft copolymers, 9, 10, 200–203, 211, 260–262, 338, 340, 348
Group contribution methods, 57–59

Index

H

Hashin's equation, 284
Heat distortion enhancers, 15, 288, 289, 327, 336, 337, 352
Heat distortion temperature, 15, 16, 290, 327, 336, 337, 340
Heat of mixing, 24–26, 51, 71, 105, 179–182, 267, 301
Heterogeneous blends, 282, 283
Hildebrand "solubility parameter," 47–63, 153
Hydrogen bonding, 28, 59, 187, 206–207, 209, 210, 217, 223, 297
α-Hydrogen groups, 210, 217

I

Ideal solution, 22
Immiscible, 2–4
Impact polystyrene, 3, 7, 10, 175, 284, 296, 322, 328, 339, 350, 351
Impact strength, 287, 289, 322, 328, 339, 340, 343, 344, 350, 351
Incompatibility, 2–4
Industrial applications, see Applications
Infrared spectroscopy, 188–189, 223
Interactions
  acid-base, 29, 30, 207, 247
  charge transfer, 30, 31, 207, 209, 242
  dipole-dipole, 26–28, 51, 52
  dipole-induced dipole, 27
  hydrogen bonding, 28, 59, 107, 173, 187, 206, 207, 209, 217, 223, 247
  ion-dipole, 28, 206
  random dipole-induced dipole, 26, 27
  specific, 29, 30, 207, 246–252, 278
Interfacial adhesion, 349–351
Interfacial properties, 36, 349–351
Interpenetrating networks, 2, 11–13, 111, 203–204, 262–264
Interphase, 179
Insulation (dielectric), 14
Internal pressure, 49, 51
Intrinsic viscosity, 176–177
Inverse gas chromatography, 96, 166–174, 206, 219
Ionic polymers, 30, 207–208, 246–252, 293, 334–336
IR, see Infrared spectroscopy

Isomers, 197–199
Isomorphic blends, 216, 264, 265
Isothermal compressibility, 119

K

Kelley-Bueche equation, 279
Kerner equation, 283, 284

L

Lattice theory, 25, 64–74
Linear viscoelastic functions, 298–300
LJD cell model, 77, 79
Loss tangent, 125
Lower critical solution temperature (lcst), 91–96, 141, 166, 226–229, 240, 266–268
Low profile additives, 342

M

Mark-Houwink relationship, 176
Maxwell's equation, 284, 301, 302
Mechanical compatibility, 2, 3, 10, 210–211, 215, 289, 339, 350
Mechanical loss, 122, 125, 201, 238, 253, 256, 260, 282, 283, 287
Mechanical properties, 248, 253, 262, 287–292, 324, 326, 328, 332, 333
Melting point depression, 182–184, 228, 240, 242, 243, 307, 309,
Membranes, 334
α-Methylstyrene co and terpolymers, 15, 17, 142, 209, 220, 222, 254, 255, 288, 336, 338, 340, 351
Microencapsulation, 335
Microheterogeneous, 282, 283, 314
Microscopy
  electron, 2, 110, 111, 137
  optical, 137
  phase contrast, 137–142
Miscible systems, listing by primary components, 215 ff.
Mixing, heat of, 24
Modulus
  loss, 122, 125, 298
  shear, 122, 298
  tensile, 125–127
Modulus-temperature behavior, 229, 241, 242, 244, 253, 254, 256, 281–287
Molecular forces of attraction, 26–31

N

Natural rubber, 49, 205, 252, 253, 297, 348
Neoprene, *see* Polychloroprene
Neutron scattering, 117, 149–155
Nitrile rubber, *see* Butadiene-acrylonitrile copolymers
Nitrocellulose, 16, 175, 231–233, 235, 237–239, 241, 282, 314, 337, 338
Normal stress, 296
Notched impact strength, 289–290
Nuclear magnetic resonance, 117, 184–187, 218, 221
Nucleation and growth, 31, 35–42, 144, 146
Nylon 6, 203, 210, 246, 261, 306, 341, 346
Nylon 6,6, 29, 245, 246, 256, 258, 341, 343

O

Oil resistance, 332, 341, 348
Olefin polymers, *see* Polyolefins
Oligomers, 14
Optical microscopy, 137
Osmium tetroxide (staining), 137
Ostwald ripening, 41, 42
Ozone resistance, 14, 288, 315, 323, 341, 348

P

Parachor, 55
Permeability, 221, 260, 278, 300–306, *see also* Gas permeability
Phase contrast microscopy, 137, 253, 254
Phase diagrams, 20, 60–62, 69–72, 75, 92–97, 118, 141, 149, 150, 163
Phase separation processes, 31–47, 140
Phenolic resins, 16, 17, 233, 243, 244, 265, 267, 339, 343
Phenoxy, 100, 102, 210, 239, 240, 241, 244, 245, 257, 258, 279, 281, 282, 315, 331, 337
Piezoelectricity, 294, 295
Placement entropy, 106, 107, 109
Plasticization, 14, 211, 219–223, 296, 322–327, 352, *see also* Polymeric plasticizer
Poly(acrylates), 233–238, *see also* Poly(ethyl methacrylate), Poly(methyl acrylate), Poly(methyl methacrylate), etc.
Poly(acrylic acid), 29, 108, 176, 188, 208, 209, 243, 244, 246–251, 281, 290, 348
Poly(amides), 245, 246, 249, 250, *see also* Amide polymers, Nylon 6, and Nylon 6,6
Poly(butadiene), 7, 139, 142, 175, 204, 205, 254, 292, 297, 339, 341, 348, 351
Poly(butylene terephthalate), 220, 222, 240–243, 255, 257–259, 281, 315, 337
Poly(butylene terephthalate)-poly(tetrahydrofuran) block copolymers, 9, 210, 325–327, 329, 332, 333, 343
Poly(butene-1), 266, 267
Poly(ε-caprolactone), 29, 96, 97, 99, 100, 102, 141, 156, 173–175, 183, 209, 217, 219, 221, 223, 227, 229, 231, 232, 239–241, 256, 257, 259, 261, 262, 279, 280, 282, 290, 308–312, 314, 316, 326, 338, 342, 343, 348
Polycarbonate, 9, 127, 141, 187, 203, 226, 229, 239–242, 256–259, 315, 322, 337, 338, 346, 350, 352
Polychloroprene, 253, 254, 265, 267, 341, 348
Poly(2,6 dimethyl 1,4 phenylene oxide), 3, 4, 14, 15, 128, 129, 133, 134, 135, 136, 145, 175, 183, 188, 211, 224, 225, 228, 229, 232, 233, 244, 245, 290, 291, 292, 293, 296, 298, 300, 304, 305, 313, 316, 322, 327–329, 351, 352
Poly(dimethyl siloxane), 72, 142, 155, 171, 187
Polydispersity, *see* Microheterogeneous
Polyelectrolyte complexes, 29, 30, 208–209, 215, 246–252, 291, 292, 293–295, 334–336, 352
Polyepichlorohydrin, 240, 241
Poly(ethyl acrylate), 141, 204, 210, 233–235, 237, 282
Polyethylene, 163, 164, 210, 297, 341, 342, 343, 348, 350
Poly(ethylene adipate), 242, 243, 256, 297, 325
Poly(ethyleneimine), 29, 188, 209, 247, 249–251, 294
Poly(ethylene oxide), 29, 176, 188, 209, 243, 244, 246, 248, 249, 256, 257, 260, 261, 281, 290, 343
Poly(ethylene terephthalate), 210, 240, 241, 243, 315, 338
Poly(ethyl methacrylate), 141, 184, 234, 236, 313, 314
Polyisobutene(polyisobutylene), 72, 141, 142, 148, 149, 150, 164, 203, 260, 261, 341, 348
Polyisoprene, 74, 142, 175, 348
Poly(isopropyl acrylate), 235, 237
Polymeric plasticizer, 3, 16, 288, 322–327

Index 369

Poly(methacrylic acid), 188, 208, 209, 236, 247–251
Poly(methyl acrylate), 141, 231–237, 260, 261
Poly(methyl methacrylate), 10, 37–41, 47, 95, 97, 98, 102, 125, 132, 133, 137, 139, 140, 141, 151, 152, 153, 154, 156, 162, 178, 183, 184, 188, 197, 198, 204, 209, 210, 220, 227, 233–238, 242, 243, 263–264, 280, 282, 287, 295, 313, 314, 315, 322, 337, 340, 341, 346, 350
Poly(α-methylstyrene), 151, 154, 155, 200, 201
Poly(olefin) blends, 297, 340–341, 350
Poly(phenylene sulfide), 342, 343
Polypropylene, 163, 164, 266, 267, 297, 322, 341, 350
Poly(n-propyl acrylate), 237
Poly(propyl methyacrylate), 235, 237
Poly(sodium styrene sulfonate), 293, 294
Polystyrene, 3, 4, 6, 14, 15, 41, 42, 46, 74, 95, 97, 98, 106, 118, 127, 128, 129, 133, 141, 142, 144, 148, 149, 150, 155, 162, 165, 166, 175, 178, 179, 183, 185, 188, 196, 200, 201, 203, 211, 224–230, 232, 233, 244, 245, 255, 258, 260, 261, 263, 264, 290, 291, 292, 293, 296, 298, 304, 305, 308, 313, 316, 322, 327–329, 341, 342, 343, 351, 352
Polystyrene block copolymers, 9, 11, 106, 107, 254, 255, 258
Polysulfide, 267
Polysulfone, 199–200, 230, 255, 256, 258, 343, 346
Poly(tetrafluoroethylene), 342, 343, 347
Polyurethanes, 175, 204, 231, 257, 259, 262, 263, 297, 326, 327, 330–332
Poly(vinyl acetate), 131, 137, 186, 211, 231, 232, 235, 237–239, 242, 243, 280, 281, 287, 292, 342
Poly(vinyl alcohol), 176, 186
Poly(vinyl benzyl trimethyl ammonium chloride), 293, 294
Poly(vinyl butyrate), 237
Poly(vinyl chloride), 3, 10, 13, 29, 96, 97, 99, 100, 102, 122, 124, 133, 156, 171, 172, 173, 174, 175, 183, 186, 197, 198, 199, 209, 210, 217–224, 231, 233, 237–239, 240–242, 257, 259, 260, 262, 279, 280, 282, 286, 288, 289, 290, 292, 297, 303, 304, 308–312, 314, 315, 316, 322–327, 331, 336, 337, 340, 341, 343, 350, 351, 352

Poly(vinyl fluoride), 266, 295
Poly(vinylidene fluoride), 125, 133, 141, 156, 178, 183, 184, 188, 234, 236–239, 264–267, 281, 282, 295, 313, 314, 337
Poly(vinyl methyl ether), 4–6, 16, 41, 42, 46, 95, 97, 98, 118, 133, 141, 162, 165, 166, 178, 179, 185, 187, 225, 226, 243–245, 292, 337–338
Poly(vinyl nitrate), 130, 131, 238, 239, 281, 292
Poly(vinyl pyrrolidone), 209, 245, 247, 249, 250, 251
Processing aids, 222, 327, 337, 348
Proteins, 208, 252
Proton
 acceptor, 28–30, 173, 207, 209, 219, 224, 231, 240, 243, 245, 268
 donor, 28–30, 173, 207, 209, 219, 224, 231, 240, 243, 245, 268
Pulse induced critical scattering (pics), 146–149
Pyroelectricity, 294, 295

Q

Quasi-lattice model, 80

R

Radial polymers, 10, 11, 111
Radioluminescence spectroscopy, 135–136
Rayleigh-Brillouin scattering, 143, 155, 156
Refractive index, 55, 140
Regular solution, 23, 24
Resilience, 126–127
Restricted volume entropy difference, 107
Rheological properties, 174–177, 295–300
 linear viscoelastic functions, 298
 viscometric functions, 297
Rheology, 225, 277, 295–300

S

Scanning electron microscopy (SEM), 140
Scattering methods, 140–155
Second-order transitions, 119, 120, 132, 278
Shellac, 16
Small-angle neutron scattering, 156
Solid plasticizers, *see* Plasticization; Polymeric plasticizers

Solubility parameter, 2, 48–59, 153, 189, 268, 344
Solution, definition, 19
Specific interactions, *see* Interactions, hydrogen bonding
Spectroscopic techniques, 184–187
Spherulitic growth rate equation, 306, 307, 309
Spinodal decomposition, 31, 42–47, 61, 62, 66, 68, 69, 73, 74, 86–91, 93, 96, 100, 102, 118, 145, 148, 226, 227
Statistical mechanics, 76, 77, 79
Star polymers, *see* Radial polymers
Steady viscometric functions, *see* Rheological properties
Strength, tensile or yield, 14, 262, 263, 287–292
Stress-cracking, 287, 344–346, *see also* Environmental stress crack resistance
Styrene/acrylonitrile copolymers, 15, 16, 37–41, 47, 95, 97, 98, 102, 140, 152, 153, 196, 197, 199, 200, 222, 227, 229, 231, 232, 237, 239, 241, 322, 340, 346
Styrene/butadiene copolymers, *see* Butadiene/styrene copolymers
Surface free energy, 108
Surface tension, 36, 53, 55
Surfactants, 343
Swelling, 49, 304

T

Tacticity, effect on miscibility, 132, 197, 235, 237
Takayanagi model, 283–285
Ternary solution methods, 157–174, 176–177, 180
Thromboresistance, 334, 335
Torsional braid analyzer, 122–124
Torsion pendulum, 122, 282, 310
Transmission electron microscopy (TEM), 137–140
Transesterification, 240
Transport behavior, 300–306
Thermal degradation, 14
Thermal-optical analysis, 134–135
Thermal pressure coefficient, 91, 92
Thermodielectric loss, 128–131
Thermoplastic polyurethane, 9, 29, 223, 257, 259, 326–327, 330–332, 338
Two-dimensional solubility parameters, 52, 268

U

Ultrahigh magnification, 137
Ultraviolet emission, 188
Ultraviolet spectroscopy, 188
Unsaturated hydrocarbon based polymers, 252–254
Upper critical solution temperature (ucst), 91–96, 102, 141, 165, 226, 259, 268

V

Van der Waals forces, 26–28
  potential, 81
Vapor sorption, 165, 181, 226, 228, 304
Varnish, 16
Vibrating reed, 122–124
Vinyl chloride/vinyl acetate copolymers, 238, 239
Viscoelastic properties, 121, 175, 277, 287, 295–300, 323
Viscoelastometer, 124, 125, 282
Viscosity
  blends, 297
  melt, 296–298
Volume of mixing, 177–179, 291

W

Water-soluble polymers, 246–252
Wear resistance, 323, 332, 333, 341, 342, 347, *see also* Abrasion resistance
Weld-line strength, 351, 352
Wettability, 349
WLF equation, 120, 277, 300, 309

X

X-ray scattering, 117, 155–156, 221, 312

Z

Zimm plot, 151–154